Energy Efficiency of Vehicles

Second Edition

Energy Efficiency of Vehicles

Second Edition

By Douglas R. Carroll

SAE INTERNATIONAL®

400 Commonwealth Drive
Warrendale, PA 15096-0001 USA
E-mail: CustomerService@sae.org
Phone: 877-606-7323 (inside USA and Canada)
             724-776-4970 (outside USA)
Fax: 724-776-0790

Library of Congress Catalog Number 2024942892
http://dx.doi.org/10.4271/9781468608458

ISBN-Print   978-1-4686-0844-1
ISBN-PDF   978-1-4686-0845-8
ISBN-epub   978-1-4686-0846-5

To purchase bulk quantities, please contact: SAE Customer Service

E-mail: CustomerService@sae.org
Phone: 877-606-7323 (inside USA and Canada)
             724-776-4970 (outside USA)
Fax: 724-776-0790

Visit the SAE International Bookstore at books.sae.org

Publisher
Sherry Dickinson Nigam

**Production and
Manufacturing Associate**
Michelle Silberman

To my wife Karla and my children Zachary and James.

Contents

Chapter 08 – Greenhouse Gases and How They Impact the Temperature of the Earth

Chapter 09 – Fundamentals of Batteries

Unit 03: Detailed Modeling of Vehicle Efficiency

Chapter 10 – Performance and Efficiency of Internal Combustion Engines

Chapter 11 – Simulation of Fuel Economy for City, Suburban, and Highway Driving

Chapter 12 – Review of the Book (Course)

Preface

I developed a new course for the mechanical engineering program at Missouri S&T titled "Energy Efficiency of Vehicles" and offered it for the first time in the spring semester of 2017. After teaching the course three times, I wrote a manuscript for students to use, and the manuscript was converted into this book published in fall 2020. The purpose of writing the book is to educate students and others about energy and the environment and the relationship between the energy we use and the environment. We are at a point in time when we need to make some very important decisions about energy in the next few decades, and I hope we can utilize our scientific knowledge to make good rational decisions. I do not claim to know what the right decisions are and, hence, have tried to make the book as nonpolitical as possible. My goal is to show how to do the calculations related to energy, power, and efficiency, and the impact of using different types of energy on the environment. None of us are willing to give up our modern lifestyle and go back to a more primitive one. Maintaining our lifestyle requires that we continue consuming large quantities of energy. The decisions that will need to be made in the coming decades are about how we will provide the energy needed for our modern lifestyle without damaging the environment.

Unit 1 of the course covers how to estimate the energy efficiency and fuel economy of many types of vehicles including bicycles, cars, trucks, trains, ships, and aircraft. In order to understand the bigger picture of energy and the environment, it is necessary to have a clear understanding of the relationship between energy and power. Engineers and scientists who have been working in the energy field for several years already have a clear understanding of energy and power and will probably want to skim the first part of Unit 1. Engineering students need to build a better foundation before they can appreciate the rest of the book. Studying vehicles is a good way to develop an understanding of energy and power and how the two quantities are important in calculating the energy efficiency of vehicles. The general public tends to view energy efficiency and consumption as magic. Myths, stories, and conspiracy theories abound about cars getting 100+ mpg. I have heard such stories throughout my entire career, and there are still many people who believe in them. Unfortunately, the stories are not true. Energy efficiency and consumption are not magic. Vehicles require power and energy, and the amounts are related to the size of the vehicle and how fast we want to travel. There are standard methods to estimate the efficiency and fuel economy of vehicles, which are taught in Unit 1.

Unit 2 is about the environmental consequences of consuming energy. The unit starts by showing the amount of energy used in the United States and the world. This is the energy required to power our modern lifestyle. We use many different sources of energy including solar, nuclear, hydro, wind, geothermal, biomass, coal, natural gas, and petroleum. There are data that show how much energy we use from each source. Data are also available for the carbon dioxide produced in the United States and the world and for the pollution generated while burning fossil fuels (carbon monoxide, nitrous oxide, sulfur dioxide, and particulate). It is straightforward to calculate the amount of pollution and carbon dioxide generated while burning fossil fuels. The book compares coal, natural gas, gasoline, and diesel fuel by calculating the amount of energy we get from the fuel and the amount of pollution and carbon dioxide generated by burning the fuel. The calculations will show clearly that some fuels are cleaner than others and that there are trade-offs in choosing which fuel to use. Most of our electricity is generated using fossil fuels, and we can calculate how much pollution and carbon dioxide are generated for each kW-h of electric energy. Electric cars are sometimes said to be zero-emission vehicles, but, in this book, we will account for the pollution and carbon dioxide used in generating electricity. Wind turbines and solar panels are usually regarded as zero-emission, but manufacturing and maintaining wind turbines and solar panels impact the environment. In Unit 2, the book covers how to calculate the environmental impact for different types of energy sources and the vehicles or other equipment that make use of the energy sources. There is a short section on

batteries included in Unit 2. Batteries are very important for electric vehicles, and it is important to understand the basics of how batteries work. Batteries for electric vehicles need to be small and lightweight. As we move in the direction of using more wind and solar power to generate electricity, we will need to develop battery systems to help us in balancing the power grid. Utility-scale batteries are still in their infancy. Weight and size are not as critical for utility-scale batteries. Utilities will need large, reliable, inexpensive batteries. We have made a lot of progress in developing batteries for vehicles and utilities but still have a long way to go.

The purpose of Unit 3 is to understand quantitatively how city driving impacts the energy efficiency and fuel economy of cars and trucks. Spreadsheet models are developed for cars and trucks driving in city traffic to calculate quantitatively how the city driving impacts fuel economy. The models will show that regenerative braking will significantly improve the range of electric vehicles in city driving. The models will also show how hybrid electric vehicles can get a much better fuel economy in the city than traditional gasoline or diesel vehicles. Thank you for purchasing this book. I hope you will gain an understanding and appreciation of our energy system and the impact it has on the environment.

Douglas R. Carroll

Unit 01: Introduction to Energy Efficiency

Unit 1 focuses on the energy consumption and power required by vehicles operating at their normal steady-state cruising speed. Vehicles spend most of their time cruising, and most of the energy consumed by vehicles occurs while the vehicle is cruising at its normal operating speed. Parameters are included in the book that will help you make reasonable estimates of the energy consumption and power requirements of many types of vehicles. At the end of Unit 1, you will be able to make a good estimate of the energy consumption and power requirements based on the size, weight, speed, and type of vehicle.

The unit starts with human and animal-powered vehicles. Average humans can provide approximately 0.1 hp of sustained power, limiting the performance of human-powered vehicles such as bicycles. Horses can provide approximately 1 hp when pulling a buggy. For human and animal-powered vehicles, we will measure the energy consumed in Cal and relate the energy to weight loss. The reader will be able to understand and calculate how exercise relates to weight loss.

Unit 1 covers the steady-state driving of cars and trucks and how to estimate fuel economy in miles per gallon (mpg). For electric vehicles, the fuel economy is measured in kW-h of energy used per mi traveled. Unit 1 does not account for hills or the stop-and-go traffic of city driving. Those topics are covered in Unit 3. Unit 1 also covers the steady-state fuel economy of trains, boats, ships, airplanes, jets, and rockets.

In my many years of teaching engineering courses, I have found that most people have trouble understanding the difference between energy and power. In the first section of the book, the reader must understand the difference, both qualitatively and quantitatively. In the basic physics class, the reader should have learned that energy is the dot product of force and distance:

$$\text{Energy} = \text{Force} * \text{Distance}$$

The force may vary with the distance traveled, so it is more precise to integrate the force along the distance traveled. The dot product is necessary because the force may not be in the same direction that the vehicle is traveling.

All vehicles have a drag force associated with their motion that opposes the motion of the vehicle. For most vehicles, aerodynamic drag is an important part of the total drag force on the vehicle. Rolling vehicles have a drag associated with the tires rolling. Pedestrians experience a drag force related to the kinematics of the walking motion. Ships experience a drag force that comes from the motion of the water around the ships and the wake formed as the ship moves through the water. Airplanes and jets experience aerodynamic drag on the wings and fuselage and induced drag that is proportional to the weight of the vehicle. Sleds are dragged across the ground and experience a frictional drag force. The total drag force on a vehicle will be the sum of all drag forces. To move the vehicle at a steady-state, there must be a thrust force equal to the drag force. The energy required to move a vehicle through a distance is equal to the total drag force multiplied by the distance traveled.

The total drag force for most vehicles will vary along the distance traveled, especially for wheeled vehicles that will travel up and down hills and accelerate and decelerate. If the force varies with the distance traveled, Equation (1.1) can be rewritten as an integral equation, integrating the force along the distance traveled. An equivalent approach to integration is to find the average drag force on the vehicle and multiply by the distance traveled. The average drag force for wheeled vehicles is approximately equal to the drag force at a steady speed on level ground. The uphill and downhill loads tend to cancel if the starting and ending points are at the same altitude. Acceleration and deceleration also tend to cancel if the vehicle does not do a lot of braking.

Assuming that the vehicle travels at a constant speed, and the altitude does not change significantly over the distance traveled, then the average drag force on the vehicle is equal to the drag force at a steady speed on level ground or level flight for aircraft. This approximation allows us to make a good estimate of the energy consumption and energy efficiency of vehicles when operating at their normal cruising speed.

Example 1.1

If the total drag force on a bicycle is 2 lb, and a person rides the bicycle 3 mi (5280 ft/mi), how much energy is required?

$$\left(2\,\text{lb}\right) * \left(3\,\text{mi}\right) * \left(5280\,\frac{\text{ft}}{\text{mi}}\right) = 31,680\,\text{ft-lb} \qquad (1.2)$$

To ride the bicycle 3 mi, the person riding would need to provide 31,680 ft-lb of energy. The energy comes from the carbohydrate fuel that the person consumes. If we can make a good estimate of the total drag force on the bicycle, we can compute how much energy is required to ride a distance, and we can calculate the number of Cal the person must consume to provide the energy.

In the metric system, newtons (N) are used for the force unit and meter (m) for the distance. Energy would be calculated in N-m. The energy unit Joule (J) is defined as 1 N-m. Electric energy is usually measured in either W-h or kW-h. Human and animal energy is usually measured in nutritional Cal. (A nutritional Cal is 1000 of the calorie unit used in basic science classes. To distinguish between the two units, it is customary to use a capital C for nutritional Cal and a lower-case c for scientific calories.) The energy in fuels is usually measured in British thermal units (BTU). Energy is such an important quantity that many different units are used. It is always possible to convert from one energy unit to another. The table will help in making unit conversions (Table 1.1).

Table 1.1 Unit conversions.

Force: 1 Pound (lb) = 4.448 Newton (N)

Distance: 1 Foot (ft) = 0.3048 meter (m)

Distance: 1 Mile = 5280 ft = 1609 m = 1.609 kilometers (km)

Energy: 1 ft-lb = 1.356 N-m = 1.356 J

Energy: 1 kW-h = 3,600,000 J = 2,655,000 ft-lb

Energy: 1 BTU = 1055 J = 778.2 ft-lb

Energy: 1 Cal = 1000 calorie = 4184 J = 3086 ft-lb

Power: 1 horsepower (hp) = 745.7 W = 0.7457 kW = 550 ft-lb/s

Power is the rate at which energy is consumed. If the person riding the bicycle above rides 3 mi in 30 min (6 mph average speed), the bicycle vehicle consumes 31,680 ft-lb of energy in 30 min. The average power required is:

$$\text{Power} = \frac{31,680\ \text{ft-lb}}{(30\ \text{min})*(60\ \text{s/min})} = 17.6\ \text{ft-lb/s} \tag{1.3}$$

One horsepower is equal to 550 ft-lb/s, so the person riding the bicycle would need to provide 17.6/550 = 0.032 hp or 23.9 W to propel the bicycle along at 6 mph. Most people could provide that much power easily. Two pounds is a typical drag force for a bicycle at 6 mph. This leads to another fundamental equation for power:

$$\text{Power} = (\text{Force})*(\text{Speed}) \tag{1.4}$$

If we know the total drag force on the vehicle and the speed of the vehicle, we can calculate the power required to push the vehicle along. Energy is the product of power and time. The aerodynamic drag force on the vehicle will increase with the square of the speed [1], and for bicycles, the aerodynamic drag is very significant. Since force is multiplied by speed, the power requirement for the rider to overcome aerodynamic drag increases with the cube of speed. Because of this relationship, increasing the average speed of the bicycle by a small amount requires a significant increase in the amount of power the rider must provide. The aerodynamic energy required to ride a distance is proportional to the square of velocity. Energy can be expressed as the product of power and time or the product of force and distance:

$$Energy = (Power) * (Time) = (Force) * (Distance) \tag{1.5}$$

For human-powered vehicles such as bicycles, rowboats, and canoes, the units of Joules and ft-lb are useful because a Joule or ft-lb is a small but significant amount of energy for a human to produce. W and ft-lb/s are small but significant amounts of power for a human to produce. Most adult humans can produce 1 hp (745.7 W) for 10–15 s. After 10–15 s, most humans will tire and must rest or reduce power output [2]. Horses can provide 1 hp for a few hours before they tire and must rest or reduce power output. A typical adult human can provide 50–100 W power for a few hours [3].

When we study electric and fuel-powered vehicles, the units of Joules and ft-lb become such tiny quantities that we use other units. The kW-h is a common energy unit for electric vehicles (1 kW-h = 3,600,000 J). Hp-h is also a common energy unit, especially for farm tractors. Power is usually measured in either ft-lb/s, hp, W, or kW.

Example 1.2

A bicycle rider faces a total drag force of 3.8 lb while riding at 10 mph on level ground. How much energy will be required to move the bicycle 1 mi (i.e., how much energy per mile traveled must the rider provide)? How much power will the rider need to provide?

Discussion: The drag force on the bicycle (and other wheeled vehicles) comes from rolling resistance in the tires and aerodynamic drag. The thin, high-pressure racing bike tires have lower rolling resistance than the wide low-pressure tires. At 10 mph and faster, most of the drag force on the bicycle will come from aerodynamics. Rolling resistance will be the dominant drag force at low speeds.

Solution: The energy and power requirement is calculated using Equations (1.6) and (1.7).

$$Energy = (3.8 \text{ lb}) * (5280 \text{ ft}) = 20,064 \text{ ft} - \text{lb/mi traveled} \tag{1.6}$$

$$Power = (3.8 \text{ lb}) * \left(\frac{10 \text{ mi}}{h}\right) * \left(\frac{5280 \text{ ft}}{\text{mi}}\right) * \left(\frac{h}{3600 \text{ s}}\right) = 55.7 \text{ ft-lb} / s \tag{1.7}$$

Recognizing that 1 hp = 550 ft-lb/s = 745.7 W, the rider must provide 0.101 hp or 75.6 W of power to push the bicycle along at 10 mph. A typical human can provide 50–100 W of sustained power; therefore, it is reasonable to expect that a human could travel a significant distance at 10 mph. A well-conditioned adult can provide 150–200 W of sustained power and can travel faster than 10 mph.

Figure 1.1 Riding a bicycle.

Example 1.3

James Watt defined the horsepower as 550 ft-lb/s. This means that a horse can produce 550 ft-lb of useful energy each second while working. It is also observed that a horse can pull a buggy along a dirt road at 7 mph (Figure 1.2).

(a) If it takes 1 hp to pull the buggy along at 7 mph, estimate the total average drag force on the buggy.

$$\left(\frac{7 \text{ mi}}{h}\right)*\left(\frac{5280 \text{ ft}}{\text{mi}}\right)*\left(\frac{h}{3600 \text{ s}}\right) = 10.27 \text{ ft / s} \tag{1.8}$$

The buggy travels 10.27 ft in 1 s, and the horse produces 550 ft-lb of energy in 1 s. Since energy is the drag force multiplied by the distance traveled:

$$550 \text{ ft-lb} = \left(\text{Drag Force}\right)*\left(10.27 \text{ ft}\right); \quad \text{Drag force} = 53.6 \text{ lb} \tag{1.9}$$

Figure 1.2 Horse-drawn carriage.

(b)   If the buggy is to be pulled by a human that can produce 0.1 hp, how fast will the human be able to pull the buggy (Figure 1.3)?

For this problem, let us use the idea that power is force times speed. These types of problems can be solved using the idea that Energy = Force * Distance or that Power = Force * Speed.

$$55 \text{ ft-lb / s} = \left(53.6 \text{ lb}\right) * \left(\text{Speed}\right); \quad \text{Speed} = 1.027 \text{ ft / s} = 0.7 \text{ mph} \qquad (1.10)$$

A human can pull with 53.6 lb and pull the buggy along the dirt road but will not be able to pull it very fast. The pulling speed is limited by the power that can be provided. The horse can pull the buggy faster because it can provide more power, not because it is pulling with a larger force. It is important to understand the difference between force, energy, and power.

Figure 1.3 Pulling a rickshaw.

© Shutterstock

Example 1.4

A small fishing boat is being pushed through the water using an electric trolling motor that provides 250 W of power. The propeller is 50% efficient in converting the rotational power of the shaft driving the propeller into thrust power for the boat. The drag force on the boat will be proportional to the velocity squared.

$$\text{Drag Force} = \left(175 \, \text{kg} \, / \, \text{m}\right) * V^2$$

How fast will the electric trolling motor pull the boat along in m/s and mph? If the battery for the trolling motor holds 1 kW-h of energy, how long will the trolling motor be able to produce this thrust force? How far (in miles) will it be able to pull the boat (Figure 1.4)?

Figure 1.4 Boat with an electric trolling motor.

The propeller will be significantly less than 100% efficient in converting the power provided to the shaft into thrust power when used to provide the thrust power. For this problem, we will assume that the electric motor is 100% efficient in providing the power to the shaft that drives the propeller. The propeller is 50% efficient in converting this power into thrust power that pushes the boat along. The electric motor will consume 250 W of power from the battery. Because of the 50% efficiency of the propeller, it will only provide 125 W of thrust power to the boat.

$$\text{Thrust Power} = \left(250\,\text{W}\right) * \left(0.50\right) = 125\,\text{W} \qquad (1.11)$$

To solve for the speed of the boat, we use the relationship that power equals drag force multiplied by speed.

$$125\,\text{W} = \left[\left(175\,\text{kg}\,/\,\text{m}\right) * V^2\right] * V \qquad (1.12)$$

$$V = 0.894\,\text{m}\,/\,\text{s} = 2\,\text{mph} \qquad (1.13)$$

The boat will be able to travel at 2 mph when the trolling motor is using 250 W power from the battery. A typical trolling motor can provide up to 500 W power, so this motor is operating at about 50% power. The battery holds 1 kW-h of energy = 1000 W-h. Using the idea that energy is power multiplied by time:

$$1000 \text{ W-h} = \left(250 \text{ W}\right) * \left(\text{Time}\right); \ \text{ Time} = 4 \text{ h} \tag{1.14}$$

The trolling motor would be able to operate for 4 h before the battery is depleted. Traveling for 4 h at 2 mph, the boat would travel 8 mi.

Example 1.5

The Gossamer Albatross was pedal-powered across the English Channel in 1979 by Bryan Allen [4]. The aircraft was designed by Paul MacCready. The distance was 22.2 mi, and it required 2 h and 49 min for the flight. Assume that Bryan is an exceptional athlete capable of providing 400 W average power during the flight. Assume that the propeller is 85% efficient in converting Bryan's power into thrust power. Estimate the drag force on the airplane in pounds (Figure 1.5).

Figure 1.5 Gossamer Albatross.

© Shutterstock.

This was a phenomenal achievement in its time. Airplanes require a lot of horsepower and designing an airplane that a human could power long enough to travel across the English Channel was an amazing accomplishment. Dr. MacCready led the design effort to design an airplane that could fly slow enough with a low enough drag force that a human could power it. Power is drag force multiplied by speed, so it was important to keep drag force and speed to a minimum. Bryan Allen was an exceptional athlete, and his ability to provide lots of power was an important part of the success of the project.

$$\text{Thrust Power} = \left(400 \text{ W}\right) * \left(0.85\right) = 340 \text{ W} = 250.8 \text{ ft-lb / s} \tag{1.15}$$

$$\text{Average Speed} = \frac{22 \text{ mi}}{2 + \frac{49}{60} \text{ h}} = 7.88 \text{ mph} = 11.56 \text{ ft / s} \tag{1.16}$$

Recognizing that power is drag force multiplied by speed:

$$250.8 \text{ ft-lb / s} = \left(\text{Drag Force}\right) * \left(11.56 \frac{\text{ft}}{\text{s}}\right); \text{ Drag Force} = 21.7 \text{ lb} \tag{1.17}$$

When the flight was started the winds were calm, but a headwind developed during the flight, and the airspeed of the vehicle was significantly higher than the 7.88 mph average land speed during the flight. The crew had planned for a 2 h flight, but the headwind significantly increased the flight time. The instruments on the aircraft recorded a maximum airspeed of 18 mph. If we were to add 5 mph to account for the headwind, the average airspeed would be 12.88 mph, and the average drag force would be 13.3 lb, which is probably closer to the actual drag force on the airplane.

Summary for Section 1.1: In the first section, we looked at different vehicles and discussed energy and power and the difference between energy and power. The energy used by a vehicle is equal to the total drag force on the vehicle multiplied by the distance traveled. For human-powered vehicles, the energy comes from the human and will relate to the number of Cal burned in powering the vehicle the distance traveled. For vehicles powered by an internal combustion engine, the energy will relate to the amount of fuel consumed over the distance traveled. For electric vehicles, the energy will relate to the electric energy drawn out of the batteries in powering the vehicle the distance traveled.

The power required for a vehicle is equal to the product of drag force and speed. The power requirement for a vehicle increases with speed, even if the drag force remains constant. The drag force for most vehicles also increases with speed, so the power requirement of the vehicle increases significantly as speed increases.

1.2.
Basics of Thermal Efficiency [5]

Power is the rate at which the engine (or human or animal) can provide energy. Power is also the rate at which the vehicle uses energy. For steady-state operation, the engine must provide the power required for the vehicle. Engines, motors, humans, and animals are not 100% efficient in converting fuel into useful energy to power the vehicle. For internal combustion engines, most of the heat energy in the fuel goes out the tailpipe as heat. The same is true for jet engines, rocket engines, the steam turbines used at the power plants, and all heat engines. Less than half of the energy in the fuel will be converted to useful energy. We refer to this as the thermal efficiency of the engine. If the engine converts 28% of the energy in the fuel to useful mechanical energy, we say the engine has a thermal efficiency of 28%. When driving on the highway, the gasoline engines in our cars and trucks have a thermal efficiency of 25 to 30%. The diesel engine typically has a thermal efficiency that is a little higher than gasoline engines. This is especially true for the diesel engines in the large trucks. To estimate the fuel consumption of a vehicle, we need to know the thermal efficiency.

Humans and animals have a thermal efficiency of about 20% [6]. About 80% of the carbohydrate fuel we consume is converted to heat, and we heat up as we exercise. Only about 20% of the carbohydrate fuel is converted into useful mechanical energy. The thermal efficiency comes from the way our muscles operate and how they use the carbohydrate fuel.

Electric motors and generators have high thermal efficiency. A typical electric motor will convert 85% of the electric energy it uses into useful mechanical energy. Only about 15% of the electric energy will convert into heat. High-efficiency electric motors can have a thermal efficiency of 95% or perhaps even a little higher. But no electric motor is 100% efficient.

In the power plant, we use a steam turbine to power a generator that is highly efficient in converting the mechanical energy from the steam turbine into electrical energy. The steam turbine/generator has a thermal efficiency of typically 35 to 40%, in converting the energy in the fuel (coal, natural gas, nuclear, etc.) into electrical energy. The transmission and distribution lines and electrical substations are very efficient in transmitting the electrical energy to the customer. The overall efficiency of our electrical system is approximately 33%, meaning that approximately one-third of the energy in the fuel used at the power plant is delivered to the customer as electrical energy. For example, a natural gas power plant might use a billion BTU of heat energy from the natural gas it burns each day. Approximately one-third of a billion BTU of electric energy would be produced and delivered to customers each day. The other two-thirds of a billion BTU of heat energy would simply add heat to the atmosphere.

When we use electricity to power our vehicles, we need to recognize that the electric energy used to charge the batteries is only about 33% of the energy in the original fuel source used to make the electricity (coal, natural gas, or nuclear). The batteries and electric motor in the car will have high efficiencies, typically 85% efficiency, but the overall efficiency relative to the original fuel source used to produce the electricity will only be 85% of 33%, which is about 28%. Internal combustion engines are typically 25 to 30% efficient when cruising on the highway. When fossil fuel is used to produce electricity, the overall efficiency in getting useful energy from the fuel source is not much different for electric vehicles and internal combustion vehicles.

Example 1.6

When we talk about diet and Cal, a Cal is 4184 J. It is common knowledge that a human at rest emits about the same heat as a 100-W light bulb. If a human is using 100-W power while at rest, how many Cal would the person consume during the day, assuming the person sat around doing nothing all day? (Cal = 1000 calorie = 4184 J)

$$100 \text{ W} = \left(\frac{100 \text{ J}}{\text{s}}\right) * \left(\frac{3600 \text{ s}}{\text{h}}\right) * \left(\frac{24 \text{ h}}{\text{day}}\right) * \left(\frac{\text{Cal}}{4184 \text{ J}}\right) = 2065 \text{ Cal / day} \tag{1.18}$$

People use less than 100 W while sleeping, so this may be a high estimate. For a very sedentary person, an estimate of 2000 Cal per day is reasonable. When we talk about weight and nutrition 1-lb of weight (fat) is equal to about 3500 Cal. A person would need to consume about 3500 Cal less than they burn to lose 1 lb. Or a person who consumes 3500 Cal more than they burn will gain 1-lb of weight. Physical activity will increase the number of Cal burned in a day but not as much as we would hope. The example problems and homework will help you gain an understanding of how exercise affects weight loss.

Energy is very important in all fields of science, and there are many different units used for energy. Google can be very helpful in looking up unit conversions. Learning to convert energy units is part of working in the energy field.

Example 1.7

The aerodynamic drag force on a car is proportional to the square of its velocity. For a typical midsize sedan, the aerodynamic drag force F is given by:

$$F = \left(0.0262\right)V^2 \tag{1.19}$$

Where F is in lb and V is in mph.

Find the extra energy required for a car to go on a 250-mi trip if the car travels at 80 mph rather than 70 mph. That is, find the aerodynamic energy required if the car travels at 70 mph and at 80 mph, and subtract the former from the latter to find the difference (Figure 1.6).

$$\text{At 70 mph, Drag Force} = \left(0.0262\right)\left(70\right)^2 = 128.4 \text{ lb} \tag{1.20}$$

$$\text{At 80 mph, Drag Force} = \left(0.0262\right)\left(80\right)^2 = 167.7 \text{ lb} \tag{1.21}$$

Figure 1.6 Midsize sedan.

Equations 1.20 and 1.21 show that the aerodynamic drag force on the car is substantially higher at 80 mph than at 70 mph. Energy consumed on a trip is drag force multiplied by the distance traveled, and the distance in this case is 250 mi. The energy required to overcome aerodynamic drag is calculated as follows:

$$\text{At 70 mph, Drag Energy} = \left(128.4\ \text{lb}\right)\left(250\ \text{mi}\right)\left(5280\ \text{ft / mi}\right) = 169.5 \times 10^6\ \text{ft-lb} \tag{1.22}$$

$$\text{At 80 mph, Drag Energy} = \left(167.7\ \text{lb}\right)\left(250\ \text{mi}\right)\left(5280\ \text{ft / mi}\right) = 221.4 \times 10^6\ \text{ft-lb} \tag{1.23}$$

$$\text{Extra energy} = 221.4 \times 10^6 - 169.5 \times 10^6 = 51.9 \times 10^6\ \text{ft-lb} \tag{1.24}$$

This is the correct answer, but we need to understand what it means. Is this a lot of energy or a small amount of energy? For humans or horses, this is a lot of energy. One horsepower is 550 ft-lb/s of power. One way to look at this is to solve for the number of hours it would take a horse to produce this much energy. Solving for the time t required:

$$\left(550\ \text{ft-lb / s}\right)\left(t\right) = 51.9 \times 10^6\ \text{ft-lb}, t = 94{,}364\ \text{s} = 26.2\ \text{h} \tag{1.25}$$

We would need to work a horse for 26.2 h to produce enough energy to provide the excess of energy required for the car to make the trip at 80 mph rather than 70 mph. Horses can be worked 4–5 h per day, so it would take several days for a horse to produce this much energy. For horses, humans, or any animal, 51.9×10^6 ft-lb represents a lot of energy.

For cars and trucks, we need to look at the energy content of the fuel. Table 1.2 shows the energy content of many types of fuel [7].

Table 1.2 Energy content of common fuels.

Fuel type	MJ/L	MJ/kg	BTU/Imp gal	BTU/US gal
Regular gasoline	34.8	~47	150,100	125,000
Premium gasoline	34.8	~46	150,100	125,000
Autogas (LPG) (60% propane, 40% butane)	25.5–28.7	~51	N/A	N/A
Ethanol	23.5	31.1	101,600	84,600
Gasohol (10% ethanol, 90% gasoline)	33.7	~45	145,200	121,000
E85 (85% ethanol, 15% gasoline)	25.2	~33	108,878	90,660
Diesel	38.6	~48	166,600	138,700
Biodiesel	35.1	39.9	151,600	126,200
Aviation gasoline	33.5	46.8	144,400	120,200
Jet fuel (naphtha)	35.5	46.6	153,100	127,500
Jet fuel (kerosene)	37.6	~47	162,100	135,000
Liquefied natural gas	25.3	~55	109,000	90,800
Liquid hydrogen	9.3	~130	40,467	33,696

© SAE International

Cars and trucks purchase fuel by the gallon, so the most important column for cars and trucks is the column that shows BTU/US gal. Gasoline has 125,000 BTU/gal, but in the Midwest, most of the gasoline we purchase has a 10% ethanol content and is called gasohol in the table. Gasohol has an energy content of 121,000 BTU/gal, which is a little less than gasoline. Many gas stations carry the E85 fuel, which has an energy content of 90,660 BTU/gal. The fuel economy of a vehicle, as measured in miles per gallon (mpg), is proportional to the energy content of the fuel. A car or truck will get slightly lower fuel economy burning gasohol compared to burning regular gaso-

line. The car will get significantly lower fuel economy burning the E85 fuel. Diesel fuel has an energy content of 138,700 BTU/gal, which is significantly higher than gasoline. Part of the reason that diesel vehicles get better fuel economy than gasoline vehicles is because diesel fuel has more energy per gallon. Diesel engines have a higher thermal efficiency than gasoline engines, which also gives them a fuel economy advantage over gasoline-powered vehicles. The result is that diesel-powered vehicles get significantly better fuel economy than gasoline-powered vehicles.

BTU Definition: Heating water is an important use of energy. The BTU is a unit of energy originally designed to support the application of heating water. A BTU is the amount of energy required to heat 1 lb of water by 1°F. A pint of water weighs about 1 lb. To raise the temperature of a pint of water 1°F requires 1 BTU of heat energy.

Example 1.8

How much energy is required to heat a quart of water from 65 to 150°F? Assume that the quart of water has a weight of 2 lb for this example (Figure 1.7).

$$\left(2\,lb\right)\left(150°F - 65°F\right)\left(\frac{1\,BTU}{lb°F}\right) = 170\,BTU \tag{1.26}$$

The stove will need to add 170 BTU of energy to the water to raise the temperature from 65 to 150°F. A lot of heat goes into the air and stove. The stove will need to provide more than 170 BTU heat energy to heat the water, but this type of analysis is a good place to start when estimating the heat energy required.

When estimating the energy required for vehicles, we estimate in units of Joules and ft-lb. The energy content of the fuel is usually given in BTU, so converting from BTU to Joules and ft-lb is common.

$$BTU = \left(1055\right)J = \left(778.2\right)ft\text{-}lb \tag{1.27}$$

Figure 1.7 Boiling a pot of water.

To continue building depth in the subject, it is necessary to have a brief discussion of thermodynamics. The first and second laws of thermodynamics are important in understanding the energy efficiency of vehicles [5].

First Law of Thermodynamics: Energy is conserved. Conservation of energy is the first law of thermodynamics. There are many different types of energy, and it is possible to convert from one type to another. The important types of energy for this subject follow:

- Chemical energy (fuel)—especially fossil fuel: This energy can be easily converted to heat by igniting the fuel. It is not (from a practical viewpoint) possible to convert the heat back into fuel. The energy is usually measured in BTU or kW-h. 1 W-h = 3.6 million Joules = 3412 BTU = 860.4 Cal = 2.655 million ft-lb.

- Electrical energy: Electricity is produced primarily from fossil fuel at power plants. Significant amounts of electricity are produced by nuclear, hydro, wind, and solar sources. Electricity can be stored in batteries as chemical energy. Electric energy is usually measured in kW-h or MW-h.

- Gravitational energy: Gravitational energy is conservative in the sense that it requires energy to push a vehicle uphill, but the energy is recovered when the vehicle goes downhill.

- Kinetic energy: Kinetic energy is also conservative, but most of the kinetic energy is converted to heat when braking. Energy is required to speed the car up and is stored as kinetic energy. When braking an electric vehicle, some of the kinetic energy can be recovered with regenerative braking, but for most vehicles, the kinetic energy is converted to heat using the brakes.

- Friction energy: A friction force absorbs energy. Friction converts mechanical energy into heat energy. It is not reversible. Braking, rolling resistance, and aerodynamic drag fall under the general category of friction energy.

Second Law of Thermodynamics and Heat Engines: The second law of thermodynamics says that when we convert from one form of energy to another, some of the energy will be converted to waste heat and (for all practical purposes) lost. All energy conversions are less than 100% efficient. Energy is conserved, but the portion of energy converted to waste heat is, for all practical purposes, lost.

Converting electric energy to kinetic energy using an electric motor is very efficient. Typically 85% of the electric energy will be converted into mechanical and ultimately kinetic energy. Converting mechanical energy to electric energy using a generator is equally efficient. The electric motor and generator will heat up in use, and the heat energy is the "lost" energy in the conversion process.

Charging and discharging a battery is converting from electrical energy to chemical energy and back to electrical energy. This is also very efficient. Lead-acid batteries are typically 80% efficient or better in storing electric energy. Lithium-ion batteries are typically 90% efficient. Batteries will heat up when being charged and discharged, and the heat energy is the energy that is "lost" in the charge/discharge cycle.

Heat engines are inefficient in converting chemical energy in fuel into mechanical energy. The steam turbines used in power plants are about 40% efficient at converting heat energy in coal or natural gas into mechanical energy. Further, inefficiencies in the generator, transmission, and distribution make the overall process about 33% efficient in converting the energy in fuel into useful electric power at your house. The internal combustion engines in our vehicles are less efficient than the steam turbine because they must operate over a range of speeds. Under load on the highway, 25 to 30% is good efficiency for a gasoline engine, and 30 to 35% efficiency is good for a diesel engine. Large trucks, trains, and ships are vehicles that use high-efficiency diesel engines.

Example 1.9

Going back to the problem of the car traveling 250 mi at 80 mph or 70 mph, it was discovered that it would require an extra 51.9×10^6 ft-lb of energy to make the trip. If it is a gasoline-powered car operating at 25% efficiency, how many extra gallons of fuel will be required for the trip? Assume the gasoline (gasohol) has 121,000 BTU of energy per gallon.

$$\left(\frac{121{,}000 \text{ BTU}}{\text{gal}} \right)\left(\frac{778.2 \text{ ft-lb}}{\text{BTU}} \right) = \frac{94.16 \times 10^6 \text{ ft-lb}}{\text{gal}} \quad \left(\text{Total Energy per gallon} \right) \tag{1.28}$$

If the gasoline engine in the car had a thermal efficiency of 100%, then each gallon of the gasohol fuel would provide 94.16×10^6 ft-lb of useful energy. However, because of the laws of thermodynamics, the engine will be much less than 100% efficient. An efficiency of 25% is typical for a gasoline engine. This means that 75% of the energy in the fuel is going out the tailpipe, into the radiator, or radiated elsewhere as heat. Only 25% of the energy in the fuel is converted to useful mechanical energy that can be used to push the car down the road.

$$\left(\frac{94.16 \times 10^6 \text{ ft-lb}}{\text{gal}} \right)(0.25) = \frac{23.54 \times 10^6 \text{ ft-lb}}{\text{gal}} \quad \left(\text{Useful Energy per gallon} \right) \tag{1.29}$$

If we divide the required energy by the useful energy per gallon, we can find the number of gallons of fuel required.

$$\frac{51.9 \times 10^6 \text{ ft-lb}}{23.54 \times 10^6 \text{ ft-lb / gal}} = 2.205 \text{ gal} \tag{1.30}$$

The final answer is that if we choose to drive the car at 80 mph rather than 70 mph, it will require that we use an extra 2.205 gal of fuel to make the 250-mi trip. The extra cost of driving the trip at 80 mph vs 70 mph would be about 2.2 gal of fuel.

The next question to be addressed is how much total fuel the car would use in making the 250-mi trip and how to calculate the fuel economy in mpg. From Equations (1.22) and (1.23), we found that the energy to overcome the aerodynamic drag for the car was 169.5×10^6 ft-lb at 70 mph and 221.4×10^6 ft-lb at 80 mph. The car will have rolling resistance too, and there will be additional energy required to overcome the rolling resistance, which is not included in the analysis.

$$70 \text{ mph} \rightarrow \frac{\left(169.5 \times 10^6 \text{ ft-lb} \right)}{23.54 \times 10^6 \text{ ft-lb / gal}} = 7.200 \text{ gal;} \quad \frac{250 \text{ mi}}{7.200 \text{ gal}} = 34.7 \text{ mpg} \tag{1.31}$$

$$80 \text{ mph} \rightarrow \frac{\left(221.4 \times 10^6 \text{ ft-lb} \right)}{23.54 \times 10^6 \text{ ft-lb / gal}} = 9.405 \text{ gal;} \quad \frac{250 \text{ mi}}{9.405 \text{ gal}} = 26.6 \text{ mpg} \tag{1.32}$$

This analysis accounts only for the aerodynamic drag. If we add the energy required to overcome rolling drag the fuel economy of the vehicle will be lower in both cases. This process illustrates how to estimate the fuel economy of the car. The next step in the analysis is to consider an electric car traveling the 250-mi trip and estimate the number of batteries required to make the trip.

Example 1.10

Going back to the problem of the car traveling 250 mi at 80 mph or 70 mph, it was discovered that it would require 169.5×10^6 ft-lb at 70 mph and 221.4×10^6 ft-lb at 80 mph to make the trip. If it is an electric car, and the motor/drive is 85% efficient, how many lead-acid batteries would be required to make the trip at 70 and 80 mph?

(a)   Assume that the lead-acid batteries weigh 80 lb and store 1000 W-h of energy each. These are typical values for a 12 V 80 amp-h battery that might be used to power a trolling motor on a fishing boat. How many lead-acid batteries are required and how much would the battery system weigh?

(b)   Suppose lithium-ion batteries are used that weigh 20 lb each and store 1000 W-h each. How many lithium-ion batteries are required and how much do they weigh? (Figure 1.8)

Figure 1.8 Lithium-ion battery.

At 70 mph the energy required to make the 250-mi trip was 169.5×10^6 ft-lb. If the battery and electric motor system are 85% efficient, we would need:

$$\frac{169.5 \times 10^6 \text{ ft-lb}}{0.85} = 199.4 \times 10^6 \text{ ft-lb energy} \quad \left(\text{Energy in batteries}\right) \tag{1.33}$$

To calculate how many batteries are required, we need to relate the energy units of ft-lb and W-h. The conversion is 2655 ft-lb = 1 W-h.

$$\frac{199.4 \times 10^6 \text{ ft-lb}}{2655 \text{ ft-lb / W-h}} = 75{,}090 \text{ W-h} \quad \left(\text{Energy in Batteries}\right) \tag{1.34}$$

Since each battery holds 1000 W-h of energy, we would need 76 batteries fully charged to provide the required energy to allow the car to travel the 250 mi. For this example, the lead-acid and lithium-ion batteries hold the same amount of energy per battery, so it is 76 batteries of either type.

Lead-acid batteries weigh 80 lb each, so the total weight of the lead-acid batteries would be 6080 lb. A typical sedan weighs 3500 lb, so the batteries would be much heavier than the car. This illustrates the problem of using lead-acid batteries to power an electric car. To provide enough energy to power the car a reasonable distance, the batteries are very heavy. As the car gets heavier, the rolling resistance energy increases. It becomes impractical to make a lead-acid-powered car that can travel 250 mi on a single charge.

Lithium-ion batteries weigh 20 lb each, so the battery weight required to provide the energy to travel 250 mi is 1520 lb. Although adding 1520 lb of batteries to a 3500-lb vehicle is a lot, it is possible to make a lithium-ion battery-powered car that can travel 250 mi on a single charge.

Gasoline has a density of 6.65 lb/gal. The weight of the 7.200 gal to make the trip is 48 lb. If we do a comparison, 6080 lb of lead-acid batteries, 1520 lb of lithium-ion batteries, or 48 lb of gasoline is required to make the 250-mi trip. Lead-acid batteries are 126 times as heavy as gasoline, and lithium-ion batteries are 62.7 times as heavy as gasoline. This is the primary challenge in developing electric cars. The batteries are very heavy compared to gasoline.

Traveling at 80 mph, the energy required increases, and more batteries are required. Similar calculations show that 99 batteries are required, and the weights for lead-acid and lithium-ion batteries are 7920 lb and 1980 lb, respectively.

Example 1.11

(This example problem was developed for the Honda 750 motorcycle. The equations and results are correct for the Honda 750). Assume that a motorcycle and rider have a total weight of 700 lb. The drag force on the motorcycle is given by the equation:

$$\text{Force}\left(\text{lb}\right) = 5.6 + \left(0.019\right)*V + \left(0.0108\right)*V^2 \quad \left(\text{where V is in ft / s}\right) \tag{1.35}$$

What is the fuel economy of the motorcycle cruising at 70 mph? Assume that the fuel is gasohol and the engine/drive system is 22% efficient in converting the chemical energy in the fuel into useful mechanical energy (Figure 1.9).

For this problem, we are assuming the motorcycle is driving on level pavement. The force equation accounts for rolling and aerodynamic drag and is a good estimate of the total drag force on the motorcycle. The motorcycle has an air-cooled engine that will have a lower thermal efficiency than a water-cooled engine. The estimate of 22% thermal efficiency is reasonable. First, we need to convert 70 mph into ft/s.

$$\left(\frac{70\,\text{mi}}{\text{h}}\right)\left(\frac{5280\,\text{ft}}{\text{mi}}\right)\left(\frac{\text{h}}{3600\,\text{s}}\right) = 102.7\,\text{ft / s} \tag{1.36}$$

Figure 1.9 Honda 750 motorcycle.

$$\text{Force} = 5.6 + \left(0.019\right) * \left(102.7\right) + \left(0.0108\right) * \left(102.7\right)^{2} = 121.4 \text{ lb} \tag{1.37}$$

In the force equation, the first two terms come from the rolling drag, and the third term comes from the aerodynamic drag. Going term by term, the drag force on the motorcycle is given by:

$$\text{Force} = 5.6 + 1.95 + 113.84 = 121.4 \text{ lb} \tag{1.38}$$

When going fast on a motorcycle, most of the energy goes into overcoming the aerodynamic drag. In this case, of the 121.4 lb of total drag force on the motorcycle, 113.84 lb comes from the aerodynamic drag. The aerodynamic drag accounts for 93.8% of the total drag force on the motorcycle. For slower driving, the aerodynamic drag term is much less, and rolling resistance is a larger part of the total drag. The same is true for bicycles and scooters, except the definition of fast would be 15 mph for bicycles and 30 mph for scooters. When going fast on a two-wheeled vehicle, most of the energy goes into overcoming the aerodynamic drag. The power required to push the motorcycle along at highway speed is force multiplied by speed.

$$\text{Power} = \left(121.4 \text{ lb}\right)\left(102.7 \text{ ft / s}\right) = 12,468 \text{ ft-lb / s} = 22.7 \text{ hp} \tag{1.39}$$

We will assume that the motorcycle is burning gasohol, which has 121,000 BTU/gal. The thermal efficiency of the engine is 22%. From these numbers, we can calculate the useful energy per gallon in the fuel, the fuel consumption for the motorcycle, and the fuel economy as shown below:

$$\text{Useful Energy} = \left(121,000\right)\left(0.22\right) = 26,620 \text{ BTU/gal} \tag{1.40}$$

$$\text{Fuel Consumption} = \left(\frac{12,468 \text{ ft-lb}}{s}\right)\left(\frac{\text{BTU}}{778.2 \text{ ft-lb}}\right)\left(\frac{\text{gal}}{26,620 \text{ BTU}}\right)\left(\frac{3600 \text{ s}}{h}\right) = 2.167 \text{ gph} \tag{1.41}$$

$$\text{Fuel Economy} = \left(\frac{70 \text{ mph}}{2.167 \text{ gph}}\right) = 32.3 \text{ mpg} \tag{1.42}$$

Notice in the final step that, if we know the speed of the vehicle in mph and the fuel consumption in gal/h (gph), the fuel economy is speed divided by fuel consumption. This approach will be used for all the fuel-powered vehicles studied in this chapter. The calculations show a fuel economy of 32.3 mpg, which is very close to what is typically achieved for a Honda 750 motorcycle. Continuing with the same motorcycle, what is the top speed of the motorcycle, assuming the engine/drive can deliver a maximum of 40 hp to the rear wheel?

$$\left(40 \text{ hp}\right)\left(550 \text{ ft-lb / s hp}\right) = 22,000 \text{ ft-lb / s} \tag{1.43}$$

And since power is force multiplied by velocity, we can set up an equation to solve for the top speed of the motorcycle:

$$22,000 \text{ ft-lb / s} = \left[5.6 + \left(0.019\right) * V + \left(0.0108\right) * V^2\right] * V \tag{1.44}$$

Solving this equation, the top speed of the motorcycle is 124.8 ft/s, which is the same as 85.1 mph. Holding the engine at full throttle on level pavement, the motorcycle would top out at about 85 mph, which is typical for a Honda 750 motorcycle. We can use this approach to plot the power vs speed curve for the motorcycle, which is illustrated in the graph (Figure 1.10).

At higher speeds and power requirements, the thermal efficiency of the engine will be near 22%, and the fuel economy analysis is reasonable. At lower speeds and power requirements, the thermal efficiency of the engine will be less than 22%, and the analysis used in this unit will yield a fuel economy estimate that is too high.

Figure 1.10 Power vs speed for Honda 750 motorcycle.

Power vs Speed Plot (Honda 750)

In Unit 1, we will focus on estimating the fuel economy of vehicles at their cruising speeds. Trains, ships, airplanes, and jets tend to get up to their cruising speed and spend most of the trip at cruising speed. Estimating the fuel economy at cruising speed is a good way to estimate the overall fuel economy of the vehicle. Cars and trucks driving on the highway spend most of their time at cruising speed. The analysis in Unit 1 is useful in estimating the fuel economy for highway driving of cars and trucks but overestimates the fuel economy in city driving. In Unit 3, we will study the details and model city driving of cars and trucks.

1.3.
Homework

To understand the material covered so far, the reader needs to solve some homework problems. Partial answers are given for the problems to help the reader ensure correct processes that are used in solving the problems. The first chapter in Unit 1 covers the relationships between drag force, speed, energy consumption, power, and fuel economy that apply to all types of vehicles. Subsequent chapters go into detail on how to estimate these quantities for different types of vehicles. At the end of Unit 1, the reader will be able to make good estimates of the drag force, power, energy consumption and energy efficiency for many types of vehicles based on the size, shape, weight and speed of the vehicle.

1. Suppose that a small ultralight is powered by a gasoline engine. The cruising speed for the ultralight is 33 mph.

   (a) Assume that it requires 25 hp of thrust to power the ultralight at 33 mph and calculate the drag force on the ultralight at 33 mph. Express your answer in lb.

   (b) Calculate the energy required for the ultralight to travel for 2 h (66 mi) at 33 mph. Express your answer in BTU.

   (c) Gasoline has an energy density of 121,000 BTU/gal. Assuming the engine and propeller are 20% efficient in converting the energy in the gasoline into thrust energy, how large should the fuel tank be to allow for 2 h of cruising? (Answer: 5.26 gal. We would probably need 6 gal to allow for takeoff and landing.) (Figure 1.11)

2. In the Iditarod, a sled is being pulled by 10 dogs each capable of producing 0.2 hp. It is observed that the sled is traveling along at 4 mph on level ground.

   (a) What is the average drag force on the sled?

   (b) How fast would the sled travel if only 8 dogs were pulling? (Answer: 3.2 mph) (Figure 1.12)

Figure 1.11 Ultralight aircraft.

© Shutterstock

Figure 1.12 Sled with sled dogs.

© Shutterstock

3. A person decides to participate in a 50-mi bicycle fundraiser as a community service project. The event starts at 8:00 in the morning, and the person finishes the event at 3:30 in the afternoon. In the event, all riders are required to take a 1-h break at the halfway point to rehydrate. Moreover, the person will have several short stops along the way for stop signs and other minor reasons. Assume an additional 30 min of stopped time to account for these minor stops.

(a) Subtracting out the break time and the minor stops, what is the average speed of the person while riding?

(b) The drag force on the person comes from aerodynamics and rolling drag. With the speed V in mph, assume that the drag force in pounds is given by the following equation:

$$F = (0.016)V^2 + 1.8 \text{ lb} \tag{1.45}$$

(c) Using the average speed while traveling above, how much energy will be required for the person to travel the 50 mi? Please express your answer in ft-lb.

(d) One Cal is 3088 ft-lb of energy. Our leg muscles are about 20% efficient in converting the sugars into useful work, so to produce the energy above, a person will need to burn about five times that much Cal energy. How many Cal will the person burn during the 50-mi bike ride?

(e) There are approximately 3500 Cal associated with 1 lb of weight. If a person were to make this 50-mi bike ride 10 times during a month, how much weight loss would be associated with riding the bike? (Answer: 3.56 lb) (Figure 1.13)

4. It is observed that a 49cc scooter gets 90 mpg while traveling at 30 mph. If we assume the engine is 22% efficient in converting the heat energy in the fuel to the energy used to propel the scooter along, what is the average drag force on the scooter? Assume the scooter is burning 10% ethanol gasoline. (Answer: 43.6 lb) (Figure 1.14)

5. Driving into a gas station, you notice that the price of the regular unleaded (10% ethanol blend) is $2.29 per gallon, and the E85 gasoline (85% ethanol blend) is priced at $1.99 per gallon. On average, your vehicle has been getting 18.3 mpg using the regular unleaded (10% ethanol) gas.

(a) What is the fuel cost per mile in driving your vehicle using regular unleaded gasoline?

Figure 1.13 Riding bicycles.

© Shutterstock.

Figure 1.14 Scooter.

(b) Assuming your car is designed to be able to use E85 (flex fuel vehicle), what would you estimate your fuel economy to be using E85 gasoline? Based on this number, what is your fuel cost per mile using E85? (Answer: 13.7 mpg)

(c) Which fuel will give you the overall lowest fuel cost per mile?

6. A cruise ship requires 45,000 hp from the engines to propel itself along at 26 mph. The propellers driving the ship are 85% efficient, so the actual power delivered to the water to provide thrust is 38,250 hp.

(a) Calculate the drag force on the cruise ship. Express your answer in pounds.

(b) If the engines are 30% efficient and the fuel has an energy density of 138,700 BTU/gal, how many gallons of fuel are required to travel 300 mi? (Answer 31,750 gal) (Figure 1.15)

Figure 1.15 Cruise ship.

© Shutterstock.

References

1. Munson, B.R. et al., *Fundamentals of Fluid Mechanics*, John Wiley & Sons, 1994. Or any good fluid mechanics textbook.

2. Wikipedia has a good article on human power. https://en.wikipedia.org/wiki/Human_power.

3. Wikipedia article on horsepower. https://en.wikipedia.org/wiki/Horsepower.

4. Wikipedia article about the Gossamer Albatross. https://en.wikipedia.org/wiki/MacCready_Gossamer_Albatross.

5. Moran, M.J. et al., *Fundamentals of Engineering Thermodynamics*, John Wiley & Sons, Hoboken, 2014. Or any good thermodynamics textbook.

6. http://mb-soft.com/public2/humaneff.html.

7. Selected information from the source. https://en.wikipedia.org/wiki/Energy_density.

The first term that needs to be defined is the thermal efficiency of internal combustion engines. The thermal efficiency of the engine is the percentage of the chemical energy in the fuel that is converted into useful work. Most of the chemical energy in the fuel will be converted to heat. A thermal efficiency of 25 to 30% is good for gasoline engines for highway driving. Car companies have made gradual improvements in the thermal efficiency of gasoline engines, and the newest cars have engines that approach 30% thermal efficiency for highway driving. Diesel engines used in passenger cars and trucks typically have a thermal efficiency of 30% for highway driving. The diesel engines used in large trucks, trains and ships can approach a thermal efficiency of 35% under typical cruising conditions and are among the most efficient internal combustion engines in service.

The thermal efficiency of an internal combustion engine is highest when the throttle is open and a lot of power is being generated by the engine. The thermal efficiency of the engine is zero at idle. Thermal efficiency is lower for city driving and higher for highway driving.

Fuel economy is related to, but not the same as thermal efficiency. Fuel economy is normally measured in miles per gallon (mpg). Most newer cars have an instantaneous fuel economy meter on the dash for the driver to view. Since the computer in the fuel-injection system knows the rate at which the car is using fuel [gallons per hour (gph)], and since the car speed is also known by the computer system (miles per hour), the instantaneous fuel economy is calculated as the car speed divided by the fuel consumption.

The instantaneous fuel economy gauge can be misleading. When accelerating, the thermal efficiency of the engine is good, but the fuel economy is poor. We push on the gas pedal (open the throttle) to accelerate the vehicle, and this increases the gph fuel consumption. Because fuel economy is speed divided by fuel consumption, the fuel economy during acceleration is low. Opening the throttle improves the thermal efficiency of the engine, so the thermal efficiency of the engine when accelerating is good.

When coasting, the thermal efficiency of the engine is poor, but the fuel economy is good. When we let off the gas pedal (close the throttle), we reduce the gph fuel consumption of the engine. Because fuel economy is speed divided by fuel consumption, the fuel economy during coasting or braking is high. Closing the throttle reduces the thermal efficiency of the engine, so the thermal efficiency of the engine when coasting or braking is low. Higher thermal efficiency will correspond to better overall fuel economy, but the instantaneous fuel economy readings are misleading.

As the car speeds up, the drag force on the vehicle increases and the engine must produce more power. Higher power output from the engine increases the thermal efficiency, and higher thermal efficiency tends to increase the fuel economy. But having the engine provide more power tends to lower the fuel economy of the vehicle. Achieving the best fuel economy requires that we find a balance between good thermal efficiency and low drag force.

At lower speeds, the improvement in thermal efficiency is more important than the increase in the drag force. For a typical car or truck, driving at 30 mph will yield better fuel economy than driving at 20 mph even though the drag force is higher at 30 mph. The engine has higher thermal efficiency at 30 mph than at 20 mph.

Most cars and trucks driving at 80 mph will yield a lower fuel economy than driving at 70 mph. The engine has higher thermal efficiency at 80 mph, but the increase in drag force at 80 mph more than offsets the improvement in thermal efficiency. Typical cars and trucks get the best fuel economy at 55–60 mph.

Electric motors have zero efficiencies at zero speed, but their efficiency increases rapidly with speed and then levels out at speeds above 15 mph. Electric cars will get their maximum economy (kW-h/mi) at much slower speeds than internal combustion engine-powered cars. For electric cars and trucks, driving slower will always increase the driving range of the vehicle, as long as the vehicle is going 15 mph or faster. If you are driving an electric vehicle and running low on charge, slowing down will allow you to go further before the car needs to be charged.

2.2.
Rolling and Aerodynamic Drag for Wheeled Vehicles

Power is the rate at which a vehicle uses energy. The common units for power are hp, ft-lb/s, W, and kW. For unit conversions, hp = 550 ft-lb/s = 745.7 W = 0.7457 kW. For internal combustion engines, the power is normally given in hp, and for electric motors, the power is normally given in kW. This chapter focuses on wheeled vehicles. The drag forces on wheeled vehicles are rolling resistance and aerodynamic drag. The engine and drive system for the vehicle must provide enough power to overcome the rolling resistance and aerodynamic drag on the vehicle.

Rolling Resistance [1]: The rolling resistance force comes from the interaction of the wheels with the pavement or driving surface. It is proportional to the weight of the vehicle. Other things being equal, the rolling resistance force will double if we double the weight of the vehicle. Rolling resistance force increases as the speed of the vehicle increases, but the primary effect is the weight of the vehicle. The basic equation to calculate the rolling resistance force is:

$$F = (W)(Crr) \tag{2.1}$$

Where F is the rolling resistance force, W is the weight of the vehicle, and Crr is the rolling resistance coefficient.

The rolling resistance coefficient is measured experimentally. For pneumatic tires rolling on pavement, Crr is a function of the type and size of tires used and the air pressure in the tires. This section will focus on using pneumatic tires on the pavement because that is the case for nearly all wheeled vehicles. When driving in sand or gravel or off-roading, the rolling resistance is much higher than on pavement. Trains have steel wheels rolling on steel rails, which results in lower rolling resistance than pneumatic tires on pavement. Trains are an efficient way to move heavy loads.

Pneumatic rubber tires have a lower rolling resistance than solid rubber tires or any other type of wheel we would consider using on cars, trucks, bicycles, motorcycles, scooters, etc. Companies that make tires are always looking for ways to improve the durability, performance, and rolling resistance. Rolling resistance is especially important for the tires used on large trucks. Lower rolling resistance tires allow trucks to get better fuel economy, which lowers shipping costs and allows trucking companies to make more profit.

As a tire rolls on the pavement, the portion of the tire in contact with the pavement is flattened slightly in the area called the contact patch. The rolling motion causes the tire to be flexed back and forth as it goes through the process of being flattened and un-flattened. Rubber has a hysteresis loop, and flexing the tire back and forth causes the tire to heat up. Some of the energy the vehicle uses is converted to heat in the tires because of the mechanics of rolling. The energy lost to heat in the tire is the most common explanation given for rolling resistance. The rolling resistance is low with properly inflated tires but not negligible. It is a significant force and energy loss for wheeled vehicles, especially for heavy vehicles like trucks and trains.

Most of us rode bicycles when we were young and remember that when the bicycle had a low tire, it took more effort to pedal the bicycle than when the tires were inflated. The same thing happens with car and truck tires, except it is more subtle because the engine will have plenty of power to overcome the increase in rolling resistance. Driving a car with a low tire will cause a lot of heat to be put into the tire as you drive and will cause the tire to overheat, fail, and blow out. It is important to keep the tires properly inflated. There are a few statements below that will help you understand why some tires have lower rolling resistance than other tires:

1. High-pressure tires have lower rolling resistance than low-pressure tires. The rolling resistance of a tire will decrease as the tire pressure is increased, even when the pressure is increased beyond the manufacturer's recommended pressure. High-pressure tires are more susceptible to damage and do not perform as well in cornering, braking, and

acceleration as low-pressure tires. Choosing the best tire for a vehicle involves more than just rolling resistance. Using higher-pressure tires will lower the rolling resistance. Racing bicycles use high-pressure tires because they have a very low rolling resistance.

2. Narrow tires tend to have a lower rolling resistance than wider tires. When we turn a vehicle to go around a corner, the turn radius is slightly different for the inside and outside edges of the tires. The difference in turn radius causes scrubbing between the inside and the outside edge of the tire, increasing rolling resistance. Narrow tires will have less scrub.

3. Large-diameter tires have lower rolling resistance than smaller-diameter tires. This has to do with the geometry of flattening the contact patch. The deformation is more severe for small-diameter tires than for large-diameter tires, which results in less energy lost in the deformation and lower rolling resistance.

To quantify the energy and power required to overcome rolling resistance, we need typical values for the rolling resistance coefficient Crr.

Table 2.1 lists typical values for different types of wheeled vehicles [1, 2, 3].

Table 2.1 Typical values for rolling resistance coefficient Crr.

Vehicle	Rolling Coefficient Crr
Racing bicycle	0.004
Electrathon or solar car	0.0055
Motorcycle street bike. The narrower, higher-pressure tires have lower rolling resistance, and the wider, high-performance tires have higher rolling resistance.	0.006–0.010
Large truck (18-wheeler). The newer-style tires have lower rolling resistance.	0.006–0.010
Passenger cars and trucks. The wider, high-performance tires tend to have higher rolling resistance.	0.010–0.012

© SAE International

These values can be used to estimate the rolling resistance at low speeds, but at higher speeds the rolling resistance coefficient increases, and we need a speed correction factor. Some textbooks define two rolling resistance coefficients, one constant and one multiplied by speed as illustrated below:

$$Crr = C_1 + (C_2)(V) \tag{2.2}$$

This approach is great from a theoretical point of view, but in practice it is difficult to get an accurate experimental value for the C_1 term and almost impossible to get an accurate value for the C_2 term. Part of the increase in the rolling resistance coefficient comes from windage, which is an aerodynamic loss in the wheels, but is normally included with the rolling resistance.

For this book, I will use an approximation that the rolling resistance coefficient varies with the term (1 + V/200), where the speed V has units of mph. This is a reasonable approximation for how the rolling resistance of pneumatic tires increases with speed. With this assumption, the rolling resistance force F_r can be calculated as:

$$F_r = (W)(Crr)\left(1 + \frac{V}{200}\right) \tag{2.3}$$

Where W is the weight of the vehicle, Crr is the rolling resistance coefficient of the tires, and V is the speed of the vehicle in mph.

Example 2.1

Suppose a large truck weighs 70,000 lb and is traveling at 70 mph. What is the rolling drag force on the truck, and how much horsepower is required to overcome rolling drag? Use Crr = 0.006.

$$F_r = (70,000\text{ lb})(0.006)\left(1 + \frac{70}{200}\right) = 567\text{ lb} \tag{2.4}$$

Power is the drag force multiplied by velocity. The power required to overcome rolling resistance for the truck is calculated as:

$$\text{Power} = (567\text{ lb})(70\text{ mph})\left(\frac{5280\text{ ft}}{\text{mi}}\right)\left(\frac{\text{h}}{3600\text{ s}}\right) = 58,212\text{ ft-lb / s} = 105.8\text{ hp} \tag{2.5}$$

A loaded 18-wheeler truck has a typical weight of 70,000 lb. The engine/drive will need to provide 105.8 hp to the wheels to overcome rolling resistance on level ground at 70 mph. The engine/drive will also have to overcome aerodynamic drag, which will be covered later in this chapter. When climbing a hill, the engine/drive must provide the power to overcome gravitational loading. The horsepower required to climb hills can be higher than rolling and aerodynamic loading combined. We will cover gravitational loading in Unit 3.

Simplified Equations: In the United States, we measure the weight of the vehicle in lb and the speed in mph. The data for the weight of vehicles and load limits on roads and bridges are usually given in pounds, kips, or tons. Speed limits are posted in mph, and most of us have a feeling for what the speed in mph means. To avoid having to go through the unit conversions every time, we will assume that the weight (W) of the vehicle is in lb and the speed (V) is in mph. Then, we can develop equations with the unit conversions all brought into one constant and simplify the analysis.

The drag force on a vehicle is normally measured in either lb or N. Equations (2.6) and (2.7) are used to get the drag force in units of lb or N. The power a vehicle needs to provide is normally given in W or kW for electric vehicles and horsepower for internal combustion vehicles. Equations (2.8) and (2.9) are used to get the power required in hp or W.

$$\text{Rolling Drag Force} = (W)(Crr)\left(1 + \frac{V}{200}\right)\text{lb} \tag{2.6}$$

$$\text{Rolling Drag Force} = (4.448)(W)(Crr)\left(1 + \frac{V}{200}\right)\text{N} \tag{2.7}$$

$$\text{Rolling Power} = (0.002667)(W)(V)(Crr)\left(1 + \frac{V}{200}\right)\text{hp} \tag{2.8}$$

$$\text{Rolling Power} = (1.989)(W)(V)(Crr)\left(1 + \frac{V}{200}\right)\text{W} \tag{2.9}$$

For emphasis, Equations (2.6) to (2.9) above assume that the weight W is in lb and the speed V is in mph.

Example 2.2

There is a small train ride at the mall for children. The engine weighs 280 lb and holds a 165 lb driver. Three cars weigh 140 lb each and carry up to four children. Assume that the average child weighs 40 lb. The track is 120 ft long on an oval loop. The train travels at 2.2 mph while running. The average rolling resistance is Crr = 0.02 (Figure 2.1).

(a)   Calculate the energy required for the train to make one 120 ft loop around the track.

$$\text{Total Weight} = 280 + 165 + \left(140 + 4\left(40\right)\right)\left(3\right) = 1345\,\text{lb} \tag{2.10}$$

$$\text{Drag Force} = \left(1345\,\text{lb}\right)\left(0.02\right)\left(1 + \frac{2.2}{200}\right) = 27.2\,\text{lb} \tag{2.11}$$

$$\text{Energy} = \left(27.2\,\text{lb}\right)\left(120\,\text{ft}\right) = 3263.5\,\text{ft-lb} \quad \left(\text{Energy per lap}\right) \tag{2.12}$$

Figure 2.1 Small train.

(b) Convert the energy to W-h.

$$\left(3263.5 \text{ ft-lb}\right)\left(\frac{1.356 \text{ J}}{\text{ft-lb}}\right)\left(\frac{\text{W-h}}{3600 \text{ J}}\right) = 1.23 \text{ W-h} \ \left(\text{Energy per lap}\right) \tag{2.13}$$

(c) Calculate the power the train uses while running in W.

$$\text{Power} = \left(1.989\right)\left(1345 \text{ lb}\right)\left(2.2 \text{ mph}\right)\left(0.02\right)\left(1 + \frac{2.2}{200}\right) = 119 \text{ W} \tag{2.14}$$

(d) Based on this analysis, would one large deep-cycle lead-acid battery be adequate to power the train? Assume the battery has 1000 W-h of electric energy for powering the train.

$$\text{Time} = \frac{1000 \text{ W-h}}{119 \text{ W}} = 8.4 \text{ h} \ \ \left(\text{continuous operation}\right) \tag{2.15}$$

$$\text{Laps} = \frac{1000 \text{ W-h}}{1.233 \text{ W-h / lap}} = 811 \text{ laps} \tag{2.16}$$

The answer is that a large deep-cycle lead-acid battery would be adequate to power the train.

Example 2.3

Suppose an adult person is comfortable walking at 3.5 mph for a long period. Assume that the power required for walking at this rate is 50 W, and the power varies linearly with the speed at which the person is walking.

Assume that the person is going to pull a wagon with children in it and the weight of the wagon and children is 120 lb. What is the walking speed when pulling the wagon if the person is limited to 50 W? Assume the rolling resistance coefficient for the wagon is Crr = 0.015 (Figure 2.2).

From the information given in the problem, we can develop an equation relating the walking speed for the person to the power required. The power requirement is 50 W at 3.5 mph and varies linearly with the walking speed. This approach would be accurate over a range of speeds. If the person were to walk very slow or very fast, the power requirement would not vary linearly with speed. Over a range of speeds, the walking power for the person would vary according to the following equation:

$$\text{Walking Power} = \left(50 \text{ W}\right)\left(\frac{V}{3.5 \text{ mph}}\right) \tag{2.17}$$

Figure 2.2 Person pulling wagon.

The rolling power for the wagon is given by:

$$\text{Rolling Power} = (1.989)(120\,\text{lb})(V)(0.015)\left(1 + \frac{V}{200}\right)W \tag{2.18}$$

If we assume the person has 50 W of available power, the equation can be set up and solved as follows:

$$50\,W = (50\,W)\left(\frac{V}{3.5\,\text{mph}}\right) + (1.989)(120\,\text{lb})(V)(0.015)\left(1 + \frac{V}{200}\right)W \tag{2.19}$$

Equation (2.19) is solved to yield V = 2.79 mph. Using the assumptions given in the problem, the person would be able to pull the wagon at 2.79 mph with the same level of effort as walking at 3.5 mph. I am sure the reader recognizes that the person could swing their arms walking but would have to hold on to the wagon while pulling. The walking motion is a little different from the pulling motion and would impact the analysis. The simple analysis done in this problem is a good place to start and helps quantify why it takes more effort to pull the wagon than to walk, or that you must slow down when pulling the wagon if you want to maintain the same level of effort.

Aerodynamic Drag [4]: Aerodynamics and rolling resistance are the two large drag forces on ground vehicles rolling on level ground. Aerodynamics tends to be the more significant force for lightweight, high-efficiency vehicles like bicycles. Rolling resistance tends to be the dominant force for dense, heavy vehicles like large trucks and trains. The aerodynamic drag force is given by the equation below:

$$\text{Aerodynamic Drag Force} = \frac{1}{2}\rho V^2 A Cd \qquad (2.20)$$

Where ρ is the density of air, V is the speed, A is the characteristic area, and Cd is the drag coefficient.

For ground vehicles, we normally use the frontal area for the characteristic area. The frontal area is the projected area seen while standing directly in front of the vehicle.

When doing wind tunnel tests to measure Cd, it is possible to measure the density of the air, the velocity, and the drag force. A characteristic area is assumed, and the equation is used to calculate the drag coefficient. When looking up a drag coefficient in the literature, the number does not have any meaning unless you know the characteristic area used. Drag coefficients are sometimes calculated based on the total surface area of the vehicle, which yields a much different drag coefficient than when the frontal area is used. In this book, we will use the frontal area. Most drag coefficients for ground vehicles in the literature are based on the frontal area.

The power to overcome aerodynamic drag is the drag force multiplied by the speed:

$$\text{Aerodynamic Drag Power} = \frac{1}{2}\rho V^3 A Cd \qquad (2.21)$$

Aerodynamic drag force is proportional to speed squared, but the power to overcome aerodynamic drag is proportional to speed cubed. At low speeds, the aerodynamic power is small, but at high speeds, the aerodynamic power becomes large. The aerodynamic drag coefficient Cd is related to the shape of the vehicle. The ideal shape to minimize aerodynamic drag is the teardrop or torpedo shape as illustrated in Figure 2.3.

Figure 2.3 Airflow around teardrop shape.

© SAE International

To minimize aerodynamic drag, the air should flow smoothly around the shape as illustrated in **Figure 2.3**. When the shape is less than ideal, the airflow streamlines separate from the object, which is often referred to as flow separation, as illustrated in **Figure 2.4**.

Figure 2.4 Airflow around less ideal shapes.

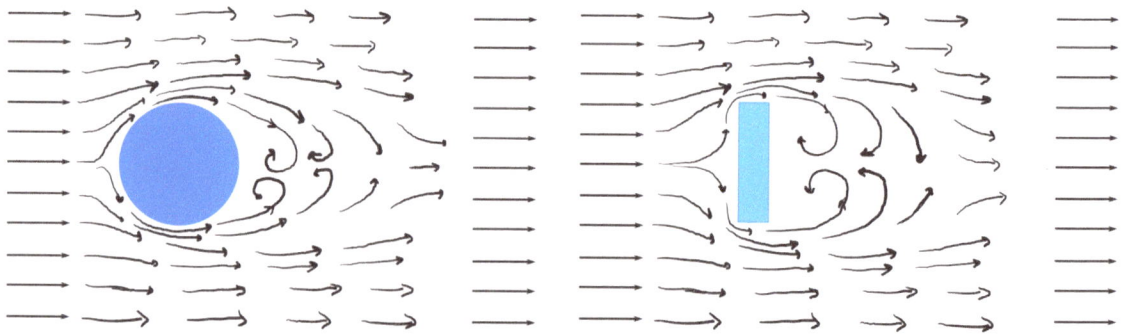

© SAE International

When air flows around a sphere or flat plate, the flow will separate from the object on the backside creating what are called eddy currents. Flow separation creates a low-pressure region behind the object that pulls backward on the object. There is an increase in pressure on the front side of the object pushing backward, and a decrease in pressure behind the object pulling it backward. The low pressure on the backside is the more significant term. Most of the drag on an object comes from the low pressure, or suction pressure, on the backside. The teardrop shape prevents flow separation on the backside and greatly reduces the aerodynamic drag on the object.

These aerodynamic principles are used when designing the shape of most vehicles. The cars with the lowest drag coefficients tend to be rounded on the front and windshield and tapered on the back. There are many considerations in designing the shape of a car. Minimizing aerodynamic drag is important, but it would be impractical to taper the back of the car to a point. A compromise is made between aerodynamic drag, the functionality, drivability, and aesthetics of the car.

For airplanes and jets, the aerodynamic drag is very important, and the shape of the fuselage and wings are designed to have a very low aerodynamic drag. The water drag on a ship hull is also a very important force, and ship hulls are designed to minimize water drag.

One way to look at the aerodynamic drag is to separate it into pressure drag and skin friction drag components. The pressure drag comes from the airflow creating high pressure on the front side of the vehicle and low pressure on the backside of the vehicle. The drag force is then the difference between the front and back pressures multiplied by the frontal area of the vehicle. Pressure drag is the dominant term for bicycles, motorcycles, cars, and trucks, and that is why we use the frontal area as the characteristic area.

Skin friction drag is caused by the frictional (or shear) force between the object and the air. The drag force is proportional to the surface area of the vehicle. Skin friction is the dominant term for vehicles that have an excellent aerodynamic shape like airplanes and jets. Solar cars and electrathon vehicles are designed to have a low aerodynamic drag coefficient, and skin friction drag is the dominant term for these vehicles. When the vehicle has an excellent aerodynamic shape with very little flow separation, the dominant term will be skin friction drag, and it makes sense to use the surface area of the vehicle as the characteristic area.

Example 2.4

Suppose the wind is blowing at 25 mph as you stand outside. The density of air is 1.225 kg/m³. You are 5′ 6″ (1.676 m) tall and your average width is 12 in. (0.3048 m), making your frontal area 0.511 m². Assume that your drag coefficient is Cd = 0.90. What is the force of the wind pushing on you? (or is it pulling on you? ☺) (Figure 2.5)

Figure 2.5 Drag force caused by wind.

© Shutterstock.

We are given all the information to make the calculation. The speed of 25 mph needs to be converted to 11.17 m/s, and then we plug into the formula for aerodynamic drag force:

$$\text{Aerodynamic Force} = \frac{1}{2}\rho\, V^2\, A\, Cd = \frac{1}{2}(1.225)(11.17)^2(0.511)(0.90) = 35.17\ N \tag{2.22}$$

The 35.17 N answer can be divided by 4.448 to yield a force of 7.90 lb. Therefore, a 25-mph wind yields a drag force of approximately 8 lb, and we would notice an 8-lb force pushing on us, but it would not knock us over. At 40 mph wind speed, the force increases to 20.2 lb, which would be more noticeable. You would be leaning into the wind to keep from falling over. If the wind speed is increased to 70 mph, the drag force becomes 61.9 lb, which would make it hard to stand up.

We think of the wind pushing on us, but it is a little more accurate to think of the wind creating a low-pressure area behind us and suction pressure pulling us backward. There is an increase in pressure in front of us pushing us backward, but the low pressure behind us is the more significant term.

Airplane wings are tilted slightly in flight, so the air is directed downward slightly as it exits the rear of the wing. As air flows over an airplane wing, there is a low-pressure region created above the wing and a high-pressure region created below the wing. The net lift on the wing is a combination of the two pressures, but the low pressure above the wing is the more significant term.

The examples and explanations given are to help the reader understand how aerodynamics works. When people first begin studying aerodynamics, they recognize that the leading edge should be rounded to allow air to flow smoothly around the object. However, in the beginning, most people think that once the air gets past the object, it can no longer cause a drag force and they miss the part that the backside of the object should be tapered. Most of the drag on an object comes from the low-pressure region on the backside, and tapering the backside to prevent flow separation is the most important part of reducing the drag on the object. Ideally, we round the leading edge and taper the back to minimize the drag on an object, but tapering the back is the more important part.

Now that we have a basic understanding of aerodynamics, we will need to develop equations to help us do energy and power analysis for cars and trucks. Recognizing that we will measure the speed of the vehicle in mph, we need to get more usable formulas for us to use in calculating the aerodynamic drag force and power.

The density of air varies a little with the temperature and pressure but is a fairly constant value. For this class, we will assume the density of air is 1.199 kg/m³.

In the literature, drag areas are typically given in m². As we use the equations, we will assume that the frontal area of the vehicle is measured in m². The drag coefficient C_d is a dimensionless quantity.

The constants in the formulas include the ½ factor, the density of air, and all unit conversions necessary to allow us to measure the speed of the vehicle in mph and the frontal area in m². The equations below are similar to the equations used to calculate the rolling drag force and power, Equations (2.6) to (2.9).

$$\text{Aerodynamic Drag Force} = (0.1198)\, V^2\, A\, Cd\ N \tag{2.23}$$

$$\text{Aerodynamic Drag Force} = (0.02694)\, V^2\, A\, Cd\ lb \tag{2.24}$$

$$\text{Aerodynamic Power} = \left(0.05357\right) V^3 A \, Cd \, W \qquad (2.25)$$

$$\text{Aerodynamic Power} = \left(7.184 \times 10^{-5}\right) V^3 A \, Cd \, hp \qquad (2.26)$$

Where V is in mph, A is the frontal area in m², and Cd is dimensionless.

For typical modern sedans, Cd is between 0.30 and 0.35. For typical sport utility vehicles (SUVs), Cd is between 0.35 and 0.45 with more rounded SUVs having a lower drag coefficient. For pickup trucks, Cd is between 0.40 and 0.50. These values will help in allowing us to estimate the aerodynamic forces and power requirements of typical vehicles.

Example 2.5

Assume a typical sedan below is loaded with people and luggage so that the total weight is 4500 lb. The aerodynamic drag coefficient is 0.32, and the car will travel on a road trip where the car will average 70 mph. Assume a coefficient of rolling resistance of 0.010 for the tires.

(a) Find the rolling drag force and the aerodynamic drag force (in lb).

(b) Find the rolling power and aerodynamic power (in hp).

(c) If the gasoline engine and drive system in the car have an overall thermal efficiency of 25%, find the fuel economy of the car in mpg.

(d) Suppose a diesel engine and drive system are used with an overall thermal efficiency of 30% and find the fuel economy of the car in mpg (Figure 2.6).

Figure 2.6 Frontal and side areas of a typical sedan.

The first order of business is to estimate the frontal area of the vehicle. There are many places on the Internet to find the width and height of a vehicle, but the true frontal area is seldom given. A reasonable estimate is to assume the frontal area is 70% of the product of the width and height of the vehicle.

The length of the car is not used in estimating the frontal area. To use the equations for aerodynamic drag force and power, the frontal area needs to be in m², and this vehicle is 1.466 m tall and 1.849 m wide. The frontal area is estimated as:

$$\text{Frontal Area} = \left(1.849\,\text{m}\right)\left(1.466\,\text{m}\right)\left(0.70\right) = 1.9\,\text{m}^2 \tag{2.27}$$

The rolling drag force is calculated using Equation (2.6), and the aerodynamic drag force is calculated using Equation (2.24):

$$\text{Rolling Drag Force} = \left(4500\right)\left(0.010\right)\left(1 + \frac{70}{200}\right) = 60.75\,\text{lb} \tag{2.28}$$

$$\text{Aerodynamic Drag Force} = \left(0.02694\right)\left(70^2\right)\left(1.9\right)\left(0.32\right) = 80.26\,\text{lb} \tag{2.29}$$

Rolling power is calculated using Equation (2.8), and aerodynamic power is calculated using Equation (2.26):

$$\text{Rolling Power} = \left(0.002667\right)\left(4500\right)\left(70\right)\left(0.010\right)\left(1 + \frac{70}{200}\right) = 11.34\,\text{hp} \tag{2.30}$$

$$\text{Aerodynamic Power} = \left(7.184 \times 10^{-5}\right)\left(70^3\right)\left(1.9\right)\left(0.32\right) = 14.98\,\text{hp} \tag{2.31}$$

The total power required to overcome rolling power and aerodynamic power is 26.32 hp. The power will need to be provided by the engine. It is convenient to convert hp into units of BTU/s as we work through the calculations. We do not normally think of engine power in units of BTU/s, but, while estimating fuel consumption and fuel economy, it is a convenient unit to use.

$$\left(26.32\,\text{hp}\right)\left(\frac{550\,\text{ft-lb / s}}{\text{hp}}\right)\left(\frac{\text{BTU}}{778.2\,\text{ft-lb}}\right) = 18.60\,\text{BTU / s} \tag{2.32}$$

The engine/drive system must provide 18.62 BTU/s to the wheels to overcome the rolling and aerodynamic drag on the car. The engine/drive system for the car is 25% efficient, so the rate of fuel consumption required to get the 18.62 BTU/s power to the wheels is obtained by dividing by 0.25.

$$\text{Fuel Consumption} = \frac{18.60\,\text{BTU / s}}{0.25} = 74.42\,\text{BTU / s} \tag{2.33}$$

The gasohol fuel (10% ethanol) has 121,000 BTU/gal energy, and there are 3600 s in an hour. The fuel consumption in gph is calculated as:

$$\text{Fuel Consumption} = \left(74.42\,\text{BTU / s}\right)\left(\frac{\text{gallon}}{121,000\,\text{BTU}}\right)\left(\frac{3600\,\text{s}}{\text{h}}\right) = 2.214\,\text{gph} \tag{2.34}$$

The fuel economy of the vehicle is then the speed (70 mph) divided by the fuel consumption (2.22 gph).

$$\text{Fuel Economy} = \frac{70\,\text{mph}}{2.214\,\text{gph}} = 31.6\,\text{mpg} \tag{2.35}$$

A gasoline-powered sedan would achieve a highway fuel economy of 31.6 mpg, which is typical for a midsize sedan. Part d of the problem is to investigate how powering the car with a diesel engine would impact the fuel economy. Diesel fuel has a higher energy content per gallon (138,700 BTU/gal), and the diesel engine will have higher thermal efficiency (30%). The combination of these two things causes the diesel-powered car to have a significantly higher fuel economy in miles per gallon. It still requires 26.32 hp = 18.60 BTU/s to power the car at 70 mph on level ground. The calculations for fuel economy of a midsize sedan are below:

$$\text{Fuel Consumption} = \frac{18.60\,\text{BTU}/\text{s}}{0.30} = 62.01\,\text{BTU}/\text{s} \tag{2.36}$$

$$\text{Fuel Consumption} = \left(62.01\,\text{BTU}/\text{s}\right)\left(\frac{\text{gallon}}{138,700\,\text{BTU}}\right)\left(\frac{3600\,\text{s}}{\text{h}}\right) = 1.61\,\text{gph} \tag{2.37}$$

$$\text{Fuel Economy} = \frac{70\,\text{mph}}{1.61\,\text{gph}} = 43.5\,\text{mpg} \tag{2.38}$$

This example illustrates the difference in using a diesel engine instead of a gasoline engine to power the car. The diesel-powered vehicle will get significantly better fuel economy than the gasoline-powered vehicle. Part of the reason for the better fuel economy is that diesel fuel has more energy per gallon than gasoline. The diesel engine will also have a higher thermal efficiency than the gasoline engine. Lost in this analysis is that the diesel engine will be heavier and more expensive than the gasoline engine. With the added weight, the diesel car will not get quite as good of fuel economy as in this example.

Example 2.6

Suppose a well-conditioned human can produce 70 W of useful power (approximately 0.1 hp). Suppose the person and bicycle together weigh 200 lb, and the rolling resistance coefficient of the tires is 0.005. Assume the frontal area is 1 m², and the drag coefficient is 0.7.

What is the steady-state speed for the person on level ground? (Figure 2.7)

The person is capable of providing 70 W power, and the power will be used to overcome rolling and aerodynamic drag.

$$\text{Power} = \text{Rolling Power} + \text{Aerodynamic Power} = 70\,\text{W} \tag{2.39}$$

$$\left(1.989\right)\left(200\right)\left(V\right)\left(0.005\right)\left(1 + \frac{V}{200}\right) + \left(0.05357\right)\left(V^3\right)\left(1\right)\left(0.7\right) = 70 \tag{2.40}$$

Figure 2.7 Person riding bicycle.

$$V = 10.8 \, \text{mph} \qquad (2.41)$$

This analysis shows that the person would be able to travel along at 10.8 mph average speed. Rolling power is 22.7 W and aerodynamic power is 47.3 W. Notice that even at a moderate speed of 10.8 mph, most of the power goes in overcoming the aerodynamic drag. The aerodynamic drag power goes with the cube of speed so trying to go faster requires a lot more power. Going slower will require significantly less power.

Producing 70 W is a moderate output for an average person. If the equation is solved for 35 W power, the average speed would be 7.92 mph. Producing 35 W is a relatively low power output for an average person. If the person is not in a big hurry, they can ride along at 8 mph without expending a lot of effort.

A well-trained, young adult man can produce approximately 200 W of sustained power. Solving the equation for 200 W yields an average speed of 16.37 mph. At that speed, 164.5 of the 200 W (82%) goes to aerodynamic power. In a bicycle race, the riders will lean over the handlebars to reduce their frontal area, reducing the aerodynamic power required. Reducing aerodynamic drag is the main consideration if you want to go fast on a bicycle. The fastest human-powered vehicles put the rider in a recumbent position to reduce the frontal area and use a fairing around the vehicle to reduce the aerodynamic drag coefficient.

Example 2.7

A team is preparing a solar car track race where the goal is to go as far as possible during an 8-h period. We will assume it is a large oval track, which, for the speeds of the solar cars, can be considered flat and level.

The rules of the race allow the car to carry batteries that will hold 3000 W-h of electric energy to power the car and a solar array that provides an average of 600 W during the 8-h period. The car and driver have a total weight of 600 lb. The tires have a rolling resistance coefficient of 0.0055. The car has a frontal area of 1.2 m² and a drag coefficient of 0.105. The electric motor and drive system have an overall efficiency of 94% in converting the energy from the batteries and solar array to the mechanical energy required to power the car.

Find the power required to drive the car at 25 mph. Using this value, how much energy will the car use during the 8-h day? Will there be enough energy in the batteries and from the solar array to drive all day at 25 mph? (Figure 2.8)

Figure 2.8 Solar car.

The total energy available is 3000 W-h in the batteries and 600 W for 8 h from the solar array, a total of 7800 W-h of energy.

$$\text{Energy Available} = 3000 \text{ W-h} + \left(600 \text{ W}\right)\left(8 \text{ h}\right) = 7800 \text{ W-h} \qquad \text{(2.42)}$$

For the first part of the example problem, we will assume the car is to drive at 25 mph. We calculate the power for driving at 25 mph from the basic parameters of the vehicle.

$$\text{Power} = (1.989)(600)(25)(0.0055)\left(1 + \frac{25}{200}\right) + (0.05357)(25^3)(1.2)(0.105) = 290 \text{ W} \qquad (2.43)$$

The motor/drive system has an efficiency of 94%, so the power to be drawn by the motor from the battery/solar array is 290 W above divided by 0.94.

$$\text{Battery / Solar Power} = \frac{290 \text{ W}}{0.94} = 308.6 \text{ W} \qquad (2.44)$$

If we drive for 8 h, the total energy used for the day can be calculated as:

$$\text{Energy Used} = (308.6 \text{ W})(8h) = 2469 \text{ W-h} \qquad (2.45)$$

Because we have 7800 W-h of energy available for the day and only need 2469 W-h, the car will be able to drive all day. We will not run out of energy, but we will have a lot of energy left over at the end of the day. In solar car racing, the goal is to travel as far as possible during the day, so this would be a poor strategy for the race. We would only travel 200 mi. We should have driven faster and used up more of the energy. The subsequent examples help the reader understand how to work toward an optimum solution to racing the solar car.

Example 2.8

Find the power (in W) required to drive the solar car at 45 mph. Using this value, how much energy will the car use during the 8-h day? Will there be enough energy in the batteries and from the solar array to drive all day at 45 mph? If not, how far will the car have gone when the batteries are dead?

$$\text{Power} = (1.989)(600)(45)(0.0055)\left(1 + \frac{45}{200}\right) + (0.05357)(45^3)(1.2)(0.105) = 977 \text{ W} \qquad (2.46)$$

$$\text{Battery / Solar Power} = \frac{977 \text{ W}}{0.94} = 1039.3 \text{ W} \qquad (2.47)$$

$$\text{Energy Used} = (1039.3 \text{ W})(8 \text{ h}) = 8314 \text{ W-h} \qquad (2.48)$$

The 8314 W-h of energy required exceeds the 7800 W-h of energy available, so it is not possible to drive all day at 45 mph. At some point during the day the batteries will run out of energy, and the car will have to stop. Because the car stops before the 8-h period is complete, the solar array will not have time to gather the 4800 W-h of energy originally calculated. The equation below can be solved to find the time T when the car runs out of energy:

$$\left(1039.3\ W\right)\left(T\right) = 3000\ \text{W-h} + \left(600\ W\right)\left(T\right) \qquad T = 6.830\ h \tag{2.49}$$

The car will need to stop with approximately 70 min remaining in the 8-h race period. It will travel a total of 307.3 mi when it runs out of energy (45 mph * 6.830 h = 307.3 mi). This is a much better strategy than driving at 25 mph all day and only traveling 200 mi, but it is not the optimum strategy. If this strategy were used, the car could sit parked beside the road for around 40 min charging the batteries and then drive for a while until the end of the 8-h driving period. The car would be able to go more than 307.3 mi, but this is still not the optimum strategy for racing.

Example 2.9

What is the ideal target speed for the solar car so it runs out of energy at the finish line and goes as far as possible on the energy available?

To get the optimum performance from the car, we would want to drive steady all day and run out of energy at the end of the 8-h period.

$$\left(\text{Driving Power}\right)\left(8\ h\right) = 7800\ \text{W-h} \tag{2.50}$$

$$\frac{\left[\left(1.989\right)\left(600\right)\left(V\right)\left(0.0055\right)\left(1+\dfrac{V}{200}\right) + \left(0.05357\right)\left(V^3\right)\left(1.2\right)\left(0.105\right)\right]\left(8\right)}{0.94} = 7800 \tag{2.51}$$

$$V = 43.8\ \text{mph} \tag{2.52}$$

The optimum strategy for the race is to drive at 43.8 mph. The car will use all the energy available during the 8-h period and will travel 350.2 mi (43.8 mph * 8 h = 350.2 mi). For an energy-limited race like this, it is the best strategy to drive the car at a steady speed during the racing period.

The next series of examples are designed to illustrate some of the fundamental problems in developing electric cars and trucks. Electric vehicles have been around for more than 100 years, and there are many cases where electric vehicles are the best choice. Electric wheelchairs are good examples. It would not make sense to use an internal combustion engine for a wheelchair. Electric forklifts and other small electric vehicles are used in manufacturing and warehouses. Electric vehicles are clean, are quiet, and have no emissions. The problem has always been the limited range of the vehicle before it needs to be recharged. Prior to the development of lithium-ion batteries the range was too small for cars and trucks. That limitation has been known for more than 100 years. The lithium-ion batteries developed in the last few decades offer the best chance for electric cars and trucks, but even those batteries have their limitations.

Example 2.10

Suppose that our project goal is to design an electric car that can drive 400 mi on one charge. That is, we will charge the batteries and drive 400 mi on the highway before having to recharge the batteries. This would give the car a range similar to most gasoline-powered cars available today. It will take longer to charge the batteries than to fill up with gasoline, but a 400-mi range would be enough for most road trips. It will not be as convenient as a gasoline car for trips longer than 400 mi, but it is a place to start.

For the first iteration, we decide to use lead-acid batteries because they are cheap and available compared to other types of batteries. We find a 12-V, lead-acid battery that has 110 amp-h of charge capacity and weighs 68.2 lb. This is the battery that we will use.

We need to make an estimate of how much the car will weigh to estimate the rolling resistance. A typical midsize gasoline-powered sedan weighs 4000 lb. Using lead-acid batteries will make the car heavier, so as a first guess we will assume a weight of 6000 lb for the car, allowing approximately 2000 lb for the batteries. Assume that low rolling resistance tires are used so that the rolling coefficient is Crr = 0.009. Assume an aerodynamic drag area of 2.68 m² and the drag coefficient is 0.30. Assume that the electric motor/drive system is 90% efficient and that the car will be driven at 70 mph. Estimate the number of batteries required and the weight of the batteries (Figure 2.9).

Figure 2.9 Midsize Sedan.

$$\text{Wheel Power} = (1.989)(6000)(70)(0.009)\left(1+\frac{70}{200}\right)+(0.05357)(70^3)(2.68)(0.30) = 24,923 \text{ W} \qquad (2.53)$$

$$\text{Battery Power} = \frac{24,923 \text{ W}}{0.9} = 27,692 \text{ W} \tag{2.54}$$

To drive 400 mi at 70 mph takes 5.714 h of driving. The total energy required for the trip is the power multiplied by the time.

$$\text{Energy for Trip} = \left(27,692 \text{ W}\right)\left(5.714 \text{ h}\right) = 158,241 \text{ W-h} \tag{2.55}$$

We will use 12-V lead-acid batteries that have a 110 amp-h capacity. We can estimate the energy stored in the battery by multiplying the voltage by the amp-h capacity. This is a high estimate for the battery. It will probably not be able to provide quite this much energy because the voltage will drop below 12 V as the energy is drawn out. If we draw the energy out slowly, the battery will be able to provide close to this much energy.

$$\text{Each Battery} = \left(12 \text{ V}\right)\left(110 \text{ amp-h}\right) = 1320 \text{ W-h} \tag{2.56}$$

$$\text{Batteries Required} = \frac{158,241}{1320} = 120 \text{ batteries} \tag{2.57}$$

At 68.2 lb per battery, this is 8176 lb of batteries to provide the necessary energy to drive the 400 mi. Adding 8176 lb to a 4000 lb car is not realistic. The lead-acid batteries are too heavy. It is not practical to design a vehicle that can travel 400 mi using lead-acid batteries.

Example 2.11

Suppose the car above weighs 4000 lb unloaded, and we plan to load 2000 lb of batteries on it. Assume that we can only use 90% of the rated capacity of the batteries without damaging them. How far can we go at 70 mph before the batteries will need to be recharged?

Because the car has a total weight of 6000 lb, the power required to drive at 70 mph is 27,692 W, as calculated in the previous example. If we can only use 90% of the energy in a battery, the energy per battery can be calculated as follows:

$$\text{The energy in one battery} = \left(12 \text{ V}\right)\left(110 \text{ amp-h}\right)\left(0.9\right) = 1188 \text{ W-h} \tag{2.58}$$

The number of batteries in 2000 lb is:

$$\text{Number of Batteries} = \frac{2000 \text{ lb}}{68.2 \text{ lb / battery}} = 29.3 \text{ batteries} \tag{2.59}$$

The amount of time the car can be driven is the total amount of energy in the battery system divided by the 27,692 W required to power the car.

$$\text{Drive Time} = \frac{\left(1188 \text{ W-h / battery}\right)\left(29.3 \text{ battery}\right)}{27,692 \text{ W}} = 1.258 \text{ h} \tag{2.60}$$

$$\text{Distance} = \left(70 \text{ mph}\right)\left(1.258 \text{ h}\right) = 88 \text{ mi} \tag{2.61}$$

For this example, 33% of the weight of the vehicle was lead-acid batteries. Even so, the vehicle can only travel 88 mi before needing a recharge. This example helps illustrate why electric cars powered by lead-acid batteries were never a viable solution.

In the late 1990s, General Motors developed a small electric car named the EV1 that used lead-acid batteries [5]. The first-generation car came out in 1997 and had a range of 55 mi using lead-acid batteries. The battery capacity was 18,270 W-h and the weight of the battery system was approximately 1000 lb. Total vehicle weight including the batteries was approximately 3000 lb. An optional nickel–metal hydride (NMH) battery system was available that had a range of 105 mi. Lithium-ion batteries suitable for cars were not available in the late 1990s. The second-generation EV1 came out in 1999 with slightly larger battery systems giving a range of 78 mi on lead-acid and 142 mi on NMH. The cars were only available on a lease basis, and in 2003 GM recalled all the EV1 cars and took them out of service.

Information about the EV1 is included to help convince the reader that the calculations are reasonable and accurate. Making 33% of the weight of the vehicle lead-acid batteries, we calculate a range of 88 mi. The EV1 was approximately 33% lead-acid batteries by weight and had a range of 55 mi on the first generation and 78 mi on the second generation.

Example 2.12

Suppose that instead of carrying 2000 lb of lead-acid batteries, we carry 2000 lb of lithium-ion batteries. Lithium-ion batteries have the highest energy density of any batteries commonly available. We find that there is a battery available with a nominal voltage of 12.8 V and 200 amp-h capacity. The battery weighs 60 lb and costs $1750.00. How far can the car go on 2000 lb of lithium-ion batteries? How much will the 5000-lb battery pack cost? Assume that we can use 90% of the theoretical energy available in the batteries. The size of the batteries is 12.75″ long, 6.5″ wide, and 8.7″ tall (Figure 2.10).

The power required is still 27,692 W because the car still weighs 6000 lb. The energy stored in one battery is:

$$\text{The energy in one battery} = \left(12 \text{ V}\right)\left(100 \text{ amp-h}\right)\left(0.9\right) = 1152 \text{ W-h} \tag{2.62}$$

$$\text{Number of Batteries} = \frac{2000 \text{ lb}}{60 \text{ lb / battery}} = 33.33 \text{ batteries} \tag{2.63}$$

Figure 2.10 Lithium-ion battery.

$$\text{Drive Time} = \frac{\left(1152\ \text{W-h / battery}\right)\left(33.3\ \text{battery}\right)}{27{,}692\ \text{W}} = 2.77\ \text{h} \qquad (2.64)$$

$$\text{Distance} = \left(70\ \text{mph}\right)\left(2.77\ \text{h}\right) = 194\ \text{mi} \qquad (2.65)$$

$$\text{Battery Cost} = \left(33.33\right)\left(\$1750.00\right) = \$58{,}333.00 \qquad (2.66)$$

The range of 194 mi is much closer to a reasonable driving range that would be acceptable to people than the 88 mi possible with the lead-acid batteries. If the batteries can be charged quickly, most people would be satisfied with a range of 194 mi. But, if it takes hours to charge the batteries, the car is not suitable for traveling on the highway. Most people will want to travel more than 194 mi in a day when traveling on the highway. The cost of purchasing the batteries is much too expensive to be practical, but car companies will be able to buy the batteries at a much lower cost. The cost will come down as the manufacturing processes are improved. Lithium-ion batteries do not have precious metals or other materials that make them inherently expensive. It is reasonable to think that the batteries can be reduced to a reasonable cost.

Example 2.13

Suppose that instead of carrying 2000 lb of batteries, we carry 2000 lb of gasohol fuel. If the engine/drive is 28% efficient at converting the chemical energy in the fuel to useful mechanical energy, how far can the car go on 2000 lb of fuel?

$$\text{Power} = \left(1.989\right)\left(6000\right)\left(70\right)\left(0.009\right)\left(1+\frac{70}{200}\right) + \left(0.05357\right)\left(70^3\right)\left(2.68\right)\left(0.30\right) = 24{,}923 \text{ W} \tag{2.67}$$

$$\text{Fuel Power} = \left(\frac{24{,}923 \text{ W}}{0.28}\right)\left(\frac{\text{BTU}}{1055 \text{ J}}\right)\left(\frac{3600 \text{ s}}{\text{h}}\right) = 303{,}733 \text{ BTU / h} \tag{2.68}$$

The gasohol fuel has 121,000 BTU/gal, and the density of the fuel is 6.085 lb/gal. 2000 lb of fuel is (2000)/(6.085) = 328.7 gal. The energy stored in the 328.7 gal of fuel is 39.77×10^6 BTU. The time the car can travel is then:

$$\text{Drive Time} = \frac{39.77 \times 10^6}{303{,}733} = 130.9 \text{ h} \tag{2.69}$$

$$\text{Distance} = \left(70 \text{ mph}\right)\left(130.9 \text{ h}\right) = 9166 \text{ mi} \tag{2.70}$$

With 2000 lb of gasohol fuel (328.7 gal), the car could drive across the country three times on one tank of fuel. This series of examples should help the reader understand one of the largest challenges of changing to electric vehicles. It is impossible to make an electric vehicle that has a driving range comparable to vehicles powered by internal combustion engines. Electric vehicles are better than internal combustion-powered vehicles in many ways. They are clean, quiet, and pleasant to drive. Lithium-ion batteries are much better than lead-acid batteries, but they are only marginally acceptable. We are still looking for a better solution.

In recent years, companies have begun experimenting with developing hydrogen-powered vehicles. If hydrogen is used as a fuel, the combustion product is water. If hydrogen is used as the fuel in a traditional internal combustion engine, there will also be some nitrous oxides produced, but it will be much cleaner than burning gasoline or diesel fuel. Currently the cost of hydrogen is too high to have practical hydrogen-fueled internal combustion engines or jet engines. The cost of producing hydrogen will need to come down significantly to make hydrogen-powered vehicles a practical reality.

Hydrogen provides more energy for its weight than gasoline or diesel fuel. But hydrogen must be stored at high pressure, and the weight of the pressure vessel will be greater than the weight of the hydrogen. The weight of the hydrogen + tank will be greater than the weight of the gasoline + tank, but the hydrogen + tank will be lighter in weight than batteries. Refilling a hydrogen tank takes the same amount of time as refilling a gasoline tank. Infrastructure is a problem. There are not many hydrogen fueling stations.

For the hydrogen fuel to make sense, we need an extremely cheap source of nonpolluting electricity. If we use fossil fuels (coal and natural gas) to generate electricity, then the hydrogen fuel idea will create more pollution and carbon emissions than running the vehicles on gas and diesel. There is a lot of discussion in the literature about different sources of hydrogen. Stripping hydrogen from natural gas or coal is less expensive than producing hydrogen by electrolysis of water, but the stripping process produces carbon dioxide as a by-product. Gray hydrogen is produced from coal, blue hydrogen from natural gas, and green hydrogen from electrolysis.

The long-term idea is that solar energy is going to become so cheap that we can afford to use electrolysis to generate hydrogen and have a clean green low-cost energy source. We are not there yet. Some people predict that by 2035 or 2040, the cost of electricity generated using solar energy will be so low that using hydrogen fuel will make sense. If the cost of hydrogen is low enough, it is possible to run internal combustion and jet engines on hydrogen and such engines would have zero carbon emissions.

Fuel cells have chemistry and operational characteristics similar to batteries, but they use up the hydrogen fuel as they produce electricity. To recharge a fuel cell, you fill the tank with hydrogen, similar to filling your gas tank. Fuel cells are typically approximately 60% efficient in converting the energy in the hydrogen to electric energy, which is much better than the 25–30% efficiency of an internal combustion engine. If we have hydrogen fuel, it makes sense to use a fuel-cell electric vehicle rather than an internal combustion engine that runs on hydrogen.

Example 2.14

Reconsider the problem of the 4000 lb car with 2000 lb of batteries or fuel. Consider using a hydrogen fuel-cell electric car with 2000 lb allowed for hydrogen fuel. Assume the fuel cell is 60% efficient in converting the energy in the hydrogen into electricity. Assume the electric motor/drive system is 90% efficient in converting the electric energy into mechanical energy. What is the range of the vehicle?

The power required from the fuel cell is 27,692 W, same as the 6000 lb electric car with a 90% efficient motor/drive system. The fuel cell is 60% efficient in converting the energy in the hydrogen fuel into electricity.

$$\text{Hydrogen Power} = \frac{27,692}{0.60} = 46,154 \text{ W} \tag{2.71}$$

Liquid hydrogen can have an energy density as high as 130 MJ/kg, but the gaseous hydrogen used in automotive applications will have an energy density closer to 115 MJ/kg. 2000 lb of hydrogen fuel will have an energy of:

$$\left(2000 \text{ lb}\right)\left(\frac{115 \times 10^6 \text{ J}}{\text{kg}}\right)\left(\frac{\text{kg}}{2.205 \text{ lb}}\right)\left(\frac{\text{W-h}}{3600 \text{ J}}\right) = 28.97 \times 10^6 \text{ W-h} \tag{2.72}$$

The amount of driving time on the fuel and the range are calculated as follows:

$$\text{Time} = \frac{28.97 \times 10^6 \text{ W-h}}{46,154 \text{ W}} = 627.8 \text{ h} \tag{2.73}$$

$$\text{Distance} = \left(70 \text{ mph}\right)\left(627.8 \text{ h}\right) = 43,945 \text{ mi} \tag{2.74}$$

From the calculations, the range on hydrogen fuel is much better than on gasohol fuel, but the calculations are misleading. The pressure tank required to hold 2000 lb of hydrogen would be very heavy. It would be more realistic to say that the weight of the tank + hydrogen is 2000 lb, and most of the 2000 lb weight would be in the tank. The Toyota Mirai is a hydrogen-fuel-cell car. The tanks on the car are reported to weigh 85 kg and store 5 kg of hydrogen. Approximately 5.56% of the total weight (tank + hydrogen) is hydrogen. A more realistic range estimate for the car above with 2000 lb of (tank + hydrogen) would be 5.56% of the 43,945 mi, which is 2441 mi. This should be compared to the 9166 mi for 2000 lb of gasohol, 194 mi for 2000 lb of lithium-ion batteries, and 88 mi for lead-acid batteries. In calculating the driving range for a car or truck, the following conclusions can be drawn from this analysis:

A.   The hydrogen (tank + fuel) is 9166/2441 = 3.75 times as heavy as gasoline.
B.   Lithium-ion batteries are 9166/194 = 47.2 times as heavy as gasoline.
C.   Lead-acid batteries are 9166/88 = 104 times as heavy as gasoline.

2.3. Trains

Trains are wheeled vehicles but need to be analyzed differently from other wheeled vehicles. Trains are very heavy, and the rolling drag is much greater than the aerodynamic drag. It is reasonable to neglect the aerodynamic drag for typical trains in the United States. Aerodynamics is important for the bullet trains used in other countries. The engine for a light passenger train will typically weigh 120 tons (240,000 lb). A large freight-train engine will typically weigh 250 tons (500,000 lb). Passenger cars typically weigh 40 tons empty and 50 tons loaded. Freight cars vary considerably in weight and can be much heavier than passenger cars.

The pure rolling resistance coefficient for steel wheels rolling on steel rails can be less than 0.001, but that is an idealistic situation. Rolling resistance increases with dirt on the track, sponginess of the ties underneath the rails, contact between the wheels and side of the rails, and corners. A realistic average value for the rolling resistance coefficient is 0.005 to 0.006 for trains, which is slightly better than the best large truck tires. Trains are the most energy-efficient way to move heavy loads over long distances.

Assume a 250-ton engine pulling thirty 100-ton cars at 60 mph for 1200 mi. The average rolling resistance coefficient is 0.006 for the train. The diesel locomotive is 35% efficient in converting the energy in the diesel fuel into useful energy to power the train. Neglect aerodynamic drag. How much horsepower is required to pull the train at 60 mph? How much fuel is required to make the 1200-mi journey?

$$\text{Weight} = \left(250\,\text{tons} + 30(100\,\text{tons})\right)\left(2000\frac{\text{lb}}{\text{ton}}\right) = 6.5 \times 10^6\,\text{lb} \tag{2.75}$$

For this analysis, we will use the horsepower required to pull the train along on the level ground. The engine will need more horsepower to pull the train up grades, but the level ground approach is best for estimating the total energy required to make the 1200-mi trip. Equation (2.8) is used to estimate the rolling power required. Aerodynamic drag is neglected for this analysis:

$$\text{Power} = (0.002667)(6.5 \times 10^6)(60)(0.006)\left(1 + \frac{60}{200}\right) = 8113\,\text{hp} \tag{2.76}$$

$$\text{Time for 1200 mi} = \frac{1200\,\text{mi}}{60\,\text{mph}} = 20\,\text{h} \tag{2.77}$$

The total energy required from the engine is the average power multiplied by the travel time.

$$\text{Energy} = (8113\,\text{hp})(20\,\text{h}) = 162,260\,\text{hp-h} \tag{2.78}$$

The energy unit of hp-h is a commonly used unit for farm tractors and trains. The diesel engine used for the train is 35% efficient in converting the chemical energy in the fuel into useful energy. Fuel energy is calculated as follows:

$$\text{Fuel Energy} = \frac{162,260\,\text{hp-h}}{0.35} = 463,601\,\text{hp-h} \tag{2.79}$$

To get fuel consumption, we need to convert the hp-h into BTU and recognize that diesel fuel has 138,700 BTU/gal. One hp-h of energy is equal to 2544.3 BTU.

$$(463,601\,\text{hp-h})\left(\frac{2544.3\,\text{BTU}}{\text{hp-h}}\right)\left(\frac{\text{gal}}{138,700\,\text{BTU}}\right) = 8504\,\text{gal} \tag{2.80}$$

The train will use 8504 gal of diesel fuel to make the 1200-mi trip. The fuel economy is 0.141 mpg, but it is more relevant to measure the ton-miles per gallon for a train. The ton-miles per gallon accounts for the fact that the train is moving a large amount of freight over the distance.

$$\frac{(3250\,\text{tons})(1200\,\text{mi})}{8504\,\text{gal}} = 459\,\text{ton} - \text{mi / gal} \tag{2.81}$$

2.3.1.
Summary for Wheeled Vehicles

The drag force on wheeled vehicles comes from rolling drag and aerodynamic drag. For light-weight vehicles such as bicycles, motorcycles, and scooters, the aerodynamic drag force is the dominant term. These vehicles have a poor aerodynamic shape, which gives them a high aerodynamic drag force for their size. Rolling drag and aerodynamic drag are both significant factors for cars and trucks. Rolling power tends to be more dominant at lower speeds, and aerodynamic power tends to be more dominant at higher speeds. For very heavy vehicles like trains, rolling resistance is the dominant term.

To get the power to the wheels to overcome rolling drag and aerodynamic drag, the energy must go through the internal combustion engine or electric motor and the drive system. To calculate the fuel power or electric power required, we divide the total of rolling and aerodynamic power by the efficiency of the motor/drive system.

$$\text{Fuel Power or Electric Power} = \frac{\text{Rolling + Aerodynamic Power}}{\text{Motor \& drive efficiency}} \qquad (2.82)$$

The motor/drive efficiency includes the efficiency of the drive system. Significant power is lost in the transmission and final drive systems of a conventional car or truck. The overall efficiency of a conventional gasoline-powered car or truck is 25 to 30% for highway driving. Diesel-powered cars and trucks are closer to 30% efficient, and the best large 18-wheeler trucks and trains have an overall efficiency that can approach 35%. These efficiency values are for highway driving. In city driving, the thermal efficiency of the internal combustion engine is much lower than on the highway.

The motor/drive efficiency for electric cars is typically 85%. Electrathon and solar cars can have efficiencies that approach 95%. Bicycle chain drive systems are essentially 100% efficient. The human powering the bicycle is typically 20% efficient in converting the carbohy-drate energy into useful energy to power the bicycle.

This unit provided the tools for analyzing wheeled vehicles and estimating the energy consumption and overall efficiency of the vehicle. For further depth, the reader should work through the homework problems below.

2.4.
Homework

1. Your shipping company manages a small fleet of 18-wheeler trucks. The tires on the trucks have an average rolling resistance coefficient of 0.0075. There is the opportunity to use a new and improved tire design that will have a rolling resistance coefficient of 0.0072. Assume that the average truck in your fleet weighs 60,000 lb and that you average 5,000,000 mi/year in truck trips. Also, assume that on average the trucks are traveling at 65 mph. Assume that the efficiency of the diesel engine and drive system is 35%.

(a) How much energy per year would you save by going with the new and improved tires? Please express your answer in ft-lb and BTUs.

(b) Assuming an engine efficiency of 35%, how many gallons of fuel per year would be saved by going to the lower rolling resistance tires? (Answer: 16,670 gal)

2. Assume a typical pickup below is loaded so that the total weight is 7000 lb. The aerodynamic drag coefficient is 0.52, and the truck is to travel on a road trip where it will average 70 mph. Assume a coefficient of rolling resistance of 0.010 for the tires.

(a) Find the rolling drag force and the aerodynamic drag force (in lb).

(b) Find the rolling power and aerodynamic drag power (in hp).

(c) If the gasoline engine and drive system in the truck have an overall thermal efficiency of 25%, find the fuel economy of the truck in miles per gallon. (Answer: 16 mpg)

(d) Suppose a diesel engine and drive system are used with an overall thermal efficiency of 30% and find the fuel economy of the truck in miles per gallon. (Answer: 22 mpg) (Figure 2.11)

3. A team is preparing an electrathon vehicle for a contest where the goal is to complete a 100-km race as quickly as possible. The race will be conducted on an oval track which, for the speeds the electrathon vehicle will travel, can be considered flat and level racing.

The rules of the race allow the car to carry batteries that will hold 1000 W-h of electric energy to power the car. The car and driver will have a total weight of 500 lb. BMX tires are used which have a rolling resistance coefficient of 0.006. The car has a frontal area of 0.8 m^2 and a drag coefficient of 0.12. The electric motor and power system have an overall efficiency of 92% in converting the energy in the batteries to the mechanical energy required to power the car.

Find the power (in W) required to drive the car at 30 mph. Using this value, how much energy will be required for the car to complete the full 100 km distance of the race? Will there be enough energy in the batteries to allow the car to complete the race at 30 mph? (Answer: Power = 374.7 W) (Figure 2.12)

Figure 2.11 Dimensions of a typical truck.

Length
5.22 m

Width
2.032 m

Height
1.882 m

© Shutterstock

Figure 2.12 Electrathon vehicle.

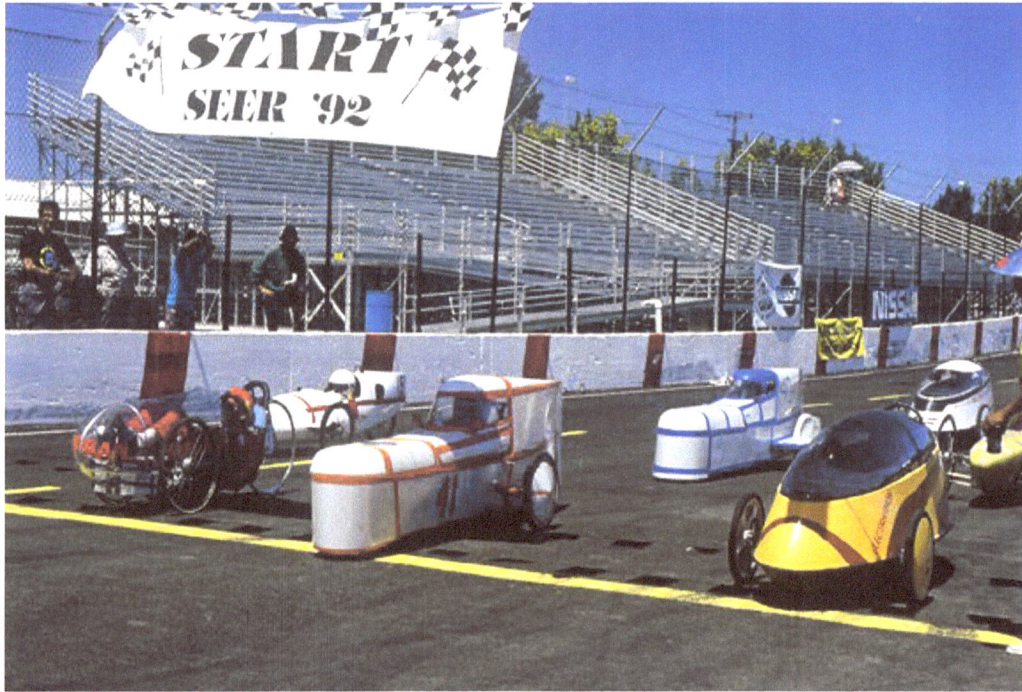

4. A team is preparing an electrathon vehicle for a contest where the goal is to complete a 100-km race as quickly as possible. The race will be conducted on an oval track, which, for the speeds the electrathon vehicle will travel, can be considered flat and level racing.

The rules of the race allow the car to carry batteries that will hold 1000 W-h of electric energy to power the car. The car and driver will have a total weight of 500 lb. BMX tires are used which have a rolling resistance coefficient of 0.006. The car has a frontal area of 0.8 m² and a drag coefficient of 0.12. The electric motor and power system have an overall efficiency of 92% in converting the energy in the batteries to the mechanical energy required to power the car.

What is the optimized speed for the car to allow it to complete the full 100 km in the least time possible, using all energy in the batteries? Assume the car travels at constant speed the entire distance. What is the race time for the 100-km race? (Answer: 38.65 mph)

5. A team is preparing an electrathon vehicle for a contest where the goal is to complete a 100-km race as quickly as possible. The race will be conducted on an oval track, which, for the speeds the electrathon vehicle will travel, can be considered flat and level racing.

The rules of the race allow the car to carry batteries that will hold 1000 W-h of electric energy to power the car. The car and driver will have a total weight of 500 lb. BMX tires are used which have a rolling resistance coefficient of 0.006. The car has a frontal area of 0.8 m² and a drag coefficient of 0.12. The electric motor and power system have an overall efficiency of 92% in converting the energy in the batteries to the mechanical energy required to power the car.

Suppose the team was too conservative in the beginning, traveling at 20 mph for the first 30 km of the race, and realizing they should have been going faster. How much energy have they used in traveling the first 30 km of the race? How much energy is left in the batteries? How fast should they travel during the last 70 km of the race to get to the finish line as quickly as possible? (Answer: 44.45 mph)

6. While testing an electrathon vehicle, it is observed that it uses 700 W when traveling at 42 mph. The vehicle weighs 525 lb, and the tires have a rolling resistance coefficient of 0.0055. The electric motor has an efficiency of 95%. The frontal area of the vehicle is estimated to be 1.16 m². From this information, calculate the aerodynamic drag coefficient for the vehicle. (Answer: 0.0810)

7. Assume that a person plans to ride an electric bicycle across the country, a distance of 3500 mi. The person can provide an average of 50 W of power to the pedals while riding. The battery on the bicycle has a useful capacity of 800 W-h energy. The person schedules 10 h/day for riding, but there will be rest breaks so that the actual riding time will be 8 h/day. Each night the person will charge the battery and then use the battery energy the next day. Assume that the electric motor is 85% efficient in converting the energy in the battery to useful mechanical energy. The rider and bicycle and battery have a total weight of

250 lb, and the big soft tires on the bicycle have a rolling resistance coefficient of 0.008. The frontal area of the bicycle and rider is 0.60 m² and the aerodynamic drag coefficient is 0.75.

(a) What is the average power available while riding the bike (human + battery)?

(b) Using the average power, what is the steady-state speed on level ground? (Answer: 14.48 mph)

(c) Based on the steady-state speed, how many days to make the 3500-mi trip? (Answer: 30.21 days)

(d) How many Cal will the rider burn to power the bicycle on the 3500-mi trip? Please assume that the human muscles are 20% efficient in converting the carbohydrate energy into useful work. Assuming 3500 Cal/lb, how many pounds does this correspond to? (Answer: 14.9 lb) (Figure 2.13)

8. An 80,000-lb truck is driving westward on I-70 across Kansas at a steady speed of 70 mph. Assume that the rolling resistance coefficient of the tires is 0.007, the frontal area of the truck is 7.8 m², and the aerodynamic drag coefficient of the truck is 0.62.

(a) Find the horsepower required to push the truck along at steady speed on level ground.

(b) Assuming the thermal efficiency of the diesel engine and transmission is 32%, find the gallons per hour fuel consumption of the truck.

(c) Calculate the amount of fuel required for the truck to travel 400 mi. If the driver paid $2.15 per gallon, what are the fuel cost per mile for the truck and the total fuel cost for the trip? (Answer: $183.35) (Figure 2.14)

Figure 2.13 Cross-country bicycle trip.

© Shutterstock.

Figure 2.14 Eighteen-wheeler truck.

© Shutterstock

9. We decide to use a small gasoline engine-powered tractor to drag a log through the field to the house where it can be chopped and split for firewood.

(a) The log has a total weight of 800 lb. To guarantee we will have the power required to pull it, we make a high estimate of 0.9 for the coefficient of friction between the logs and the ground. Find the power required to drag the log at 10 kph. Neglect air resistance. (Answer: 8.897 kW)

(b) Assume that for driving through the field the rolling resistance coefficient for the tractor is 0.1 and that the tractor has a weight of 3000 lb. Find the power required to drive the tractor at 10 kph pulling the logs (i.e., add the rolling resistance power for the tractor to the power in (a) above). Neglect air resistance. (Answer: 12.72 kW) (Figure 2.15)

10. An electric vehicle is to be developed to allow people to drive around inside a large warehouse facility. The vehicle will average a speed of 5 mph while driving, and the rolling resistance coefficient for the tires is 0.020. The electric motor is 85% efficient in converting the electrical energy stored in the batteries to useful mechanical energy. Air resistance is negligible for this problem. The ten deep-cycle lead-acid batteries that power the vehicle weigh 65 lb each and have a useful energy capacity of 1 kW-h each. The vehicle weighs 250 lb without batteries, people, or cargo. The vehicle is designed to

Figure 2.15 Tractor pulling log.

carry up to 500 lb of people and cargo load, not counting the batteries.

(a) What is the total weight of the vehicle plus batteries plus a maximum load of people and cargo?

(b) What percentage of the total weight [above in (a)] comes from the batteries?

(c) Assuming the vehicle is fully loaded with 500 lb of people and cargo, calculate the power in W that must be drawn from the batteries to power the vehicle along at 5 mph. (Answer: 335.8 W)

(d) Based on the total energy capacity of the batteries, how long can the vehicle travel around the warehouse before it needs to have the batteries recharged? (Answer: 29.8 h) (**Figure 2.16**)

11. The Toyota Mirai is a fuel-cell vehicle. The car is 195.8 in. long, 74.2 in. wide, and 57.9 in. tall and has a curb weight of 4335 lb. The car has a good aerodynamic design, so assume the drag coefficient is 0.32 based on frontal area. Assume the rolling resistance of the tires is 0.010. The three fuel tanks hold a total of 5.6 kg of hydrogen gas.

Assume that the fuel cell is 60% efficient in converting the energy in the hydrogen fuel into electricity. The electric motors are 90% efficient in converting the electric energy into useful work. The manufacturer rates the car with a range of 357 mi, but does not say whether it is city or highway mi.

Figure 2.16 Small electric vehicle.

© Shutterstock.

Loaded, the car will weigh approximately 5000 lb. Calculate the driving range of the car if it travels at 70 mph. (Answer: 325 mi, Frontal Area is 1.94 m²) (Figure 2.17)

12. Assume a 220-ton engine is pulling 18 cars that weigh 110 tons each at 55 mph for 2000 mi. The average rolling resistance coefficient is 0.005 for the train. The diesel locomotive is 34% efficient in converting the energy in the diesel fuel into useful energy to power the train. Neglect aerodynamic drag. How much horsepower is required to pull the train at 55 mph? How much fuel is required to make the 2000-mi journey? (Answer: 8075 gal) (Figure 2.18)

Figure 2.17 Toyota Mirai.

Figure 2.18 Train.

© Shutterstock.

References

1. Carroll, D., *The Winning Solar Car* (Warrendale: SAE, 2003).

2. https://www.engineeringtoolbox.com/rolling-friction-resistance-d_1303.html.

3. https://en.wikipedia.org/wiki/Rolling_resistance.

4. Munson, B.R. et al., *Fundamentals of Fluid Mechanics*, John Wiley & Sons, Hoboken, 1994. Or any good fluid mechanics textbook.

5. https://en.wikipedia.org/wiki/General_Motors_EV1#cite_note-nimhcurrent-2.

The drag force on a boat or ship is more complicated than for an object moving within a fluid because the boat or ship is riding on the surface of the water. There exists a normal friction drag on the boat hull due to the water flowing against the hull. A poorly designed hull can produce a significant amount of pressure drag on the hull pushing the water up in front of the boat and allowing the flow to separate on the back of the boat. The drag force from water flow around the hull is the dominant drag term. The portion of the boat or ship above the water will experience aerodynamic drag, which is significant but smaller in magnitude than the water drag force.

Boats and ships create waves as they pass through the water, and it takes a significant amount of energy to create the waves. This was studied extensively by William Froude, who developed the Froude number that goes with the drag on the boat or ship. The Froude number was found to be linearly proportional to the speed of the ship in the water and inversely proportional to the square root of the length of the ship. Froude's original work did not include the acceleration of gravity in the definition of the Froude number, but gravity was added later to make the Froude number a dimensionless quantity as is common in the subject of fluid mechanics. The modern definition for the Froude number (Fr) is [1]:

$$Fr = \frac{V}{\sqrt{gL}}$$

(3.1)

Where V is the speed of the boat in the water, g is the acceleration of gravity, and L is the length of the boat or ship at the waterline.

The Froude number is a dimensionless quantity. Boats and ships with similar Froude numbers will create similar waves, assuming a good hull design. The hull should be designed to allow water to flow smoothly around the boat or ship.

3.1.1.
Brief Derivation of the Froude Analysis for Ships

It has been known for at least 150 years that the drag force on a ship is proportional to the weight of the ship and the square of the Froude number. The reader could accept this as fact and skip this derivation. The derivation is included in the book to give the reader a deeper understanding of where the Froude analysis comes from. The basic water drag force on a ship is given by the equation below:

$$\text{Drag Force} = \frac{1}{2}\rho V^2 A\, Cd$$

(3.2)

Where ρ is the density of water, V is the speed of the ship in the water, A is the surface area of the hull in contact with the water, and Cd is the drag coefficient of the hull (based on the surface area of the hull in contact with the water).

For a ship of weight mg, the volume of water displaced must have a weight equal to the weight of the ship. The volume of water displaced is given by the following:

$$\text{Volume Displaced} = \frac{mg}{\rho g} = \frac{m}{\rho}$$

(3.3)

Where m is the total mass of the ship and ρ is the density of water. Seawater has a higher density than freshwater.

The assumption made in the Froude analysis is that there is a relationship between the surface area of the hull in contact with the water (A) and the volume of water displaced by the hull (volume displaced). A shape factor (SF) is defined so that the area (A) is proportional to the volume divided by the length of the hull in contact with the water (L). The equation for the assumption is shown below:

$$A = (SF)\frac{(\text{Volume Displaced})}{L} = (SF)\frac{\left(\frac{m}{\rho}\right)}{L} = (SF)\left(\frac{m}{\rho L}\right)$$

(3.4)

Equation (3.4) would be true for any given ship, but the equation would be of little value if the SF constant is different for different sizes of ships. Equation (3.4) and the Froude analysis are only valuable if the SF is constant for ships that have a similar hull shape.

It is difficult to calculate the SF for real hulls, but we can get an idea of the correct value by looking at an idealized shape. The shape to be considered is the hemisphere. Suppose the hull of the ship in question is a hemisphere. A more precise way of saying this is that the area of the ship in contact with the water is a hemisphere. The area of a hemisphere is $A = 2\pi r^2$, the volume is given by Volume = $(2/3)\pi r^3$, and the length is $L = 2r$. The SF for a hemisphere is 6. That is:

$$SF = \frac{AL}{\text{Volume}} = \frac{(2\pi r^2)(2r)}{\frac{2}{3}\pi r^3} = 6 \qquad (3.5)$$

The SF is constant for all sizes of hemispheres. It does not vary with the radius of the hemisphere. While this is not proof, it can be shown mathematically that the SF is constant for a given hull shape. Once a hull shape is defined, making the hull larger or smaller does not change the SF.

Different hull shapes will have different SFs, but if we assume the ship has a good hull shape, the SF will be fairly constant from ship to ship.

Equation (3.4) is substituted into Equation (3.2) to yield the following:

$$\text{Drag Force} = \frac{1}{2}\rho\, V^2\,(SF)\left(\frac{m}{\rho L}\right)Cd \qquad (3.6)$$

The constants can be gathered at the beginning of the equation. Using the definition of the Froude number in Equation (3.1), Equation (3.6) can be rewritten as:

$$\text{Drag Force} = \left[\left(\frac{1}{2}\right)(SF)(Cd)\right]mg\, Fr^2 \qquad (3.7)$$

Equation (3.7) says that the drag force on a ship will be linearly proportional to the weight of the ship (mg) and proportional to the square of the Froude number. The constant term is determined empirically. Typical ship hulls have a SF of about 40, and the drag coefficient for perfectly smooth ship hulls is typically 0.0045 [2]. Using SF = 40 and Cd = 0.0045 the constant term in Equation (3.7) would be 0.09 for a perfectly smooth hull. In looking at data for ships, I adjusted the constant to 0.113 to be more realistic for ships with a little fouling on the hull. It is a simple model allowing for an estimate of the drag force on a ship, but the reader should understand it is a rough estimate.

A detailed discussion of the hydrodynamics of boats is beyond the scope of the book. The water drag force on a boat or ship is proportional to the weight of the boat or ship, so in some ways, it is similar to rolling resistance on a wheeled vehicle. For lower speeds, the water drag force increases with the square of the speed, making it similar to aerodynamic drag. As the boat speeds up, it begins to plane on top of the water and the drag force no longer increases with the square of speed. There is a transition region as the boat approaches and surpasses the hull speed, as defined below:

$$\text{Hull Speed} = V_{Hull} = (0.3985)\sqrt{gL}$$ (3.8)

Boats and ships that travel at a speed below approximately 85% of the hull speed will have a drag force that can be approximated by Equation (3.7). The large ships on the ocean always travel at speeds well below their hull speed. Motor boats on a lake and most fishing boats and yachts on the ocean travel at speeds well above the hull speed, and Equation (3.7) overestimates the drag on the boat. The model below will be used to estimate the water drag force on boats and ships in this book. The model assumes the boat or ship has a good hydrodynamic shape so that the water flows smoothly over the hull. For barges with square noses and sterns, the model will underestimate the drag force. For most boats and ships, the model and process below will allow the reader to make a reasonable estimate of the energy and power required for the vessel.

1. Calculate the Froude Number. The first step is to calculate the Froude number for the boat or ship. In the literature, it is common to list the length of the boat or ship but not common to list the waterline length. The waterline length is typically about 90% of the length, but you may need to use your judgment for the boat or ship in question. If you know the waterline length, use it in calculating the Froud number. If not, assume the waterline length is 90% of the length of the boat or ship. Use the waterline length for the boat or ship as it sits statically at rest in the water. Do not use the waterline length of the boat as it is planing on top of the water.

2. Calculate Hull Speed [Equation (3.8)]. The hull speed is the speed where the bow and stern of the boat or ship are riding on the crests of the transverse waves generated by the boat or ship. The waterline length of the boat or ship is equal to the wavelength of the waves generated when the boat or ship is traveling at hull speed. The hull speed is a limiting speed for most human-powered boats and large ocean ships. Achieving hull speed requires a large amount of power compared to traveling at 85% of hull speed. Olympic athletes, in specially designed boats, can exceed the hull speed but for most human-powered boats, the human cannot provide enough power to achieve hull speed. For large ships, the power to achieve hull speed is so high that it is impractical and cost-prohibitive for ships to travel at hull speed. Motorboats and jet skis routinely travel much faster than the hull speed because they have the horsepower to do it.

3. $0 < V < (0.85)\,V_{hull}$ is low-speed operation. Drag is proportional to the weight of the boat or ship and the Froude number squared. Most of the ships and boats used for transportation of freight and people operate in this range of speeds. It is the most important part of the model as far as energy consumption and energy efficiency of boats and ships. Estimate the drag force and power using the equations below.

$$\text{Drag Force} = (0.113)\,mg\left(Fr^2\right)$$ (3.9)

$$\text{Drag Power} = (\text{Drag Force})V \qquad \qquad (3.10)$$

If the ship or boat is operating at a speed below 85% of the hull speed, Equations (3.9) and (3.10) give a reasonable estimate of the drag force and power required. For speeds above 85% of hull speed, the boat will begin to rise in the water and plane out. As the boat planes out, the SF in Equation (3.5) changes significantly. The mass of water displaced by the hull is less than the mass of the boat. The assumptions made in deriving Equations (3.9) and (3.10) are invalid, and the equations overestimate the drag force and power required.

4. $V > V_{hull}$. Motorboats operate at speeds well above the hull speed. The analysis below is a method for estimating the drag force and power requirement for boats operating at speeds above the hull speed, after the boat has planed and is traveling on top of the water. Boat hulls are designed to operate over a range of speeds. It is assumed that the boat is being operated within the speed range for which it was designed. There is a relationship between the weight of the boat, the speed, and the shaft horsepower (SHP). This relationship is shown graphically in Figure 3.1.

Graphical estimations can be made using Figure 3.1. If any two of the parameters are known, the third can be estimated from the figure by drawing a line connecting the two known parameters. The example in the figure is for a boat that has a weight of 3000 lb and is traveling at a speed of 36 knots. From the graph, the engine will need to deliver 90 hp to the shaft(s) driving the propeller(s) to push the boat along. It is assumed the boat is using a propeller or propellers that have been designed for the boat and speed. Everything about the analysis assumes that the boat and power system have been properly designed for the operating speed and that the boat is cruising at a constant speed.

In Figure 3.1, SHP is the power provided to the shaft driving the propeller (or screw) and must be expressed in hp. Weight (W) is the weight of the boat in lb. Speed (S) is the speed of the boat in knots. Equation (3.11) is convenient for doing the calculations.

Figure 3.1 Planing hull speed, displacement, and shaft horsepower relationship.

Planing Hull Speed, Displacement and Engine SHP Relationship

$$\text{SHP} = \left(8.1785 \times 10^{-6}\right) W^{1.04} S^{2.20} \tag{3.11}$$

This model is simple and useful in making an estimate of the power required for the boat. It can be used to study how power varies with the speed and weight of the boat. The power from the engine must go through a drive system to get to the shaft driving the propeller. There will be a small amount of power lost in the drive system. Engine power will need to be higher than SHP. The shaft drives a propeller (screw), and the propeller will be less than 100% efficient. Some propellers are more efficient than others. The model does not account for propeller efficiency.

The model will give a rough estimate of how SHP varies with the weight and speed of the boat.

In some cases, the manufacturer will give the top speed for the boat, and if that information is available, the model can be improved. Set SHP to the maximum power from the engines, W the weight of the boat as tested for maximum speed, and S the maximum speed of the boat. Solve for the constant C in Equation (3.12) to get the constant that works for the boat in question. This process is illustrated in problem 4 at the end of the chapter.

$$\text{Constant} = C = \frac{\text{SHP}}{W^{1.04} S^{2.20}} \tag{3.12}$$

Each boat will have a different constant, though there should not be a lot of difference from boat to boat. If the constant C can be developed for the boat in question, then Equation (3.13) will yield

an accurate model for the boat over a wide range of speeds. The model assumes that the boat is going fast enough to plane out on top of the water, that is, above the hull speed for the boat.

$$\text{SHP} = C \, W^{1.04} S^{2.20} \tag{3.13}$$

Model Summary: For boats and ships that travel below 85% of hull speed Equations (3.9) and (3.10) are used to estimate the drag force and power required to push the vessel through the water. For boats traveling above the hull speed, Equation (3.11) is used to estimate the shaft hp required to push the boat through the water.

Boats traveling near the hull speed do not fit either model, as illustrated in the example below. The models presented in this chapter are simple and give a good estimate of the power requirements for boats and ships. They provide a good starting point for people interested in the energy efficiency of marine vessels.

Example 3.1

The Olympic record for a single person rowing 2000 m is 5 min, 36.6 s. The single scull boat used was 27 ft long and had a weight of about 70 lb. The rower weighed about 180 lb. Use the method above to estimate the drag force on the boat and the power required (Figure 3.2).

An Olympic athlete in a specially designed boat like this can provide enough power to exceed the hull speed of the boat. The average speed of the boat is calculated as:

$$\text{Speed} = \frac{2000 \, \text{m}}{5\left(60\right) + 36.6 \, \text{s}} = 5.94 \, \text{m} / \text{s} \tag{3.14}$$

The waterline length of the hull will be approximately 90% of the 27 ft = 24.3 ft = 7.407 m. The hull speed can be calculated from Equation (3.8) as:

$$\text{Hull Speed} = \left(0.3985\right)\sqrt{\left(9.81\right)\left(7.407\right)} = 3.4 \, \text{m} / \text{s} \tag{3.15}$$

Figure 3.2 Rowing a single scull.

The speed of the scull is greater than 85% of the hull speed, and the weight and speed of the boat are known. Equation (3.11) is used to estimate the power the rower must provide. The weight of the boat and rower is 250 lb. The speed is 5.94 m/s, which must be converted to knots (1 m/s = 1.94384 knots). The speed is 11.55 knots. Going through the calculations yields:

$$\text{SHP} = \left(8.1785 \times 10^{-6}\right)250^{1.04}\,11.55^{2.20} = 0.555\,\text{hp} = 414\,\text{W} \tag{3.16}$$

In this example, there is no motor, and shaft power is being interpreted as rowing power. The reader should recognize that this is a good way to estimate the power the rower needs, but it is a rough estimate. For this example, the rower would need to provide 414 W for 5 min and 36.6 s. It would take an Olympic athlete to provide that much power. Providing 414 W is a lot of power for a human being, but it is possible. Assuming the efficiency of the oars in the water is comparable to a propeller, this is a good estimate of the power the rower must provide. The drag force on the scull is equal to the power divided by the speed. For the Olympic athlete, the drag force is:

$$\text{Drag Force} = \frac{414\,\text{W}}{5.94\,\text{m / s}} = 69.7\,\text{N} \tag{3.17}$$

As a follow-up to this question, we could explore how long it would take a more typical athletic person to row the scull through the 2000 m course. Most young men can train up to providing about 200 W power. Assuming a weight of 250 lb and a power of 200 W, Equation (3.11) can be used to estimate the speed of the scull. Going through the calculations, the speed is 8.3 knots = 4.27 m/s, which is still above the hull speed. It would take a more typical athlete 7 min, 48.4 s to travel the 2000 m course.

A less athletic person would be able to provide about 70 W power. Assuming 250 lb weight and 70 W power, Equation (3.11) yields a speed of 5.15 knots = 2.65 m/s, which is below the hull speed of 3.4 m/s. The model is not valid when the boat is traveling below the hull speed, so the result is invalid. Equations (3.9) and (3.10) can be applied if the speed is below 85% of the hull speed. The mass is 250/2.205 = 113.4 kg.

$$70\,\text{W} = \left(0.113\right)\left(113.4\right)\left(9.81\right)\left(\frac{V^2}{\left(9.81\right)\left(7.407\right)}\right)V \tag{3.18}$$

Solving Equation (3.16) yields a speed of 3.43 m/s, which is just above the hull speed of 3.4 m/s. Since the speed is above 85% of hull speed, the result for this model is invalid. Neither model works for this case. The speed of the boat is probably between the two results of 2.65 and 3.43 m/s. This example is presented to illustrate that when the speed is near the hull speed neither of the models presented in this chapter yields a valid solution. It is difficult to estimate power requirements for a vessel traveling near the hull speed. Ships travel at speeds well below the hull speed, so Equations (3.9) and (3.10) yield good estimates for ships. Motor boats and yachts travel at speeds well above the hull speed, so Equation (3.11) yields good estimates for motor boats and yachts. The models presented in this chapter work for the vast majority of boats and ships, but do not work for vessels traveling near the hull speed.

Example 3.2

The cruise ship Oasis will carry 6298 passengers and has a reported fuel economy of 14.4 passenger mpg of diesel fuel. The ship is 360 m long and weighs 100,000 metric tons. It is reported to have 60 MW of power for propulsion. The cruising speed is 26 mph. Assume that the engines and propeller (screw) system is about 25% efficient in converting the chemical energy in the fuel into useful mechanical thrust for the ship.

(a) Use Equations (3.9) and (3.10) to estimate the drag force and cruising power for the ship.

(b) What is the fuel economy in mpg? In feet per gallon? Assume the fuel has 138,700 BTU/gal.

(c) Most of the fuel energy is used to provide electricity for passengers and amenities. Based on 6298 passengers and 14.4 passenger mpg, what fraction of the fuel is used for propulsion? (Figure 3.3)

Large ships travel at speeds below 85% of the hull speed with few exceptions, but it is necessary to calculate the ship speed and hull speed to ensure the ship is traveling at a speed below 85% of the hull speed. A speed of 26 mph converts to 11.62 m/s. The waterline length is estimated at 90% of the total length of the ship.

$$\text{Waterline Length} \, L = (0.9)(360) = 324 \, \text{m} \tag{3.19}$$

$$\text{Hull Speed} = (0.3985)\sqrt{gL} = (0.3985)\sqrt{(9.81)(324)} = 22.46 \, \text{m/s} \tag{3.20}$$

Figure 3.3 Cruise ship Oasis.

The ship's speed of 11.62 m/s is less than 85% of the hull speed. Equations (3.9) and (3.10) will be used to calculate the drag force and power for the ship. The Froude number is calculated as:

$$Fr = \frac{11.62}{\sqrt{(9.81)(324)}} = 0.20611 \qquad (3.21)$$

The mass of the ship is 100,000 metric tons, which is 100×10^6 kg. The drag force and power required are calculated as:

$$\text{Drag Force} = (0.113)(100 \times 10^6)(9.81)(0.20611^2) = 4.709 \times 10^6 \text{ N} \qquad (3.22)$$

$$\text{Drag Power} = (4.709 \times 10^6)(11.62) = 54.7 \times 10^6 \text{ W} = 54.7 \text{ MW} \qquad (3.23)$$

According to the specifications, the ship has 60 MW of power available for propulsion. The analysis in this chapter yields a reasonable estimate for the drag force and propulsion power at cruising speed. Based on this estimate, we can calculate the fuel consumption and fuel economy of the ship. If the product of engine/drive efficiency and propeller efficiency is 25% as stated in the problem, the fuel consumption can be calculated as:

$$\text{Fuel Usage} = \frac{54.7 \times 10^6 \text{ W}}{0.25}\left(\frac{3.412 \text{ BTU / h}}{W}\right)\left(\frac{gal}{138,700 \text{ BTU}}\right) = 1300 \text{ gph} \qquad (3.24)$$

$$\text{Fuel Economy} = \frac{26 \text{ mph}}{1300 \text{ gph}} = 0.0200 \text{ mpg} = 106 \text{ ft / gal} \qquad (3.25)$$

In the literature, large ships are reported to have fuel economy in the range of 50 to 150 ft/gal for propulsion depending on the size and weight of the ship. The Oasis is a cruise ship and must provide electricity for lights, air conditioning, and entertainment for the passengers. The ship also spends long periods docked in a port using no energy for propulsion, but still consuming energy for the passengers. It turns out that propulsion is less than half of the energy consumption for the ship, as is true for a typical cruise ship.

The literature on the Oasis reports that it averages 14.4 passenger mpg of fuel consumed. The ship holds 6298 passengers when fully loaded. If we divide the passenger mpg by the number of passengers, we can obtain the average fuel economy for the ship.

$$\text{Average Fuel Economy} = \frac{14.4}{6298} = 0.002286 \text{ mpg} = 12.07 \text{ ft / gal} \qquad (3.26)$$

In traveling a mile, the ship would use (5280/106) = 50 gal of fuel for propulsion. The total fuel consumption, on average, for 1 mi traveled would be (5280/12.07) = 437 gal because of the fuel being used for things other than propulsion. A large fraction of the energy is being consumed while the ship is in port. Overall, the ship uses about 11.4% of the fuel for propulsion and 88.6% for other things.

For a 300-mi journey at 26 mph, the ship would use (300 mi) (50 gal/mi) = 15,000 gal of fuel for propulsion. In 2024, the average price for bunker fuel was approximately $2.50 per gallon. The cost for a 300-mi trip would be $37,500.00. Dividing by the 6298 passengers yields a cost of $5.95 per passenger. The cost of propulsion is an insignificant part of the cost of taking a cruise.

If we factor in an average of 437 gal/mi, the total fuel cost is $327,750.00. Dividing by the 6298 passengers on board, the cost is $52.04 per passenger. That is still a minor part of the cost of taking the cruise.

Example 3.3

A Formula 240 Bowrider is powered by a 320-hp gasoline engine. The boat is 31 ft long and has an empty weight of 5000 lb. The cruising speed for the boat is 32 mph and is rated to use 11 gph of fuel at that speed. Assume that the boat is carrying 1500 lb of people and gear.

(a)   Calculate the shaft power to move it along at 32 mph.

(b)   Compare the calculations to the fuel consumption rate of 11 gph. Assume that the engine/drive and propeller are 20% efficient overall in converting the energy in the fuel to useful thrust. Assume the gasoline has 121,000 BTU/gal (Figure 3.4).

The speed of 32 mph is converted to 14.30 m/s. The waterline length is estimated as 90% of the total 21 ft length of the boat, which is 8.504 m. The hull speed is calculated as 3.64 m/s, which means the boat speed is greater than 85% of the hull speed. Equation (3.11) will be used to analyze the boat. The total weight of the boat is 6500 lb and the speed is 14.30 m/s = 27.80 knots.

$$SHP = \left(8.1785 \times 10^{-6}\right) 6500^{1.04} \, 27.80^{2.20} = 113.5 \, hp \qquad (3.27)$$

Figure 3.4 Formula 240 Bowrider.

The calculations show that the shaft power is 113.5 hp is required for the boat to cruise at 32 mph. The engines are rated at 320 hp, which is significantly higher than the 113.5 hp calculated to power the boat. There will be power lost in the drive system, but the engines have enough power for the boat to cruise at 32 mph. The propellers attached to the engines will be less than 100% efficient, so the actual thrust power for the boat will be less than 113.5 hp.

There is another way to estimate the power requirement of the engines from the information given. The information says the boat will use 11 gph gasoline when cruising at 32 mph. If we assume the engine/drive and propeller have an efficiency of 20% and that the fuel has an energy content of 121,000 BTU/gal:

$$\text{Power} = \left(11\,\text{gph}\right)\left(121{,}000\,\frac{\text{BTU}}{\text{gal}}\right)\left(0.20\right)\left(\frac{\text{hp}}{2545\,\dfrac{\text{BTU}}{\text{h}}}\right) = 104.6\,\text{hp} \tag{3.28}$$

The answers of 113.5 and 104.6 hp are in reasonably good agreement when you consider the assumptions made in the analysis. If we had assumed the engine/drive and the propeller was 21.7% efficient instead of 20% efficient, the answers would be in exact agreement.

3.1.2.
Propellers

Propellers accelerate the flow of fluid. The speed of the air or water approaching the propeller is less than the speed of air or water exiting the propeller. Conservation of mass requires that the intake area be larger than the exhaust area. Referencing Figure 3.5, $V_1A_1 = V_2A_2$, so if $V_2 > V_1$, then $A_1 > A_2$. V_1 is the speed of the vehicle.

Figure 3.5 Flow through a propeller.

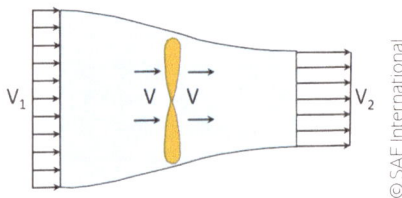

© SAE International.

Figure 3.5 illustrates how the intake area must be greater than the exhaust area because the exhaust velocity is greater than the intake velocity. There is no physical boundary around the propeller. The boundary illustrates the control volume and streamlines.

The derivation by Froude starts with conservation of mass. It is assumed that the area at the propeller is πR^2, where R is the radius of the propeller. V_1 is the intake velocity on the left side of the control volume, V is the velocity at the propeller, and V_2 is the velocity as the flow exits the control volume. The fluid is assumed to be incompressible. Conservation of mass says that the mass flow rate must be the same at all three points, and for an incompressible fluid mass flow is the product of density, area, and velocity of the flow.

$$\rho V_1 A_1 = \rho V_2 A_2 = \rho V\left(\pi R^2\right) \tag{3.29}$$

For the conservation of momentum equation, there is a thrust force F acting on the propeller. The thrust force F is what propels the airplane or ship. Applying the conservation of momentum principle to the fluid it follows:

$$F = -V_1 \rho V_1 A_1 + V_2 \rho V_2 A_2 \tag{3.30}$$

Substituting the conservation of mass equation into the conservation of momentum equation yields the following equation for the thrust force F:

$$F = \rho V \left(\pi R^2 \right) \left(V_2 - V_1 \right) \tag{3.31}$$

At this point in the derivation, an assumption or approximation must be made to develop a useful formula. Froude assumed that the velocity immediately before and after the propeller was equal to the flow velocity through the propeller. This leads to the equation below, though an equivalent way to do the derivation is to simply assume the equation below is correct:

$$V = \frac{V_1 + V_2}{2} \tag{3.32}$$

Equation (3.32) can be substituted into Equation (3.31) to yield the thrust equation. V_1 is the velocity of the airplane or ship in the fluid. V_2 is the velocity of the fluid exiting the control volume relative to the airplane or ship. V_2 is not an absolute velocity. It is the velocity relative to the vehicle (airplane or ship).

$$F = \frac{1}{2} \rho \left(\pi R^2 \right) \left(V_2^2 - V_1^2 \right) \tag{3.33}$$

Equation (3.33) can be used to calculate the thrust force the propeller provides to the vehicle. The power provided to the vehicle is the thrust force multiplied by the vehicle velocity, which is V_1 in this case.

$$\text{Power Provided to Vehicle} = F\, V_1 \tag{3.34}$$

The power being consumed by the propeller is the thrust force F multiplied by the velocity of the fluid through the propeller V. Since V is always greater than V_1, the power required by the propeller is always more than the power provided to the vehicle. A thrust efficiency (e_{prop}) for the propeller is defined as the ratio of the power provided to the vehicle divided by the power provided to the propeller.

$$e_{prop} = \frac{F\, V_1}{F \dfrac{V_1 + V_2}{2}} = \frac{2\, V_1}{V_1 + V_2} \tag{3.35}$$

The efficiency definition in Equation (3.35) is useful once the vehicle gets to cruising speed. It gives a way to calculate the efficiency of the propeller discussed in previous sections. Unfortunately, the definition says that the propeller efficiency is zero when the velocity of the vehicle is zero, such as when an airplane begins accelerating down the runway on takeoff. Mathematically, the efficiency is zero when the velocity of the vehicle is zero, but the definition does not fully capture what we mean by efficiency in that case. The propeller provides the thrust necessary to accelerate the vehicle to speed and is very important in making the vehicle functional. To say the efficiency is zero when the vehicle velocity is zero does not fully capture what is happening with the vehicle. But this is the definition that is used for propeller efficiency.

Example 3.4

An 18-ft-long, 2500-lb bass boat is traveling at 37 mph. Use the method from the previous section to estimate the drag force on the boat. The drive system is 92% efficient in transmitting the engine power to the shaft driving the propeller. The propeller has a diameter of 14 in. Find the power that the engine must provide to push the boat along at 37 mph. Also find the thrust power for the boat. Assume the density of water is 1000 kg/m³ (Figure 3.6).

Figure 3.6 Bass boat.

© Shutterstock.

The 37 mph speed of the bass boat is greater than 85% of the hull speed. The weight and speed of the boat are given, and we want to calculate the engine power required. Equation (3.11) is used to estimate the shaft power required. The speed of 37 mph is converted to 32.15 knots. If the weight of the boat is 2500 lb, and the speed is 32.15 knots, SHP is calculated as:

$$SHP = \left(8.1785 \times 10^{-6}\right) 2500^{1.04} \; 32.15^{2.20} = 57.85 \, hp \qquad (3.36)$$

Since the drive system is 92% efficient in transmitting the engine power to the shaft driving the propeller, the engine power is calculated as:

$$Engine \; Power = \frac{57.85 \, hp}{0.92} = 62.88 \, hp \qquad (3.37)$$

The thrust power provided to the boat is less than the SHP because the propeller will not be 100% efficient. The engine will need to provide 57.85 hp to the propeller, but the thrust power provided to the boat will be less than 57.85 hp. The problem is solved below using metric units. The speed V_1 = 37 mph is converted to 16.5369 m/s. The propeller radius of 7 is converted to 0.1778 m. The shaft power is equal to the drag force F on the boat multiplied by the velocity of the fluid at the propeller:

$$\left(57.85 \, hp\right)\left(\frac{745.7 \, W}{hp}\right) = F\left(\frac{16.5369 + V_2}{2}\right) \qquad (3.38)$$

Equation (3.33) is used to calculate the drag force F:

$$F = \frac{1}{2}\left(1000\right)\left(\pi\right)\left(0.1778^2\right)\left(V_2^2 - 16.5369^2\right) \qquad (3.39)$$

Equations (3.38) and (3.39) are solved to find V_2 = 17.994 m/s and F = 2499 N. The thrust power that the propeller is providing to the boat is the drag force F multiplied by the speed V_1:

$$Thrust \; Power = \left(2499 \, N\right)\left(16.5369 \, m \, / \, s\right) = 41,318 \, W = 55.4 \, hp \qquad (3.40)$$

In summary, the engine produces 62.88 hp. Loss in the drive system causes 57.85 hp to be delivered to the shaft driving the propeller. Because of inefficiencies in the propeller the thrust power provided to push the boat through the water is 55.4 hp. The efficiency of the propeller is 55.4/57.85 = 95.8%.

Example 3.5

Suppose we are planning to use a pair of 7000-hp diesel engines to power a ship that will cruise at 25 mph, and want to estimate the size of the screw (propeller) that will be required. A higher efficiency screw will allow the ship to burn less fuel, which reduces the operating cost. We estimate that it will take 500 kN thrust to push the ship through the water at 25 mph. Assume that the engines/drives are 32% efficient and the fuel cost is $1.50 per gallon. The ship is to cruise from Los Angeles to Hawaii (2500 mi), from Hawaii to Midway (1500 mi), and from Midway to Tokyo (2600 mi), refueling at each stop. The density of seawater is 1025 kg/m³.

(a) Size the screw assuming a 75% efficiency at the cruising speed. Calculate the cost of the fuel for a one-way trip.

(b) Explore the cost difference using a 70% efficient screw and an 80% efficient screw (Figure 3.7).

The speed of the ship is V_1 = 25 mph. Equation (3.35) is used to find the velocity V_2 of the water relative to the ship:

$$0.75 = \frac{2(25)}{25 + V_2}; \quad V_2 = 41.67 \text{ mph} \hspace{2cm} (3.41)$$

Figure 3.7 Small ship.

It will be convenient to work the problem in the metric system, so the velocities need to be converted to units of m/s. V_1 = 25 mph = 11.17 m/s, and V_2 = 41.67 mph = 18.62 m/s. Ships normally refer to the propeller as the screw. The radius of the screw can be calculated using the thrust formula [Equation (3.33)]. There are two screws for this ship.

$$500,000\,\text{N} = \frac{1}{2}\left(1025\right)\pi R^2\left(18.62^2 - 11.17^2\right) * 2 \ \left(\text{two screws}\right) \tag{3.42}$$

Solving Equation (3.42):

$$R = 0.8364\,\text{m}, \quad D = 1.673\,\text{m} \tag{3.43}$$

The power required on the shafts driving the screws is equal to the drag force multiplied by the average velocity through the screw:

$$\text{Shaft Power} = \left(500,000\right)\left(\frac{11.17 + 18.62}{2}\right) = 7.45\,\text{MW} = 9990\,\text{hp} \tag{3.44}$$

The power would be divided between two 7000-hp engines, so the engines will have enough power to push the ship along. The engines are not operating at full throttle. The total time required to travel from Los Angeles to Tokyo can be calculated as:

$$\text{Time} = \frac{2500\,\text{mi} + 1500\,\text{mi} + 2600\,\text{mi}}{26\,\text{mph}} = 264\,\text{h} \tag{3.45}$$

The energy required is power multiplied by the time. This should be converted to units of BTU to calculate the number of gallons of fuel required.

$$\text{Energy} = \left(7.45 \times 10^6\,\text{W}\right)\left(264\,\text{h}\right)\left(\frac{3.412\,\text{BTU}}{\text{W-h}}\right) = 6.708 \times 10^9\,\text{BTU} \tag{3.46}$$

The engine/drive system is 32% efficient in converting the energy in the fuel to power driving the shafts that power the screws. Diesel fuel has 138,700 BTU/gal, and the cost is assumed to be $1.50 per gallon. The number of gallons and cost are calculated below:

$$\text{Fuel} = \frac{6.708 \times 10^9\,\text{BTU}}{\left(0.32\right)\left(138,700\,\text{BTU / gal}\right)} = 151,147\,\text{gal} \tag{3.47}$$

$$\text{Cost} = \left(151,147\,\text{gal}\right)\left(\$1.50\,/\,\text{gal}\right) = \$226,721.00 \tag{3.48}$$

These calculations were done assuming a 75% efficient screw. The same process can be repeated for the 70% efficient screw. The V_2 velocity is 46.43 mph = 20.75 m/s. The diameter of the screw required is 1.504 m. Power required on the shafts driving the screws is 7.98 MW, versus 7.45 MW for the 75% efficient screw. Making the trip from Los Angeles to Tokyo would require 161,953 gal of fuel for a cost of $242,930.00, versus $226,721.00 for the 75% efficient screw. The lower efficiency screw would add approximately $16,209.00 in fuel cost.

The same process can be repeated for the 80% efficient screw. The V_2 velocity is 37.5 mph = 16.76 m/s. The diameter of the screw required is 1.995 m. Power required on the shafts driving the screws is 6.98 MW versus 7.45 MW for the 75% efficient screw. Making the trip from Los Angeles to Tokyo would require 141,709 gal of fuel for a cost of $212,564.00, versus $226,721.00 for the 75% efficient screw. The higher-efficiency screw would save approximately $14,157 in fuel cost.

The analysis shows the advantage of using a more efficient propeller (screw). Generally, the more efficient propellers will be larger in diameter than the less efficient propellers. The fuel savings and cost savings that go with using a more efficient propeller are very significant. The design of a propeller is beyond the scope of the book, but the readers should see that using a larger and more efficient propeller is almost always worth the extra cost.

3.2. Homework

1. The small aluminum fishing boat in Figure 3.8 is to be powered by a 5-hp gasoline motor. When cruising across the lake, the motor is at partial throttle such that it is producing 3 hp. The drive system is 90% efficient in transmitting the engine power to the shaft that drives the propeller (SHP = 2.7 hp). The total weight of the boat, two fishermen, and their gear is 1050 lb. The boat is 14 ft long, and the waterline length is 90% of the boat length. Assume that the engine has a thermal efficiency of 22% and the gasoline has an energy density of 121,000 BTU/gal.

    (a) Calculate the hull speed for the boat. Express your answer in ft/s. (Answer: 8.03 ft/s)

    (b) Assume that the boat will be cruising at more than 85% hull speed and calculate the cruising speed for the boat when the engine is producing 3 hp. Verify that the boat is traveling at more than 85% hull speed. (Answer: 12.03 knots = 13.85 mph = 20.3 ft/s)

    (c) Find the thrust force on the boat assuming the propeller has a 14-in. diameter and an efficiency of 92%. Density of water is 1000 kg/m^3. Express your answer in lb. (Answer: 719 N = 219.2 lb)

    (d) With a 3-gal tank of fuel, how far can the boat travel at the cruising speed? (Answer: 0.2867 gph, 145 mi)

2. The large ship in Figure 3.9 has a total weight of 54,500 metric tons and cruises at 28 mph on the ocean. The waterline length on the ship is 325 m. The density of seawater is 1025 kg/m^3.

    (a) Calculate the drag force for the ship at cruising speed. (Answer: 2.97 MN)

Figure 3.8 Fishing boat.

© Shutterstock.

Figure 3.9 Large ship.

© Shutterstock.

(b) Assume that the ship is powered by three screws that are 5.5 m in diameter. The drive system is essentially 100% efficient in transmitting the power from the engines to the shafts driving the screws. Calculate the required engine horsepower for the screw system. (Answer: 41.45 MW)

(c) Assume that the diesel engines are 30% efficient and the fuel has 138,700 BTU/gal. Calculate the fuel economy in mpg and feet per gallon. How much fuel is required to make a 300-mi trip? (Answer: 36,400 gal)

3. Assume that the density of seawater is 1025 kg/m^3. The 15,000-lb fishing boat in Figure 3.10 is powered by twin 120-hp motors and traveling at 18 knots. The static waterline length of the boat is 30 ft. The propellers have a diameter of 18 in.

(a) Use Equation (3.11) to estimate the shaft power required to push the boat along at 18 knots. (Answer: 104 hp)

(b) Find the drag force on the boat and the efficiency of the propellers being used. Assume that the drag force is shared equally by the two propellers. (Answer: Drag force = 7508 N, V_2 = 11.42 m/s, efficiency = 89.6%)

(c) Assume that the engines burn gasoline that has an energy density of 125,000 BTU/gal and that the engines are 25% efficient in converting the energy in the fuel into useful power. The drive system is 95% efficient in delivering the engine

Figure 3.10 Fishing boat.

power to the shaft driving the propeller. The cost of fuel is $5.18 per gallon at the marina. How many gallons per hour does the boat use while cruising at 18 knots? What is the fuel cost to power the boat at this speed for 3 h? (Answer: 8.99 gph, $139.71)

4. This problem illustrates how to adjust the constant in Equation (3.11) to fit a boat when enough information is given. The 540 Sports Cruiser Yacht specifications are that it has an empty weight of 42,000 lb, two diesel engines that provide 600 hp each, and a top speed of 30 knots. The fuel tank holds 300 gal of diesel fuel.

(a) Assume that the drive system is 95% efficient in transmitting the engine power to the shafts driving the propellers so that maximum SHP = 1200 hp * 0.95 = 1140 hp for the boat. Assume the boat is loaded to 45,000 lb and that the top speed is 30 mph. Calculate the appropriate constant for the boat. (Answer: 9.2875×10^{-6})

(b) Calculate the engine power when cruising at 18 knots. (Answer: SHP = 370 hp, engine power = 390 hp)

(c) Suppose the boat is loaded to a total weight of 50,000 lb and is cruising at 18 knots. Calculate the power the engines must provide. (Answer: 435 hp)

(d) Suppose the boat is loaded to a total weight of 50,000 lb and is cruising at 18 knots. The efficiency of the engines is 29% and the engines burn diesel fuel which has 137,800 BTU/gal. Estimate the distance the boat could travel using 300 gal of fuel. (Answer: 195 nautical mi = 224 mi)

References

1. https://en.wikipedia.org/wiki/Froude_number.

2. https://www.diva-portal.org/smash/get/diva2:1449680/FULLTEXT01.pdf.

The drag force acting on airplanes and jets is due to aerodynamic drag. Propulsion comes from the propellers on airplanes and from the jet engines on jets. For steady-state flight, the drag force on the aircraft must equal the propulsion force from the propellers or jet engines.

Airplanes and jets must produce enough lift to offset their weight. For steady-state flight, the lift force on the aircraft must equal the weight. The wings of the aircraft must be tilted upward slightly to generate the required lift force. Tilting the wings increases the aerodynamic drag force. The increase in aerodynamic drag due to tilting the wings and generating the required lift is known as induced drag.

The drag force acting on the aircraft is broken into two parts: basic drag and induced drag. The aircraft is designed for a specific cruising speed to minimize the total drag force on the aircraft at cruising speed. This allows the aircraft to get the best fuel economy possible. Aircraft burn a lot of fuel, and the weight of the fuel is a significant portion of the total weight of the aircraft. The cost of the fuel is a significant part of the operating cost. Optimizing for fuel efficiency at the cruising speed is an important design consideration.

Glide Ratio and Lift to Drag Ratio: If the power is cut, an airplane or jet will glide at a downward angle θ, as illustrated in   [1].

Figure 4.1 Glide angle.

The downward vertical component of velocity is Vsin(θ). Assuming a constant-speed gliding, the gravitational power (mg Vsin(θ)) must equal the power to overcome aerodynamic drag. The conservation of energy principle is used to formulate an equation for the drag force on the aircraft:

$$mg\ V\sin(\theta) = (\text{Drag Force})V; \quad \text{Drag Force} = mg\sin(\theta) \tag{4.1}$$

If we know the glide angle θ of the airplane or jet in flight, the equation above can be used to find the drag force on the airplane or jet. The glide angle θ varies from aircraft to aircraft and varies with the speed of the aircraft. The glide angle will be smallest at the designed cruising speed for the aircraft.

Rather than reporting a glide angle, aircraft manufacturers report the glide ratio for the aircraft, which is the cosecant of the glide angle. A glide ratio of 10 means that the aircraft travels 10 m for every 1 m it drops in altitude. The glide angle for a glide ratio of 10 is arcsine(1/10) = 5.74°.

The glide ratio is equal to the lift-to-drag ratio for the aircraft. An aircraft with a glide ratio of 10 will generate 10 lb of lift force for every 1 lb of drag force. Knowing the glide ratio allows us to make good estimates of the drag force on an aircraft, which allows us to make good estimates of the power requirements and fuel economy. The following statements are helpful in estimating the glide ratio for aircraft.

(a) Small single-engine airplanes with fixed landing gear typically have a glide ratio of about 10.

(b) Airplanes with retractable landing gear typically have a glide ratio of about 15.

(c) Older jets like the Boeing 747 typically have a glide ratio of about 16.

(d) Newer jets typically have a glide ratio of about 20.

These glide ratios are typical when the aircraft is operating at the cruising speed. The glide ratio will be lower at other speeds. This chapter covers how to estimate the fuel consumption and fuel economy of airplanes and jets operating at their designed cruising speeds. An important part of the analysis is to make a good estimate of the drag force on the aircraft, and this requires a good estimate of the glide ratio. Use the guidance given in the above paragraph to estimate the glide ratio for the airplane or jet. The drag force and thrust power required to keep the aircraft in flight are calculated from Equations (4.2) and (4.3):

$$\text{Drag Force} = \frac{\text{Aircraft Weight}}{\text{Glide Ratio}} \tag{4.2}$$

$$\text{Thrust Power} = (\text{Drag Force})(V) \tag{4.3}$$

4.2.
Fuel Economy of Airplanes

Equations (4.2) and (4.3) are used to estimate the drag force and thrust power required for airplanes and jets. Airplanes are driven using propellers and jets use jet engines. The examples below illustrate how to use the propeller formula to estimate the fuel economy of airplanes.

Example 4.1

A single-engine Cessna 350 Corvalis airplane with fixed landing gear has a mass of 1300 kg and cruises at 200 mph. Estimate the drag force on the airplane and the cruising power (Figure 4.2).

Since this is a small airplane with fixed landing gear, the glide ratio will be approximately 10. Drag force and thrust power are calculated as follows:

$$\text{Drag Force} = \frac{(1350)(9.81)}{10} = 1275\,\text{N} \tag{4.4}$$

$$\text{Thrust Power} = (1275)(200)\left(\frac{1609}{3600}\right) = 114.0\,\text{kW} = 152\,\text{hp} \tag{4.5}$$

The propeller will need to generate 1275 N of thrust to keep the plane in flight. The propeller will be less than 100% efficient, so the engine will need to provide more than 152 hp to the propeller shaft to get the required 1275 N thrust. There is also a drive system between the engine and the propeller shaft, and there will be a small amount of power lost in that system. For most airplanes, the engine directly drives the propeller shaft and the power lost in the drive system is small. For airplanes, it is reasonable to assume that the engine power is essentially the same as the power delivered to the propeller shaft. If a gear reduction is used in the drive system between the engine and propeller shaft, then there will be a loss of a few percent in the drive system. For a direct drive, the only losses are in the bearings, and bearing losses are small. For the examples in the book, I have assumed zero power losses in the drive system for airplanes.

The next step in estimating fuel consumption is to calculate the power the engine must provide. Assume that a 6-ft-diameter propeller is used to power the Columbia 350 airplane and estimate the engine power required. Assume that the density of air is 1.2 kg/m³. The speed of the airplane is 200 mph, which is converted to 89.39 m/s. The radius of the propeller is 3 ft, which is converted to 0.9144 m.

The thrust force from the propeller must be 1275 N. The thrust force formula [Equation (3.33)] can be used to find the required value for V_2, which is the velocity of the air leaving the propeller with respect to the velocity of the airplane.

$$1275\,\text{N} = \frac{1}{2}(1.2)\pi\left(0.9144^2\right)\left(V_2^2 - 89.39^2\right); \quad V_2 = 93.81\,\text{m/s} \tag{4.6}$$

Figure 4.2 Cessna 350 Corvalis.

Courtesy of Roberto Bianchi.

The engine power is the thrust force multiplied by the average velocity through the propeller.

$$\text{Engine Power} = \left(1275\right)\left(\frac{89.39 + 93.81}{2}\right) = 116.8 \text{ kW} = 156.6 \text{ hp} \qquad (4.7)$$

The engine must provide 156.6 hp to the propeller shaft. The propeller provides 152 hp of thrust power to the airplane. Approximately 4.6 hp is lost due to the inefficiency of the propeller. The 6-ft-diameter propeller has a high efficiency at the cruising speed for this airplane. The efficiency formula [Equation (3.35)] yields:

$$\text{Propeller Efficiency} = \frac{2\left(89.39\right)}{89.39 + 93.81} = 97.9\% \qquad (4.8)$$

The propeller formulas used in the analysis are a bit idealistic, and the propeller would probably not be quite this efficient, but this method is a good first estimate for power and efficiency of the propeller. Once the required horsepower of the engine is known, the fuel consumption and fuel economy are estimated using the same method as for wheeled vehicles. Assume that the engine is 25% efficient in converting the energy in the fuel into power and that the fuel has 120,200 BTU/gal. Calculate the fuel economy of the airplane in mpg.

$$\text{Fuel Power} = \frac{116,790 \text{ W}}{0.25} = 467,160 \text{ W} \tag{4.9}$$

$$\left(467,160 \text{ W}\right)\left(\frac{3.412 \text{ BTU}}{\text{W}-\text{h}}\right)\left(\frac{\text{gal}}{120,200 \text{ BTU}}\right) = 13.26 \text{ gph} \tag{4.10}$$

$$\text{Fuel Economy} = \frac{200 \text{ mph}}{13.26 \text{ gph}} = 15.1 \text{ mpg} \tag{4.11}$$

The Columbia 350 has a fuel tank of 102 gal, and the manufacturer reports a range of 1500 mi. The result of the calculations in this example (15.1 mpg) (102 gal) = 1540 mi is in good agreement with the actual performance of the airplane.

Example 4.2

A Bombardier Dash 8 passenger plane has a mass of 16,466 kg and cruises at 332 mph. It has retractable landing gear, so the glide ratio will be approximately 15. Assume that the propellers are 8 ft in diameter and that the density of air is 1.2 kg/m³. The aircraft has an 835-gal tank, and the manufacturer says it has a range of 1300 mi. Assume the engine has a thermal efficiency of 25%.

(a) Estimate the drag force on the airplane when cruising.

(b) Use the propeller formula to estimate the engine power required for cruising.

(c) Assume that the engine has a thermal efficiency of 25% and that it burns aviation fuel which has an energy density of 120,200 BTU/gal. Calculate the fuel economy of the airplane in mpg.

(d) Based on the 835-gal tank and fuel economy when cruising, estimate the range of the airplane. How does this compare with the manufacturer's estimate of 1300 mi? (Figure 4.3)

The solution to the problem is worked out using the metric system below.

$$\text{Drag Force} = \left(\frac{\left(16,466 \text{ kg}\right)\left(9.81 \text{ m}/\text{s}\right)}{15}\right) = 10,756 \text{ N} \tag{4.12}$$

The velocity of the airplane V_1 is 332 mph, which is converted to 148.4 m/s. The radius of the propellers is 4 ft, which is converted to 1.219 m. The thrust force formula [Equation (3.33)] is used to find the velocity of the air behind the propellers relative to the airplane V_2. Since this is a twin-engine airplane, the thrust needs to be multiplied by 2.

$$10,756 \text{ N} = \frac{1}{2}\left(1.2\right)\pi\left(1.219^2\right)\left(V_2^2 - 148.4^2\right)\left(2 \text{ propellers}\right); \quad V_2 = 154.7 \text{ m}/\text{s} \tag{4.13}$$

Figure 4.3 Bombardier Dash 8 airplane.

$$\text{Engine Power} = \left(10,756\right)\left(\frac{148.4 + 154.7}{2}\right) = 1.630 \times 10^6 \text{ W} \tag{4.14}$$

$$\text{Fuel Power} = \left(\frac{1.630 \times 10^6 \text{ W}}{0.25}\right)\left(\frac{3.412 \text{ BTU}}{\text{W} - \text{h}}\right)\left(\frac{\text{gal}}{120,200 \text{ BTU}}\right) = 185.1 \text{ gph} \tag{4.15}$$

$$\text{Fuel Economy} = \frac{332 \text{ mph}}{185.1 \text{ gph}} = 1.79 \text{ mpg} \tag{4.16}$$

The fuel tank holds 835 gal according to the manufacturer, so the range can be calculated as:

$$\text{Range} = \left(835 \text{ gal}\right)\left(1.79 \text{ mpg}\right) = 1498 \text{ mi} \tag{4.17}$$

The manufacturer says the range is 1300 mi, which is less than calculated. The calculations above assume the miles are spent cruising at the operating speed. Allowing for takeoff and landing will require extra fuel and reduce the range, and the plane will need to have a reserve left in the tank when it lands for safety reasons. All things considered, the method above makes a good estimate for the fuel economy of the aircraft in flight.

In addition to being able to power the engine during flight, the engines and propellers must be able to provide enough thrust to accelerate the airplane up to speed on the runway. The maximum power from the engines will be used during takeoff. As the airplane accelerates down the runway, there is a thrust force from the engines and propellers and an aerodynamic drag force on the airplane that increases with the square of the speed. Assuming the airplane has a mass m and acceleration a, Newton's second law can be written as:

$$\text{Thrust Force} - \text{Aerodynamic Drag Force} = ma \qquad (4.18)$$

Where m is the mass of the airplane and a is the acceleration on the runway. Thrust force and aerodynamic drag are functions of velocity of the airplane. Acceleration is the derivative of velocity with time. It is possible to set up a first-order differential equation with velocity as the unknown, but the nonlinear nature of the functions for thrust force and aerodynamic drag makes it difficult to find an analytic solution. An approximate solution can be developed using a spreadsheet approach. The solution method is illustrated best with an example.

Example 4.3

A turboprop airplane must accelerate to 160 mph on the runway to take off. The plane has a total mass of 60,000 kg. The two engines and 3 m diameter propellers push the air backward at a speed of V_2 = 500 mph relative to the aircraft when at full throttle. Assume that the aerodynamic drag force on the airplane varies with the square of speed, aerodynamic drag = 5.5 V^2, where V is in m/s and the force is in N. Develop a spreadsheet model for acceleration of the plane down the runway at full throttle. Assume the density of air is 1.2 kg/m³.

Develop the spreadsheet model for the plane going down the runway with a time increment of 0.1 s. The conclusion should be that it takes about 10.8 s for the airplane to take off, a distance of about 400 m down the runway (Figure 4.4).

Aerodynamic drag is proportional to the square of the velocity of an object. When we include aero-dynamic drag in the model, we always end up with a nonlinear differential equation to solve for the velocity and position of the object as a function of time. A schematic of the forces acting on the aircraft during takeoff is shown in Figure 4.5.

Recognizing that V_2 is given as 500 mph = 223.5 m/s and that the acceleration is a = dV_1/dt:

$$\frac{\rho\pi R^2}{2}\left(223.5^2 - V_1^2\right)(2) - (5.5)V_1^2 = m\frac{dV_1}{dt} \qquad (4.19)$$

Figure 4.4 Turboprop airplane.

© Shutterstock

Figure 4.5 Thrust and drag forces on airplane.

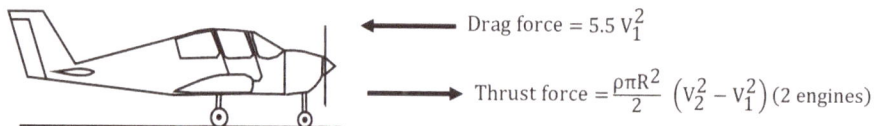

Drag force $= 5.5\, V_1^2$

Thrust force $= \dfrac{\rho \pi R^2}{2}\left(V_2^2 - V_1^2\right)$ (2 engines)

© SAE International.

In this case, it is possible to separate variables and integrate to get V_1 as an analytical function of time, but the result is difficult to work with. For this set of parameters, the analytical solution is:

$$V_1 = \left(174.078\ \text{m/s}\right)\left(\frac{e^{\frac{t}{12.3253}} - 1}{e^{\frac{t}{12.3253}} + 1}\right) \qquad (4.20)$$

As the analysis becomes more complex, it becomes impossible to find an analytical solution to the differential equation. We will develop an approximate solution using Excel. In the first column use a time increment of 0.1 to 30 s. The approximation used is that the acceleration is constant over a 0.1-s increment in time.

It is difficult to present spreadsheet development in book form. The equations for the subsequent columns in the spreadsheet below are listed. The reader should work on the spreadsheet until it matches the first several lines printed:

(a) For the first line in the spreadsheet, V1 is zero. In subsequent lines, the velocity is acceleration multiplied by the time increment plus the velocity at the beginning of the time step. That is, for line i+1, $V1_{i+1} = V1_i + a_i(0.1)$. The approximation used in this model is that the acceleration is constant over the time step. The units for V1 in this column are m/s.

(b) In the next column, the speed V1 is converted to mph by multiplying by 3600/1609.

(c) In the next column, V2 is a constant value for this model. V2 = 500(1609/3600) to convert it into m/s.

(d) The thrust force is calculated from Equation (3.39) using $\rho = 1.2$ kg/m³ and R = 1.5 m, and V2 and V1 are from the fourth and second columns in the spreadsheet, respectively. There are two propellers for this plane, so there is a factor of 2 on the thrust formula.

(e) The drag force is from the equation in the figure using V1 from column 2.

(f) The net force is the thrust force minus the drag force.

(g) The acceleration is the net force divided by the mass of 60,000 kg.

(h) For the first line in the spreadsheet, the distance the airplane has traveled on the runway is zero. In subsequent lines, the distance is the average velocity during the time step multiplied by the time increment, plus the distance at the beginning of the time step. With the assumption of constant acceleration during the time step, the average velocity is the average of the initial and final velocities during the time step. For line i+1, $d_{i+1} = d_i + (V1_i + V1_{i+1})(0.1)/2$.

(i) Propeller efficiency is 2V1/(V1 + V2).

(j) The altitude column is included for the next part of the problem where we will model the airplane taking off. For this part of the problem, the altitude is zero since the airplane is on the runway.

The first few lines of the spreadsheet are shown in Table 4.1.

Table 4.1 Spreadsheet model example.

Time (s)	V1 (m/s)	V1 (mph)	V2 (m/s)	Thrust (N)	Aero-dynamic drag (N)	Net force (N)	Accele-ration (m/s²)	Distance (m)	Propeller efficiency	Alti-tude (m)	Speed (mph)
0	0	0	223.4722	423,604.7	0	423,604.7	7.06007772	0	0	0	0
0.1	0.706008	1.579632	223.4722	423,600.4	2.741458	423,597.7	7.05996156	0.070601	0.006299	0	1.579632
0.2	1.412004	3.159238	223.4722	423,587.8	10.96565	423,576.8	7.0596131	0.211801	0.012558	0	3.159238
0.3	2.117965	4.738766	223.4722	423,566.6	24.67177	423,541.9	7.05903236	0.423598	0.018777	0	4.738766
0.4	2.823868	6.318164	223.4722	423,537	43.85828	423,493.2	7.05821942	0.705985	0.024957	0	6.318164
0.5	3.52969	7.897381	223.4722	423,499	68.52293	423,430.5	7.05717436	1.058954	0.031098	0	7.897381
0.6	4.235408	9.476363	223.4722	423,452.5	98.66274	423,353.8	7.05589732	1.482494	0.0372	0	9.476363
0.7	4.940998	11.05506	223.4722	423,397.6	134.274	423,263.3	7.05438845	1.976594	0.043264	0	11.05506
0.8	5.646436	12.63342	223.4722	423,334.2	175.3523	423,158.9	7.05264793	2.541238	0.049288	0	12.63342
0.9	6.351701	14.21139	223.4722	423,262.5	221.8926	423,040.6	7.05067599	3.176408	0.055274	0	14.21139
1	7.056769	15.78892	223.4722	423,182.3	273.8889	422,908.4	7.04847288	3.882085	0.061222	0	15.78892

Follow the first column down to 10.8 s and the speed in the third column is V1 = 160 mph. At this point, the plane has traveled about 400 m down the runway. The reader may want to compare the analytical solution [Equation (4.20)] to the approximate solution. At 10.8 s, the analytical solution yields a speed V_1 = 71.73538 m/s and the approximate solution yields

V_1 = 71.73618 m/s. This comparison is included to help convince the reader that the spreadsheet approach yields an accurate solution to the problem. At 160 mph, the airplane takes off and gravity becomes part of the analysis. Gravity can be included easily in the spreadsheet model, but it is more difficult to include it in the analytical solution.

Example 4.4

Assume that once the airplane in the previous problem reaches 160 mph it takes off and climbs at a 10% grade. Because the plane is climbing, there is an additional gravitation force opposing the motion of 10% of the airplane weight. Include the gravitational force and plot the altitude of the plane with time.

To modify the spreadsheet to account for takeoff, go into the net force column at 10.8 s and subtract 10% of the weight of the airplane from the net force to account for the plane traveling upward against gravity. The altitude is estimated assuming that the acceleration is approximately constant for the time step. The change in altitude is 10% of the average velocity multiplied by the time step. For line i+1, altitude$_{i+1}$ = altitude + (0.10) (V1$_i$ + V1$_{i+1}$) (0.1)/2. The changes in the spreadsheet must be copied down so that subsequent lines are changed. The first few lines in the transition area of the spreadsheet are shown in Table 4.2.

There are many plots that can be obtained from the data. Plotting the airplane speed (V1), thrust force, aerodynamic drag force, net force, acceleration, distance on the runway, or propeller efficiency versus time are all interesting charts. I have chosen to include the plot of altitude versus time to illustrate the airplane going down the runway and taking off, the first 30 s. The airplane takes off and climbs to 208 m in the first 30 s (Figure 4.6).

Table 4.2 Spreadsheet transition to takeoff.

Time (s)	V1 (m/s)	V1 (mph)	V2 (m/s)	Thrust (N)	Aerodynamic drag (N)	Net force (N)	Acceleration (m/s²)	Distance (m)	Propeller efficiency	Altitude (m)	Speed (mph)
10.5	69.26585	154.9764	223.4722	382,908.6	26,387.67	350,635	5.84391625	378.4191	0.473227	0	154.9764
10.6	69.85024	156.2839	223.4722	382,219	26,834.81	349,498.2	5.82497063	385.4042	0.476269	0	156.2839
10.7	70.43274	157.5872	223.4722	381,525.9	27,284.24	348,355.7	5.80592804	392.4474	0.479289	0	157.5872
10.8	71.01333	158.8863	223.4722	380,829.3	27,735.91	347,207.4	5.78679033	399.5488	0.482287	0	158.8863
10.9	71.59201	160.181	223.4722	380,129.3	28,189.79	346,053.6	5.76755938	406.708	0.485264	0.71592	160.181
11	72.16876	161.4714	223.4722	379,426	28,645.82	344,894.2	5.74823703	413.9248	0.488219	1.43760	161.4714
11.1	72.74359	162.7576	223.4722	378,719.5	29,103.96	343,729.5	5.72882516	421.1992	0.491153	2.16504	162.7576
11.2	73.31647	164.0393	223.4722	378,009.7	29,564.18	342,559.5	5.70932561	428.5309	0.494065	2.89820	164.0393
11.3	73.8874	165.3168	223.4722	377,296.8	30,026.42	341,384.4	5.68974026	435.9196	0.496957	3.63708	165.3168
11.4	74.45638	166.5898	223.4722	376,580.9	30,490.64	340,204.3	5.67007095	443.3652	0.499827	4.38164	166.5898
11.5	75.02338	167.8584	223.4722	375,862	30,956.8	339,019.2	5.65031953	450.8676	0.502677	5.13188	167.8584

Figure 4.6 Plot of altitude versus time for the first 30 s.

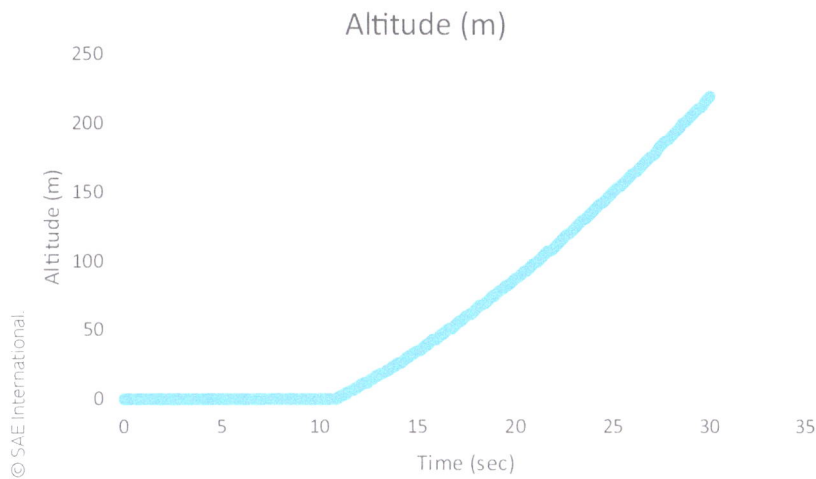

4.3.
Fuel Economy of Turbojet and Turbofan Aircraft

For turbojet and turbofan engines, we will need to analyze the mass flow rate through the engine. Air flows into the engine, fuel is added and burned, and the combustion products flow out of the engine. There is more mass flowing out of the engine than flowing in because of the fuel added. For the turbojet engine, all the air flows through the combustion chamber. For turbofan engines, some of the air flows around the combustion chamber and is added and heated with the exhaust gasses. Having a high air/fuel ratio improves the efficiency of the engines. Turbofan engines tend to be more efficient than turbojet engines. The design of an engine is complex and there are many considerations. In this chapter, we study the overall effect of the engine assuming it is well designed (Figure 4.7).

Air enters the engine with a mass flow rate of $\dot{m}_a$. Fuel is injected into the combustion canisters at a mass flow rate of $\dot{m}_f$. The exhaust gasses will have a mass flow rate of $(\dot{m}_a + \dot{m}_f)$. The velocity of the air entering the engine is the velocity of the jet aircraft, symbolized as V. The velocity of the exhaust gasses relative to the jet aircraft will be symbolized as V_e. The exhaust velocity V_e depends on the energy density of the fuel and the fuel/air mixture. The first order of business is to develop an equation for the thrust force provided by the jet engine. For steady-state flight, the thrust force of the engines must equal the drag force on the aircraft (Figure 4.8).

To apply the principle of conservation of momentum, we will need the absolute velocities of the air entering the jet engine and the exhaust gasses exiting the engine. The absolute velocity of the air entering the engine is zero (still air). The absolute velocity of the exhaust gasses exiting the engine is $(V - V_e)$.

Figure 4.7 Turbofan engine.

inlet air

primary exhaust

Figure 4.8 Mass flow and velocity relative to the aircraft.

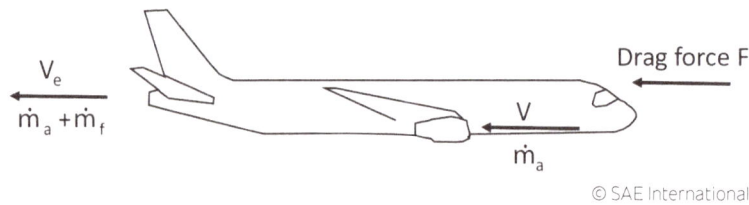

Assume that M is the mass of the aircraft and conservation of mass is being done for a small time period dt. The mass of air used during the small time period is $\dot{m}_a dt$ and the mass of fuel used during the small time period is $\dot{m}_f dt$. At the beginning of the time period, the aircraft (M) and the fuel $(\dot{m}_f dt)$ have a velocity of V. The air $(\dot{m}_a dt)$ has a velocity of zero.

$$\text{Initial Momentum} = \left(M + \dot{m}_f dt \right) V + \left(\dot{m}_a dt \right)(0) \tag{4.21}$$

At the end of the time step, the fuel and air are consumed and have a velocity $(V - V_e)$. The aircraft still has a velocity of V. The drag force (F) acted on the aircraft a time dt. The momentum equation for the end of the time step is:

$$\text{Final Momentum} = \left(M \right) V + \left(\dot{m}_a dt + \dot{m}_f dt \right)\left(V - V_e \right) + \left(F \right) dt \tag{4.22}$$

Setting the two equations equal, the dt divides out and the MV terms cancel. For steady-state flight, the drag force must equal the thrust and it follows:

$$\text{Thrust} = \text{Drag} = \left(\dot{m}_a + \dot{m}_f \right) V_e - \left(\dot{m}_a \right) V \tag{4.23}$$

A similar approach can be used for conservation of energy. At the beginning of the time step, the aircraft (M) and fuel ($\dot{m}_f dt$) are traveling at speed V. The air has zero velocity. Let E_f be defined as the energy density of the fuel (E_f = 46 to 47 MJ/kg

for jet fuel). Let e be the thermal efficiency of the engine in converting the chemical energy in the fuel to kinetic energy of the exhaust gasses. (e is typically 45% for a modern jet engine.) The energy at the beginning of the time step is:

$$\text{Initial Energy} = \frac{1}{2}(M + \dot{m}_f dt)(V^2) + \frac{1}{2}(\dot{m}_a dt)(0^2) + eE_f\dot{m}_f dt \tag{4.24}$$

At the end of the time step, the aircraft (M) continues at speed V and the exhaust gasses exit at a speed $(V - V_e)$. The drag force F acts through

a distance Vdt, so the energy consumed is F(Vdt). The energy at the end of the time step can be written as:

$$\text{Final Energy} = \frac{1}{2}(M)V^2 + \frac{1}{2}(\dot{m}_a dt + \dot{m}_f dt)(V - V_e)^2 + F(Vdt) \tag{4.25}$$

The derivation gets a little messy at this point. Set the initial and final energy equal and expand the terms. Many terms will cancel. Then

substitute for the drag force F from Equation (4.23) and simplify. The result is a formula for the exhaust velocity V_e:

$$V_e = \sqrt{\frac{2eE_f\dot{m}_f + \dot{m}_a V^2}{\dot{m}_a + \dot{m}_f}} \tag{4.26}$$

The thrust equation [Equation (4.23)] and the exhaust-velocity equation [Equation (4.26)] are the most important results from the derivation. The air/fuel ratio is important for the efficiency of the engine. To have complete combustion, it requires about 16 kg of air for each kg of fuel burned in the engine. The engine does not operate efficiently at the low stoichiometric ratio of 16:1. Older jet engines use about 70 kg of air for each kg of fuel burned, and the newer turbofan engines use 100 kg of air for each kg of fuel burned. The excess air reduces the exhaust temperature, and less of the energy goes out the back of the engine as heat. That is, a higher percentage of the chemical energy in the fuel is converted to kinetic energy of the exhaust

gasses. Mathematically, it would be desirable to go to higher air/fuel ratios, but there are practical limitations in the combustion and distribution of the heat inside the engine.

For modern jet engines, the thermal efficiency is typically about 45%. There is also a thrust efficiency of the engine with a similar definition as for the propellers $2V/(V + V_e)$. The thrust efficiency is lower for a jet engine than for a propeller, but the thermal efficiency of a jet engine is higher than for the engines used to power propellers. The two factors tend to balance and yield approximately the same overall efficiency for propeller and jet-powered aircraft. The overall efficiency of a jet engine is the product of the thermal efficiency and the thrust efficiency.

$$\text{Overall Efficiency} = \frac{2eV}{V + V_e} \tag{4.27}$$

Example 4.5

A Boeing 747 can weigh as much as 900,000 lb at takeoff. To accelerate up to speed on the runway to take off requires an initial acceleration of about 0.25 g. Assume that the engines are 45% efficient and estimate the fuel consumption during takeoff. Also, assume that the engines use 70 kg of air for each kg of fuel (Figure 4.9).

The mass of the 900,000 lb aircraft is 409,091 kg. Starting from rest, the airplane would have zero velocity and no aerodynamic drag. There would be some rolling drag, but the main drag force is the dynamic drag required for acceleration. The required thrust for the engines is mass times acceleration.

$$\text{Required Thrust} = \left(409{,}091\,\text{kg}\right)\left(0.25\right)\left(9.81\,\text{m}\,/\,\text{s}^2\right) = 1.033\,\text{MN} \tag{4.28}$$

Figure 4.9 Boeing 747 on runway.

© Shutterstock.

The exhaust velocity can be calculated from Equation (4.26). The energy density of the fuel is assumed to be 46.6 MJ/kg and the thermal efficiency is assumed to be 45%. The air/fuel ratio is 70:1.

$$V_e = \sqrt{\frac{2\left(0.45\right)\left(46.6 \times 10^6\right)\dot{m}_f + 70\dot{m}_f\left(0^2\right)}{70\dot{m}_f + \dot{m}_f}} = 768.57\,\text{m}\,/\,\text{s} \tag{4.29}$$

The required thrust value and the exhaust velocity can be substituted into Equation (4.23) to solve for the required mass flow rate of the fuel.

$$1.033 \times 10^6 = \left(70\dot{m}_f + \dot{m}_f\right)\left(768.57\right) - \left(70\dot{m}_f\right)\left(0\right) \tag{4.30}$$

Solving Equation (4.30) yields $\dot{m}_f = 18.93$ kg/s. Jet fuel has a density of approximately 3.0 kg/gal. The mass flow rate of fuel can be converted to a volumetric flow rate of 6.31 gal/s. A fully loaded 747 will use about 6.31 gal of fuel per second when accelerating down the runway. The required takeoff speed for the aircraft is about 200 mph, which is 89.4 m/s. At an acceleration of 0.25 g, it will take 36.4 s for the aircraft to accelerate to the 89.4 m/s (200 mph) takeoff speed. The aircraft will need 1629 m or 5344 ft of runway to get to the takeoff speed. This is a little over a mile of runway length. Boeing recommends that the runway be at least 11,000 ft, or about 2 mi, in length to ensure the aircraft has plenty of space for takeoff and landing.

Example 4.6

Assume a Boeing 747 has a weight of 700,000 lb while cruising at 550 mph. This would be an average cruising weight for the aircraft. The aircraft has a glide ratio of 16. Assume that the engines are e = 45% efficient and use 70 kg of air for each kg of fuel.

(a)   Calculate the drag force on the aircraft using the glide ratio.

(b)   Calculate the exhaust velocity V_e and the fuel consumption while cruising.

(c)   The fuel tank on the 747 holds 55,000 gal of fuel. Calculate the range of the aircraft.

(d)   Assume at the airport the cost of fuel is $4.00 per gallon. What is the cost of filling the tank?

(e)   If the aircraft holds 600 passengers, what is the fuel cost per passenger? What are the mpg and passenger-mpg for the aircraft? (Figure 4.10)

The drag force on the aircraft is estimated from the glide ratio:

$$\text{Drag Force} = \frac{\left(700,000\,\text{lb}\right)\left(4.448\,\text{N}\,/\,\text{lb}\right)}{16} = 194,600\,\text{N} \tag{4.31}$$

Figure 4.10 Boeing 747 in flight.

The speed of the aircraft in flight is 550 mph = 245.8 m/s. The exhaust velocity can be calculated as:

$$V_e = \sqrt{\frac{2(0.45)(46.6 \times 10^6)\dot{m}_f + 70\dot{m}_f(245.8^2)}{70\dot{m}_f + \dot{m}_f}} = 806.4 \text{ m / s} \tag{4.32}$$

The thrust equation [Equation (4.23)] is used to calculate the fuel consumption:

$$194,600 = (70\dot{m}_f + \dot{m}_f)(806.4) - (70\dot{m}_f)(245.8) \tag{4.33}$$

The rate of fuel consumption is $\dot{m}_f = 4.859$ kg / s, and since jet fuel has a density of 3.0 kg/gal, the aircraft will consume 1.62 gal/s in steady-state flight. In comparison to the 6.31 gal/s required for takeoff, you can see the aircraft uses a lot less fuel in flight. The aircraft has a 55,000-gal fuel tank, so the aircraft can fly 55,000/1.62 = 33,951 s = 9.43 h. The range of the aircraft is (550 mph) (9.43 h) = 5187 mi.

Los Angeles to Tokyo is 5479 mi, which is further than the range calculated. Seattle to Tokyo is 4783 mi. Flying across the Pacific requires all the fuel the aircraft can hold. The density of 3 kg/gal is 6.6 lb/gal. 55,000 gal of fuel would weigh 363,000 lb. The maximum takeoff weight for the aircraft is 900,000 lb, so the fuel represents 40% of the total weight of the aircraft on takeoff.

If the cost of fuel is $4.00 per gallon, the cost to fill the tank is (55,000 gal)($4.00/gal) = $220,000.00. With 600 passengers the cost is $366.67 per passenger. If an airline ticket to fly across the Pacific Ocean costs $2000.00, the fuel cost is a significant part of the cost of the ticket but is less than half the cost.

One way to calculate the fuel economy of the aircraft is to take the range of 5187 mi and divide by the fuel consumed which is 55,000 gal. The fuel economy is 0.0943 mpg. The efficiency of large jets is typically rated in passenger-mpg, which in this case is (600) (0.0943) = 56.6 passenger-mpg. Putting two people in a car that averages 28.3 mpg on the highway yields about the same fuel economy per person as flying on a Boeing 747. Jet fuel is approximately twice the cost of gasoline. In terms of cost, one person riding in a car that averages 28.3 mpg on the highway will be about the same fuel cost per person flying on a Boeing 747.

Example 4.7

We are developing a jet airplane that will have a maximum takeoff weight of 130,000 lb and requires the engines to provide a 0.25 g initial acceleration down the runway. How much thrust is required from the engines? Assuming the engine efficiency is e = 45% and that it uses 70 kg of air for each kg of fuel burned, find the mass flow rate of fuel during takeoff (Figure 4.11).

The mass of the 130,000 lb aircraft is 59,091 kg. The required thrust for a 0.25 g acceleration is:

$$\text{Thrust} = (59,091)(0.25)(9.81) = 144,920 \text{ N} \tag{4.34}$$

The exhaust velocity V_e is:

$$V_e = \sqrt{\frac{2(0.45)(46.6 \times 10^6)\dot{m}_f + 70\dot{m}_f(0^2)}{70\dot{m}_f + \dot{m}_f}} = 768.57 \text{ m / s} \tag{4.35}$$

Figure 4.11 Jet on runway.

© Shutterstock.

Mass flow for the fuel is calculated from the thrust formula:

$$144,920 = \left(70\dot{m}_f + \dot{m}_f\right)\left(768.57\right) - \left(70\dot{m}_f\right)\left(0\right) \tag{4.36}$$

Equation (4.36) is solved to yield a mass flow rate of m_f = 2.656 kg/s for the fuel.

When cruising at 500 mph the weight of the aircraft will be reduced to an average of about 106,000 lb because some of the fuel will have been consumed. Assume a glide ratio of 20 and calculate the fuel consumption while cruising.

$$\text{Drag Force} = \frac{\left(130,000\ \text{lb}\right)\left(4.448\ \text{N} / \text{lb}\right)}{20} = 23,574.4\ \text{N} \tag{4.37}$$

The cruising speed for the aircraft is 500 mph = 223.5 m/s. The exhaust velocity is calculated as:

$$V_e = \sqrt{\frac{2\left(0.45\right)\left(46.6 \times 10^6\right)\dot{m}_f + 70\dot{m}_f\left(223.5^2\right)}{70\dot{m}_f + \dot{m}_f}} = 799.97\ \text{m} / \text{s} \tag{4.38}$$

Mass flow for the fuel is calculated from the thrust formula:

$$23,574.4 = \left(70\dot{m}_f + \dot{m}_f\right)\left(799.97\right) - \left(70\dot{m}_f\right)\left(223.5\right) \qquad (4.39)$$

Solving Equation (4.39), the mass flow rate while cruising is $\dot{m}_f = 0.573$ kg / s. If the plane is to have a range of 3500 mi what should the size of the fuel tank be?

$$\text{Cruising time} = \frac{3500\,\text{mi}}{500\,\text{mph}} = 7\,\text{h} = 25,200\,\text{s} \qquad (4.40)$$

$$\text{Fuel Tank Size} = \left(25,200\,\text{s}\right)\left(\frac{0.573\,\text{kg}}{\text{s}}\right)\left(\frac{\text{gal}}{3.0\,\text{kg}}\right) = 4811\,\text{gal} \qquad (4.41)$$

The fuel tank for the aircraft would need to be at least 4811 gal to give the aircraft a 3500 mi range. Fuel economy would be (3500 mi/4811 gal) = 0.729 mpg. An aircraft of this size would be able to carry about 100 passengers, so it would yield approximately 72.9 passenger-mpg.

4.4.
Fuel Economy of Rockets

Rockets are a very small part of our transportation sector. In the grand scheme, the energy and fuel spent on rocket transportation is insignificant compared to the other methods of transportation. This section is included because many students find rockets and space exploration interesting.

Rockets must carry the fuel and oxygen to burn the fuel, which reduces the energy density of the fuel + oxygen mixture. Kerosene has an energy density of about 47 MJ/kg, but when the required oxygen is added, the energy density decreases to about 10.7 MJ/kg. Liquid hydrogen has an energy density of about 130 MJ/kg, but it takes about

8 kg of oxygen to burn 1 kg of hydrogen so, after adding the oxygen the energy density drops to about 14.4 MJ/kg.

The velocity of the fuel exiting the rocket engine depends on the energy density of the fuel, and the expression can be derived using conservation of energy and conservation of momentum. Let M be the mass of the rocket and $\dot{m}_f dt$ be the small amount of fuel burned during the small amount of time dt. In this case, $\dot{m}_f$ is the mass flow rate of the fuel + oxygen mixture. E_f is the energy density of the fuel + oxygen mixture, which is 10.7 MJ/kg for the kerosene fuel and 14.4 MJ/kg for liquid hydrogen. The thermal efficiency of the rocket engine is e. The exhaust velocity V_e is the velocity relative to the rocket. The equations for thrust and exhaust velocity are below:

$$\text{Thrust} = \left(\dot{m}_f\right)\left(V_e\right) \qquad (4.42)$$

$$V_e = \sqrt{2eE_f} \qquad (4.43)$$

In Equation (4.43) the E_f term is the energy density of the fuel + oxygen mixture. In addition to accelerating the rocket to speed, the rocket must overcome gravity to get into orbit. The universal gravitational constant is $G = 6.674 \times$ 10^{-11} m³/kg s². The mass of the earth is $M = 5.9736 \times 10^{24}$ kg. The radius of the earth is $R = 6.371 \times 10^6$ m. Moving a mass m to a height h above the surface requires that we integrate mgh.

$$\text{Gravitational Energy} = \int_{R}^{h+R} \frac{GMm}{r^2} dr = \frac{GMm}{R} - \frac{GMm}{R+h} \tag{4.44}$$

If h is much larger than the radius of the earth, the second term will be insignificant, and the energy required to raise a mass m = 1 kg is equal to GM/R = 62.6 MJ/kg. This would be the energy required to get beyond the gravitational pull of the earth. Since the energy density of the fuel + oxygen is 10.7 to 14.4 MJ/kg, it should be clear that it takes many kg of fuel to lift 1 kg of mass into deep space. The general rule of thumb is that the rocket should be 90% fuel by weight. In many ways, a rocket is a giant fuel tank with a small rocket engine at the bottom. To have enough energy to get into space at least 90% of the total weight of the rocket must be fuel.

Example 4.8

To dock with the International Space Station, a spacecraft must be lifted 400 km (250 mi) above the surface of the earth and must be accelerated to the orbital speed of the space station which is 27,600 kph (17,200 mph). This requires gravitational energy and kinetic energy. The Space Shuttle had a mass of 100,000 kg (220,000 lb). How much energy is required to lift the Space Shuttle and dock it with the International Space Station? (Figure 4.12)

The gravitational energy required can be computed from Equation (4.44):

$$\text{Gravitational Energy} = \frac{\left(6.674 \times 10^{-11}\right)\left(5.9736 \times 10^{24}\right)\left(100,000\right)}{6.371 \times 10^6}$$

$$-\frac{\left(6.674 \times 10^{-11}\right)\left(5.9736 \times 10^{24}\right)\left(100,000\right)}{6.371 \times 10^6 + 400,000} = 3.696 \times 10^{11} \text{ J} \tag{4.45}$$

The kinetic energy is ½ m V² and the orbital speed is 27,600 kph = 7666.7 m/s. Kinetic energy is calculated as:

$$\text{Kinetic Energy} = \frac{1}{2}\left(100,000\right)\left(7666.7^2\right) = 2.939 \times 10^{12} \text{ J} \tag{4.46}$$

Figure 4.12 International Space Station.

The total energy required is the sum of the gravitational and kinetic energy, which is 3.308×10^{12} J. Notice that the kinetic energy is about eight times as much as the gravitational energy. Most of the energy from the rocket engine will go toward accelerating the rocket to orbital speed for the space station. Only about 1/8 of the energy will go toward overcoming gravity. We neglected the aerodynamic drag energy required to push the rocket through the atmosphere, because aerodynamic drag energy will be small compared to gravitational and kinetic energy.

If the rocket engine was 100% efficient in converting the chemical energy in the fuel to kinetic energy of the exhaust gasses, we can estimate the amount of fuel required as:

$$\frac{3.308 \times 10^{12} \text{ J}}{10.7 \text{ MJ / kg}} = 309,200 \text{ kg fuel} \left(\text{for the kerosene fuel}\right) \tag{4.47}$$

$$\frac{3.308 \times 10^{12} \text{ J}}{14.4 \text{ MJ / kg}} = 229,700 \text{ kg fuel} \left(\text{for the hydrogen fuel}\right) \tag{4.48}$$

Please notice that the mass of the Space Shuttle is 100,000 kg and the mass of the fuel is two to three times as much as the mass of the Space Shuttle. The fuel will not be burned instantaneously as the rocket takes off. A portion of the fuel will need to be lifted and accelerated with the Space Shuttle. As more fuel is added to lift the weight of the fuel, it leads to the "rule of thumb" that 90% of the total weight on the launchpad needs to be fuel. When you see photographs of the Space Shuttle on the launchpad, it is attached to a much larger rocket. Approximately 90% of the total weight on the launchpad must be fuel. That is what is required to have enough energy to lift the Space Shuttle into orbit and accelerate it to orbital speed so it can dock with the International Space Station.

The rocket engine will not be 100% efficient. Around 40% efficiency is more realistic. A large portion of the fuel will be needed to lift and accelerate as the rocket takes off. The fuel will be used up gradually as the rocket is in flight, and the mass of the rocket will decrease as it is in flight. We will develop a spreadsheet model of a rocket blasting off. To accelerate the rocket to the required speed, it will be necessary for the mass of the rocket to be 90% fuel at takeoff. Anything less and we will not be able to get the rocket up to the required orbital speed and it will fall back to earth.

The last homework problem in the unit asks the reader to think about space travel in the practical sense. Neptune is about 2.7 billion mi from the earth. Traveling at 35,000 mph, about twice as fast as orbital speed for the International Space Station, it would take 8.8 years to get to Neptune. The fastest vehicles we have ever made were the two Voyager missions. We estimate that the spacecraft got to a speed of about 40,000 mph. To explore our solar system in a reasonable amount of time, we need to go much faster. Kinetic energy goes with the square of speed, so it takes much more energy and fuel to go faster. It is not possible to reach the speeds that we need with the current rocket fuel. We need a fuel that has a higher energy density, that is, more energy per kg. We need a lot more energy per kg. In the last homework problem, the reader is to consider the possibility of developing a nuclear rocket engine using a fuel that has a much higher energy density (hydrogen fusion). The reader will discover that even hydrogen fusion does not have enough energy to allow us to travel fast enough for human exploration of our solar system and planets. The kinetic energy required to go fast enough to get a human to Neptune and back in a reasonable period of time is more than can be provided by nuclear fusion.

Example 4.9

Develop a spreadsheet for a rocket going straight up. Assume that the energy density of the fuel is 10.7 MJ/kg, that the efficiency is 40%, that the total weight of the rocket is 500,000 kg, and that 450,000 kg of the total weight is fuel (Figure 4.13).

The mass of the rocket will decrease significantly with time as the fuel is burned. Assuming a mass flow rate of $\dot{m}_f$ for the fuel, the mass of the rocket at a time t is:

$$\text{Mass} = m = 500,000 \text{ kg} - \dot{m}_f t \qquad (4.48)$$

Figure 4.13 Rocket blasting off.

The acceleration of gravity g varies with height h above the surface of the earth according to:

$$g = \frac{GM}{(R+h)^2} = \frac{(6.674 \times 10^{-11})(5.9736 \times 10^{24})}{(6.371 \times 10^6 + h)^2} \tag{4.49}$$

The exhaust velocity V_e and thrust are calculated using Equations (4.42) and (4.43). The net lift force for the rocket is the thrust force minus the weight. Acceleration is the lift force divided by the mass.

In the spreadsheet, the approximation made is that the acceleration is constant over the time step interval chosen, which in this case is 1 s. The velocity of the rocket is incremented such that $V_{i+1} = V_i + a(\Delta t)$, where Δt is the time step. The height of the rocket is incremented according to the average velocity during the time step: $h_{i+1} = h_i + (V_{i+1} + V_i)(\Delta t)/2$.

For the spreadsheet below, it was assumed that the mass flow rate for the rocket engine was 2500 kg/s. Once the fuel is gone the rocket is done. Burning 450,000 kg of fuel at a rate of 2500 kg/s takes 180 s. At 180 s the speed of the rocket is 4933 m/s and the rocket is 232 km above the earth. This is not high enough or fast enough to dock with the space station. In this example, the rocket would fall back to earth after burning its fuel. Working with the variables it is necessary to achieve a rocket efficiency of 80% for this rocket to be able to dock with the space station. If hydrogen fuel is used, the rocket engine will need to be 60% efficient.

I hope that working through this example will illustrate to the reader the energy challenge of space flight. The fuel that we have is very marginal in providing the energy per kg mass required for space travel. Because of the limitation of the fuel, nearly all the mass of the rocket must be fuel, and the rocket engines must have a very high thermal efficiency to achieve orbit. The two Voyager missions are the only two objects that have been able to push beyond our solar system, and it took decades to get them there. It was a phenomenal achievement. Flying a rocket and docking with the space station is a phenomenal achievement. It has been done and the public sometimes gets the impression that space travel is routine, but space travel is difficult.

One of the major problems associated with space travel is the limited energy density of the fuel. Table 4.3 is provided to help the reader develop the spreadsheet for launching a rocket straight up into space. Setting it up like this the parameters can be changed, and the reader can explore many possibilities (Table 4.3).

Table 4.3 Rocket spreadsheet.

E_f	1.44E+07	J/kg			m_f		2500	kg/s			
e	0.6										
Time (s)	Mass (kg)	Mass flow (kg/s)	Gravity (m/s²)	V_e (m/s)	Thrust (N)	Lift (N)		Acceleration (m/s²)	Velocity (m/s)	Height (m)
0	500,000	2500	9.822163	4156.922	10,392,305	5,481,223	10.96245	0	0
1	497,500	2500	9.822146	4156.922	10,392,305	5,505,787	11.06691	10.96245	5.481223
2	495,000	2500	9.822095	4156.922	10,392,305	5,530,368	11.17246	22.02936	21.97712
3	492,500	2500	9.82201	4156.922	10,392,305	5,554,965	11.27912	33.20182	49.59271
4	490,000	2500	9.82189	4156.922	10,392,305	5,579,579	11.3869	44.48093	88.43409
5	487,500	2500	9.821735	4156.922	10,392,305	5,604,209	11.49581	55.86783	138.6085
6	485,000	2500	9.821546	4156.922	10,392,305	5,628,855	11.60589	67.36364	200.2242
7	482,500	2500	9.82132	4156.922	10,392,305	5,653,518	11.71714	78.96953	273.3908
8	480,000	2500	9.821058	4156.922	10,392,305	5,678,197	11.82958	90.68666	358.2189
9	477,500	2500	9.820761	4156.922	10,392,305	5,702,892	11.94323	102.5162	454.8203
10	475,000	2500	9.820426	4156.922	10,392,305	5,727,602	12.05811	114.4595	563.3082

4.5.
Homework

1. The small airplane in Figure 4.14 has a total weight of 5200 lb and cruises at 120 mph. Assume that the glide ratio for the airplane is 10 at the cruising speed. The density of air is 1.17 kg/m³ at the altitude and temperature in question. The propeller has a diameter of 75 in.

Assume that the engine burns aviation fuel that has an energy density of 120,400 BTU/gal and that the thermal efficiency of the engine/drive is 25%.

(a) Find the drag force on the airplane and the power required from the engine. (Answer: 185 hp)

(b) Find the fuel consumption in gallons per hour. If the aviation fuel has a cost of

$4.35 per gallon at the airport, what is the fuel cost per hour to fly the plane? (Answer: $68.00 per hour)

(c) The fuel tank holds 65 gal. What is the range of the airplane? Please use judgment in your analysis and account for some extra fuel used in takeoff and landing.

Figure 4.14 Small airplane.

© Shutterstock.

2. The small airplane in Figure 4.15 must accelerate to 80 mph on the runway to take off. The plane has a total mass of 1200 kg. The engine and 2-m-diameter propeller push the air backward at a speed $V_2 = 150$ mph relative to the plane when at full throttle. Assume that the aerodynamic drag force on the plane varies with the square of speed such that Aero Drag $= (0.659)\ V_1^2$, where V is in m/s and the aerodynamic drag force is in Newtons. Develop a spreadsheet model for acceleration of the plane down the runway at full throttle. Assume that the density of air is 1.2 kg/m^3.

How long does it take for the airplane to reach a takeoff speed of 80 mph? (Answer: 5.9 s)

3. Assume that once the plane in problem 2 reaches 80 mph, it takes off and climbs at a 10% grade. Because the plane is climbing, there is an additional gravitational force opposing the motion of 10% of the weight of the airplane. Include the gravitational force and plot the altitude of the plane with time.

4. The jet in Figure 4.16 has a maximum takeoff weight of 23,500 lb. The engines are capable of providing an acceleration on the runway of 0.35g during takeoff. Assume that the engines

Figure 4.15 Small airplane on runway.

© Shutterstock

Figure 4.16 Learjet in flight.

© Shutterstock.

have a thermal efficiency of 45% and estimate the fuel consumption during takeoff. Assume that the engines use 65 kg of air for each kg of fuel. The energy density of the fuel is 46.6 MJ/kg, and the fuel density is 3.0 kg/gal. Calculate the gallons per second of fuel used during takeoff. (Answer: 0.232 gal/s)

5. The jet in Figure 4.16 has a weight of 22,000 lb while cruising at 484 mph. The aircraft has a glide ratio of 15 at the cruising speed. Assume the engines have a thermal efficiency of e = 45% and consume 65 kg of air for each kg of fuel. The density of the fuel is 3.0 kg/gal.

(a) Calculate the drag force on the jet using the glide ratio.

(b) Calculate the exhaust velocity V_e and the fuel consumption while cruising.

(c) Calculate the fuel economy in mpg for the jet. (Answer: 2.499 mpg)

(d) The fuel tank on the jet holds 1200 gal. Calculate the range of the jet. (Answer: 480 mi)

(e) If the maximum takeoff weight is 23,500 lb, what percentage of this weight is fuel?

6. We are developing a jet, that is, to have a maximum takeoff weight of 220,000 lb, and require that the initial acceleration down the runway must be 0.25 g. Energy density of the fuel is 46.6E6 J/kg and the mass density is 3 kg/gal.

(a) Calculate the thrust required to provide the initial acceleration.

(b) Assuming the thermal efficiency of the engine is e = 0.45 and that the engines use 60 kg of air for each kg of fuel, calculate the mass flow rate of fuel during takeoff. (Answer: 4.84 kg/s)

(c) When cruising at 450 mph the weight is 200,000 lb because some of the fuel has been burned. The engines use 90 kg of air for each kg of fuel while cruising. Assume a glide ratio of 20 for the aircraft and calculate the fuel economy while cruising at 450 mph. If the plane is to have a range of 3000 mi, how large should the fuel tank be, and what is the weight of the fuel? (Answer: 0.390 mpg, 7685 gal, 50,800 lb, or 23.1% of the vehicle weight is fuel)

7. Assume that we plan to fire a rocket to lift a spacecraft into space using a rocket fuel/oxygen mix that has an energy density of 14.4 MJ/kg. The spacecraft is approximately the size of the Space Shuttle, which had a mass of approximately 100,000 kg. Assume 45% thermal efficiency for the rocket engines.

(a) The "rule of thumb" is that a rocket + spacecraft should be 90% fuel by mass. Assume that the rocket itself is 95% fuel by mass and that it must lift the 100,000 kg spacecraft. How much fuel will be in the rocket, and what is the total weight of the rocket and spacecraft? (Answer: fuel mass = 1,710,000 kg, total mass = 1,900,000 kg)

(b) Develop the spreadsheet like the one developed in the book for this rocket going straight up (Figure 4.17).

8. Space travel with conventional rocket fuel is difficult. The energy density of the fuel is much too low to go very far into space. We made it to the moon and back, and we might make it one way to Mars, but beyond that we will need a fuel with a higher energy density.

For this problem, assume that we develop the technology to control a nuclear fusion reaction converting deuterium-deuterium hydrogen into helium. The energy density of the fuel is

Figure 4.17 Space Shuttle launch.

250×10^{12} J/kg. Assume that we have a space-craft that has a total mass of 500,000 kg and is 50% fuel by mass. The thermal efficiency of the rocket engines is 45%. Blasting this rocket off would be a larger energy release than all the hydrogen bombs ever created, and it might destroy life on a large part of the earth. But overlooking that problem, use the spreadsheet approach to calculate the speed of the rocket when it has burned half of its fuel supply. How does that speed compare with the speed of light? (This is a thought problem to help you understand that even with the most powerful fuel we know of, nuclear fusion, there is not enough energy to accelerate the spacecraft to speeds necessary for space exploration. We would need to discover a fuel with a much higher energy density than nuclear fuel.)

Reference

1. https://en.wikipedia.org/wiki/
    Lift-to-drag_ratio#Glide_ratio.

Unit 02: Energy and the Environment

As a society, we have come a long way since colonial times. The Industrial Revolution started in 1760, and the inventions and technology developed since then have dramatically improved the quality of our lives. We have hot and cold running water. Our homes and buildings are climate-controlled. We have electric power available for the multitude of electronic devices we use. We have a safe food supply and can buy almost any food we want at any time of the year. We have a transportation system that allows us to travel long distances with ease and comfort. We have modern medicine that can often heal us when we have a medical problem. Our lives are better and easier than at any time in history.

The modern lifestyle we lead has an impact on the environment. Maintaining our lifestyle requires that we use a large amount of energy. When quantifying the amount of energy used, the numbers are so large that it is difficult to understand what the numbers mean. The first part of Unit 2 will help the reader comprehend and make sense of the amount of energy used in the United States and in the world. Burning fossil fuels adds carbon dioxide to the atmosphere, and carbon dioxide in the atmosphere contributes to global warming. In Unit 2, the book will show the reader how to calculate carbon footprint and carbon emissions from fossil fuels. The book also illustrates how to estimate other pollutants that come from burning fossil fuels including sulfur dioxide, nitrous oxides, carbon monoxide, and particulate emissions. The reader will learn to do the calculations and will be able to show quantitatively that burning coal generates more pollution and carbon emissions than burning natural gas. Burning propane, diesel fuel, gasoline, or jet fuel is cleaner than burning coal, but not as clean as natural gas. The focus in the first part of Unit 2 (Chapters 5 and 6) is on how to do the calculations associated with burning fossil fuels.

The reader will learn how to calculate carbon footprint per mile traveled for cars and trucks, trains, ships, aircraft, and other fuel-powered vehicles.

In order to reduce our carbon footprint, it will be necessary to produce the electricity we use from carbon-free sources: hydro, wind, solar, and nuclear. Chapter 7 shows how to estimate the amount of electric energy that can be produced from renewable sources, so the reader can estimate how many wind turbines, solar panels, dams, and lakes are necessary to produce the amount of electricity we will need. Chapter 7 also includes a discussion of the pros and cons of nuclear energy. Generating the electricity we need from carbon-free sources is not going to be easy and is not going to happen in the near future.

In Chapter 8, a relatively simple radiative heat transfer model is developed for the earth, so the reader can calculate how increasing the atmospheric content of carbon dioxide will increase the average temperature of the earth. The reader will learn how to make a quantitative estimate of how increasing the carbon dioxide content of the atmosphere will increase the average temperature of the earth.

Batteries are an essential part of making the renewable energy system work, and the batteries we have today are not adequate. Vehicles will require lightweight batteries that can store a lot of energy. Utilities will require huge batteries to allow them to keep the electric grid balanced and handle peak loads. The last part of Unit 2 (Chapter 9) discusses the batteries that are currently available and their strengths and weaknesses. Some companies are considering using hydrogen fuel and fuel cells to reduce carbon emissions. Added to this edition of the book is a brief unit showing how hydrogen is produced and how fuel cells work.

Human innovation, machines, and other technology allow us to live the modern lifestyle we have today. We tend to think about inventions and technology because that is what we see and what impacts our daily life. But in the background, it takes a lot of energy to power everything. Before the Industrial Revolution, most of our energy came from human and animal power. Hydropower (water wheels) was used to grind grain. Horses and dogs were important, because they took some of the workload off humans and made our lives better than what we had before. The steam engine and electric power plants gave us more energy to work with and made our lives much easier. The Lawrence Livermore National Laboratory keeps track of the energy used in the United States, and the results are summarized in   . A quad is defined as a quadrillion BTU or 1×10^{15} BTU.

In the United States, we have used approximately 100 quads annually for each of the last 10 years.

In   , the left column is the primary sources of energy we use. The first primary source of energy is solar energy, and it is important to note that the Department of Energy only considers solar voltaic panels and solar hot-water heaters in this category. Solar energy is also used in agriculture to grow the food we eat. If agriculture is included, solar is our largest and most important energy source, and we use more solar energy than all the others combined.

We often think of electricity as an energy source, but there is no naturally occurring electricity we can harness. Electricity must be generated from primary energy sources. Of the 100.3 quads of energy used in 2022, 37.7 quads were used to generate electricity. Generating electricity is the largest single use of energy.

Figure 5.1 Estimated US energy consumption in 2022.

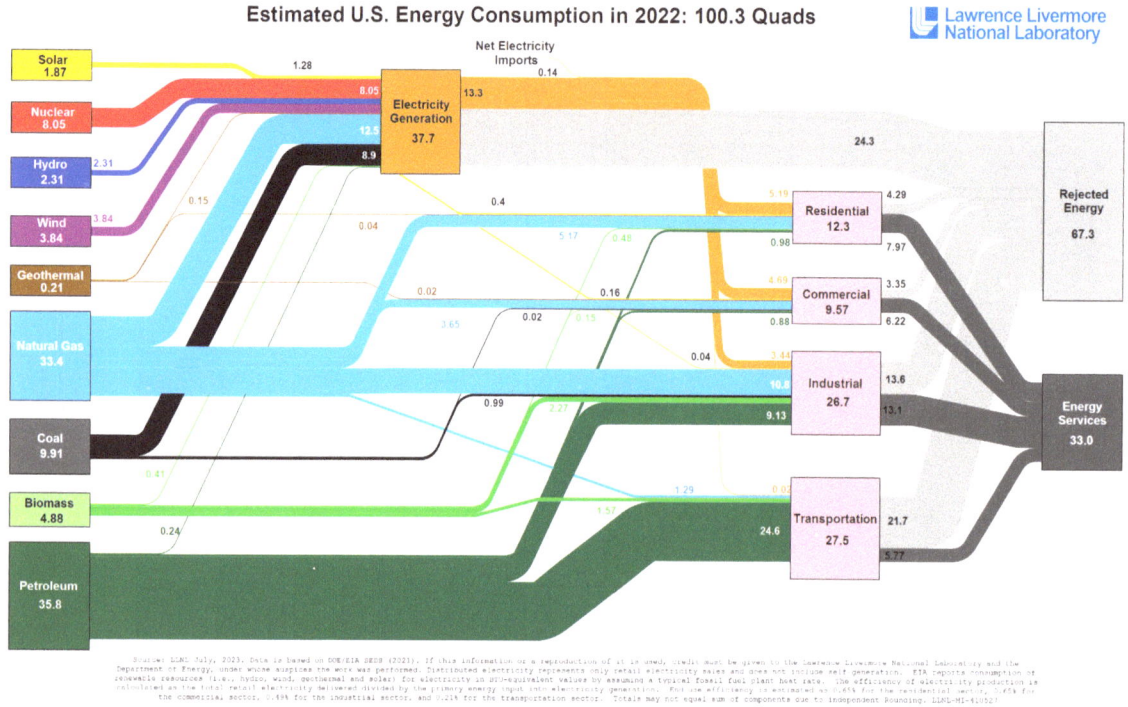

Estimated U.S. Energy Consumption in 2022: 100.3 Quads

Electricity is important because it is the most efficient way to transmit and distribute energy. We can transmit electricity over long distances and distribute it to the customers who need electricity with very little energy loss. It is the energy efficiency associated with the transmission and distribution of energy that makes electricity so valuable to us.

Early powerhouses had shafts and pulleys inside a building, and equipment was powered using a belt drive system from one of the overhead pulleys. It was dangerous and had low energy efficiency. One could envision a mechanical energy distribution system for a city that had shafts, gears, and universal joints on top of telephone poles through the city. The shaft system would go to houses and power overhead shafts and pulleys that could drive equipment in the houses using a belt system like in the old powerhouses. A mechanical distribution system like this would be very expensive and inefficient. There would be large energy losses in all gears, universal joints, pulleys, and belts. It is a ridiculous idea but visualizing a mechanical distribution system helps you appreciate the value of the electrical distribution system. It is not that electric energy is better than other forms of energy, but that we can transmit and distribute electricity very efficiently and at a low cost.

Looking at Figure 5.1, we can draw some conclusions about the energy used to generate electricity. All the primary sources of energy are used in generating electricity. A question that always comes up is "What is biomass?" Biomass includes alcohol mixed with gasoline, biodiesel, and burning of wood or other organic materials.

It is a fuel made from plants. Table 5.1 summarizes the different types of primary energy that are used to generate electricity.

If we divide electricity production by types of energy, fossil fuel contributes 23.34 quads or 61.0% of the electricity produced. Renewables contribute 6.46 quads or 16.9%, and nuclear contributes 8.44 quads or 22.1%. Some people may want to put nuclear energy in with renewables since we have an almost unlimited amount of nuclear fuel, and it is a carbon-free source of energy. Compared with the 2016 data, renewables have increased by about 2% and fossil fuels have decreased by about 2% in the mix of generating electricity. Nuclear has been steady.

Table 5.1 Primary energy used to generate electricity in the United States.

Primary energy	Amount (quads)	Percentage (%)	Type
Solar	1.28	3.4	Renewable
Nuclear	8.05	21.4	Nuclear
Hydro	2.31	6.1	Renewable
Wind	3.84	10.2	Renewable
Geothermal	0.15	0.4	Renewable
Natural gas	12.5	33.2	Fossil fuel
Coal	8.9	23.6	Fossil fuel
Biomass	0.41	1.1	Renewable
Petroleum	0.24	0.6	Fossil fuel
Totals	37.68	100.0	

© SAE International

As most of the electric energy comes from heat engines, the second law of thermodynamics says that most of the energy will become heat that goes into the atmosphere, which is called "Rejected energy" in Figure 5.1. The "Rejected energy" and "Energy services" divisions on the right side of the chart can be used to assess our overall energy efficiency as a nation. Of the 100.3 quads of energy used, 67.3 quads (67.1%) are expelled to the atmosphere as wasted heat and 33.0 quads (32.9%) become useful energy for us. Our overall efficiency is 32.9% in using our primary energy.

Figure 5.1 divides the energy we use into categories of residential, commercial, industrial, and transportation. Residential, commercial, and industrial customers use a lot of energy by heating our homes and buildings in the winter, commercial cooking and industrial canning, and the melting and processing of plastics, metals, and ceramics. When we need heat, we can use fossil fuel, especially natural gas, very efficiently to provide the heat. Modern gas furnace systems are more than 90% efficient in using the heat energy in natural gas to heat our homes. Using natural gas directly is much more efficient overall than using electricity for heat. If we use natural gas in a power plant to make electricity, only about one-third of the energy in the natural gas will be converted to electricity. The remaining two-thirds will be lost as heat in the atmosphere. An electric furnace may be 100% efficient in converting the electricity into heat, but, when you look broadly, we only recover about 33% of the heat energy originally in the natural gas. It is much more efficient to use the natural gas directly in the furnace. This is true for almost any application where we need heat; it is most efficient to use natural gas (or other fossil fuel) directly to generate the heat. There are cases where we only need a small amount of heat and using an electric heater makes sense, but in general using electricity to make heat is inefficient and wasteful. Electricity should be used to power electric motors and other electrical and electronic devices where we need electricity.

Figure 5.1 also provides information about the efficiency in generating electricity and in providing energy for residential, commercial, industrial, and transportation applications. Table 5.2 shows a summary.

Table 5.2 Summary of energy use and efficiency in electricity, residential, commercial, industrial, and transportation.

	Useful energy (quads)	Total energy (quads)	Efficiency (%)
Electricity	13.3	37.7	35.3
Residential	7.97	12.3	64.8
Commercial	6.22	9.57	65.0
Industrial	13.1	26.7	49.1
Transportation	5.77	27.5	21.0

© SAE International

The efficiencies for residential, commercial, and industrial customers are much higher than for generating electricity and transportation. In residential, commercial, and industrial applications, a significant percentage of the energy in the fuel is used to heat our homes, offices, and buildings. Heating using natural gas can be done at a high efficiency. For generating electricity and transportation, the energy in the fuel is used in a heat engine (internal combustion engine or steam turbine), which has a much lower efficiency than using the energy in the fuel directly as heat. The overall efficiency in transportation is only 21%, which means that 79% of the energy in the fuel (gasoline, diesel, and jet fuel) is dumped into the atmosphere as heat. The steam turbines used in power plants are significantly more efficient than the internal combustion engines used in transportation.

The next step is to gain a deeper understanding of how much energy is used in the United States.

In 2022, we used 100.3 quads of energy annually, but most people have a difficult time understanding the magnitude of that amount of energy. There were 333.3 million people in the United States in 2022, so each person would use 300.9 million BTUs annually. Dividing by 365 days/year, on average each person uses 824,466 BTUs daily. This was the average daily energy consumption for people in the United States in 2022.

Before the Industrial Revolution, horses were used to provide some of the energy that was used. We might consider a horse that could produce 1 hp for 8 h/day. This would be a very hard-working horse. Convincing a horse to work this hard for 8 h/day would be a challenge. But, in an ideal situation, a horse could produce 8 hp-h of energy in a day. We need to convert the energy into BTUs for comparison.

$$(8\text{ hp-h})\left(\frac{550\text{ ft-lb}/\text{s}}{\text{hp}}\right)\left(\frac{3600\text{ s}}{\text{h}}\right)\left(\frac{\text{BTU}}{778.2\text{ ft-lb}}\right) = 20,355\text{ BTU} \tag{5.1}$$

Dividing the 824,466 BTU the average person uses each day by the 20,355 BTU a horse could provide in a day yields 40.5, which means we would need 40.5 horses per person in the United States to provide the energy that we use. We would need to figure out a way that each

person could work 40.5 horses for 8 h per day. For our nation, we would need about 13.5 billion horses to provide the energy that we need. It would not be possible for us to grow enough food to feed them, and I am not sure what we would do with the manure.

Another way to look at the energy we use is to consider how we could replace it with human power. A well-conditioned human can provide about 10% of a horsepower. It would be hard for any person to do that for 8 h per day, but even in the most idealistic case, it would take 400 humans to provide the energy necessary for one person to live the modern lifestyle that we have. If all our energy came from human and animal power, we would have to live a much simpler life.

The Industrial Revolution brought us many inventions and technologies that improved the quality of our lives. As the inventions came, we gradually increased our energy consumption. Today we use a tremendous amount of energy, an amount that would be unimaginable to people who lived before the Industrial Revolution. To maintain our modern lifestyles, we will need to continue using these large amounts of energy and improve the energy efficiency of the devices we use. Using large amounts of energy is creating environmental problems, and there is no easy answer to addressing these problems. None of us want to give up our modern lifestyle. We have made gradual improvements in energy efficiency, which helps reduce our energy consumption and the associated environmental consequences. But as population increases and our level of automation increases, it seems likely that energy consumption will continue to increase. Most of our energy comes from fossil fuels, and it will have to continue that way in the near future. We are working on developing energy sources that are more environmentally friendly.

Burning fossil fuels generates carbon dioxide which is released into the atmosphere, and the carbon dioxide is causing a greenhouse effect that causes our planet to warm. The Environmental Protection Agency (EPA) tracks greenhouse gasses (GHGs) for the United States [2]. Figure 5.2 is from their 2024 report, which shows results through 2022. Carbon emissions for the United States peaked 15 or 20 years ago and we have gradually reduced our carbon emissions since. The COVID-19 year 2020 was our historically lowest carbon emissions year, but that was because we greatly reduced our travel that year. Our carbon emissions increased in 2021 and 2022 compared to 2020, but are still lower than in 2019. Each year we have increased the amount of wind and solar energy used to produce electricity, which helps reduce our carbon footprint. We have phased out many of the coal plants in the United States. The newer cars and trucks get better fuel economy on average than the older cars and trucks they replace, and there are a significant number of electric and hybrid-electric cars and trucks sold each year. Reducing the number of coal plants generating electricity, increasing the amount of wind power used to generate electricity, and improving the fuel economy of our cars and trucks are the most significant things we have done in the United States to reduce our carbon emissions.

A metric ton is 1000 kg, and according to the EPA report, the United States generated a net of 5053 million metric tons (MMT) of carbon dioxide in 2022. In 2022, 4699 MMT, or 93% of the carbon dioxide emissions, came from burning fossil fuels (natural gas, coal, and petroleum). The other major sources of GHGs include methane, nitrous oxide, and hydrofluorocarbons. The EPA converts emissions of these gasses to an equivalent amount of carbon dioxide, and the equivalent amounts are shown in Figure 5.2. For example, the natural gas industry emitted 0.401 MMT of methane into the atmosphere, but methane is about 28 times more potent GHG than carbon dioxide, so the EPA reports it as 11.2 MMT carbon dioxide equivalent. Agriculture and landfills emit a significant amount of methane and nitrous oxides into the atmosphere, which are included in Figure 5.2.

Figure 5.2 US GHG emissions and sinks by gas.

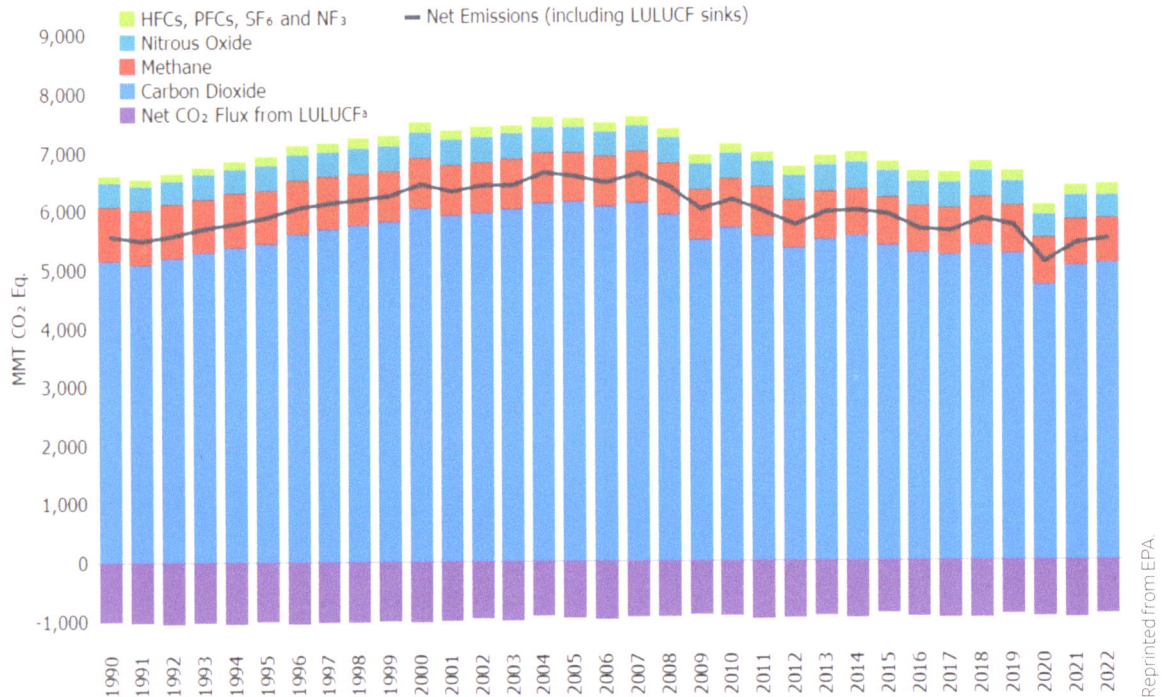

Starting in the 2021 report, EPA began including analysis of GHGs removed from the atmosphere (LULUCF in the report). Most of the removal comes from forestry (trees), though all plants removed carbon dioxide. Plants absorb carbon dioxide through the photosynthesis process, but they add carbon dioxide back to the atmosphere when they decompose. Trees are the most effective plant in removing carbon dioxide from the atmosphere because only a small percentage of the tree (leaves) decompose each year. The amount of carbon dioxide removed depends on the type of tree and where it is planted. The analysis used in the EPA report is complex. In 2022 the United States had 281 million hectares of forest that removed 787 MMT of CO_2 [2]. A hectare is 2.471 acres. Based on those numbers,

one acre of forest removes approximately 1.13 metric tons of CO_2 from the atmosphere each year. The United States generated a total of 6343 MMT (equivalent) of carbon dioxide in 2022. 854 MMT was removed from the atmosphere, yielding a net of 5489 MMT carbon dioxide (equivalent) GHG put into the atmosphere by the United States. This includes all sources of GHG emissions.

Of the 5489 MMT carbon dioxide (equivalent) emissions from the United States in 2022, 4699 MMT (85.6%) is the carbon dioxide generated from combustion of fossil fuels. The rest is from other GHGs. Of the 4699 MMT generated by combustion of fossil fuels, 898.8 MMT (19.1%) came from coal, 1706.8 MMT (36.3%) from natural gas, and 2093.4 MMT (44.5%) from petroleum. Numbers from the Department of Energy indicate we used 9.91 quad of coal, 33.4 quad of natural gas, and 35.8 quad of petroleum in 2022. Ratios can be taken to calculate the carbon footprint of the three fossil fuels as the amount of carbon dioxide generated (MMT) for each quad of heat energy in the fuel.

(a) Coal: 898.8/9.91 = 90.7 MMT CO_2/quad

(b) Natural Gas: 1706.8/33.4 = 51.5 MMT CO_2/quad

(c) Petroleum: 2093.4/35.8 = 58.5 MMT CO_2/quad

These numbers illustrate that coal will have the highest amount of carbon emissions for the amount of heat energy in the fuel. Natural gas has the lowest carbon emissions, and petroleum is in the middle, though much closer to natural gas than coal. Combustion of fossil fuels is covered in Chapter 6, and we will develop the numbers above using the basic chemistry of combustion for fossil fuels.

The EPA report is comprehensive and divides the carbon emission data in many ways. The 2022 carbon emissions from burning fossil fuels in the different sectors are: 1531.7 MMT (32.6%) in production of electricity, 334.1 MMT (7.1%) in residential, 258.7 MMT (5.5%) in commercial, 801.1 MMT (17.0%) in industrial applications, and 1751.3 MMT (37.3%) in transportation.

[The EPA reports 22.6 MMT (0.5%) was generated by US territories.] This information can be combined with the information from Lawrence Livermore National Lab in Table 5.2 to estimate the carbon footprint intensity (MMT CO_2/quad) of the different sectors. Table 5.3 summarizes the information.

Transportation produces the most carbon emissions. The engines used in powering our transportation sector generate a lot of carbon emissions for the amount of useful energy they provide, as indicated by the carbon footprint intensity. Generating electricity produces the second most carbon emissions. The carbon footprint intensity for generating electricity is much less than for transportation, but higher than in residential, commercial, and industrial applications. Residential, commercial and industrial applications use a lot of natural gas to make heat, which is an energy-efficient process, and a process that has a low carbon footprint intensity. If we were to start using electricity to heat our homes and offices, our carbon emissions would increase. It would be more expensive and generate more carbon emissions. In the future, if we are able to generate all electricity from renewable sources, then electric heat will make sense.

The numbers in Table 5.3 illustrate that in order for the United States to reduce its carbon emissions, we must reduce carbon emissions in

Table 5.3 Summary of carbon footprint for electricity, residential, commercial, industrial, and transportation.

	Useful energy (quads)	Carbon emissions (MMT)	Carbon footprint intensity (MMT CO_2/quad)
Electricity	13.3	1531.7	115.1
Residential	7.97	334.1	41.9
Commercial	6.22	258.7	41.6
Industrial	13.1	801.1	61.2
Transportation	5.77	1751.3	303.5

transportation and in generating electricity. Those are the two largest sources of carbon emissions and the two highest carbon footprint intensities. This is why there is such a strong push for renewable energy and electric vehicles. Generating electricity has a much lower carbon footprint intensity than transportation, so electric vehicles will have a lower carbon footprint than petroleum-powered vehicles. Batteries and electric motors are not 100% efficient in using the electricity, so the improvement is not as dramatic as comparing the carbon intensity factors of 303.5 and 115.1 in Table 5.3, but there is a significant improvement. Chapter 6 illustrates how to calculate the carbon footprint of electric and petroleum-powered vehicles.

The next step is to look at the energy used in the world. Figure 5.3 illustrates world energy consumption since 1990. The world used about 590 quads of energy in 2022. There was a time when the United States used about one-fourth of the energy in the world, but that has gradually been reduced to about 17% as the rest of the world has modernized. The United States represents about 4.2% of the world's population using 17% of the energy, so on a per-person basis we use a lot of energy. Data for Figures 5.3, 5.4, and 5.5 were obtained from Enerdata's Global Energy Statistical Yearbook [3] at https://www.enerdata.net/publications/world-energy-statistics-supply-and-demand.html.

Figure 5.3 World and regional energy consumption.

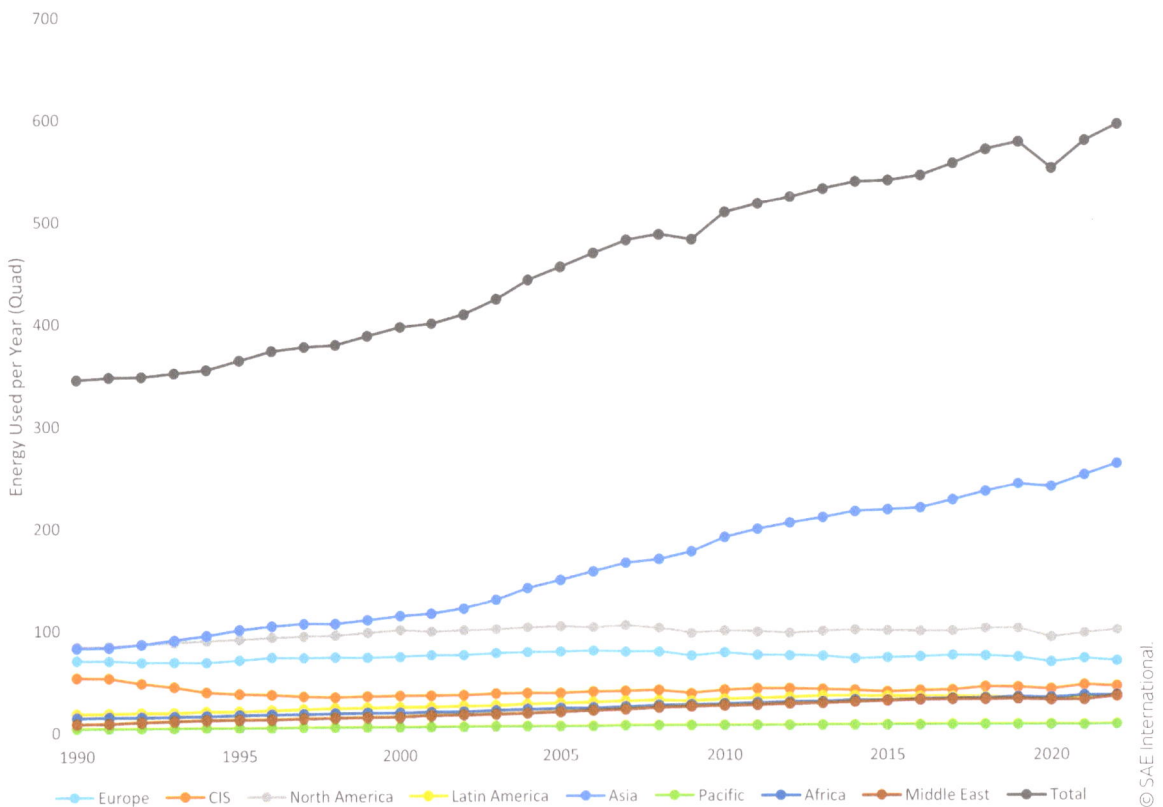

Figure 5.4 Energy consumption in Commonwealth of Independent States (CIS), Latin America, Africa, the Middle East, and Pacific.

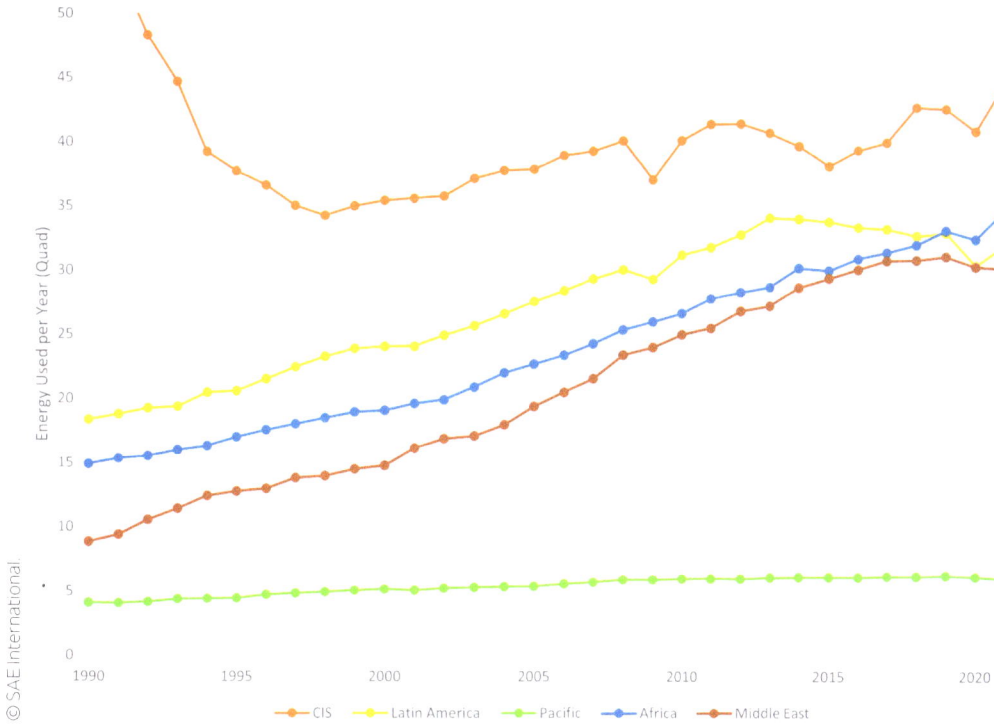

Figure 5.5 World consumption of petroleum, natural gas, and coal.

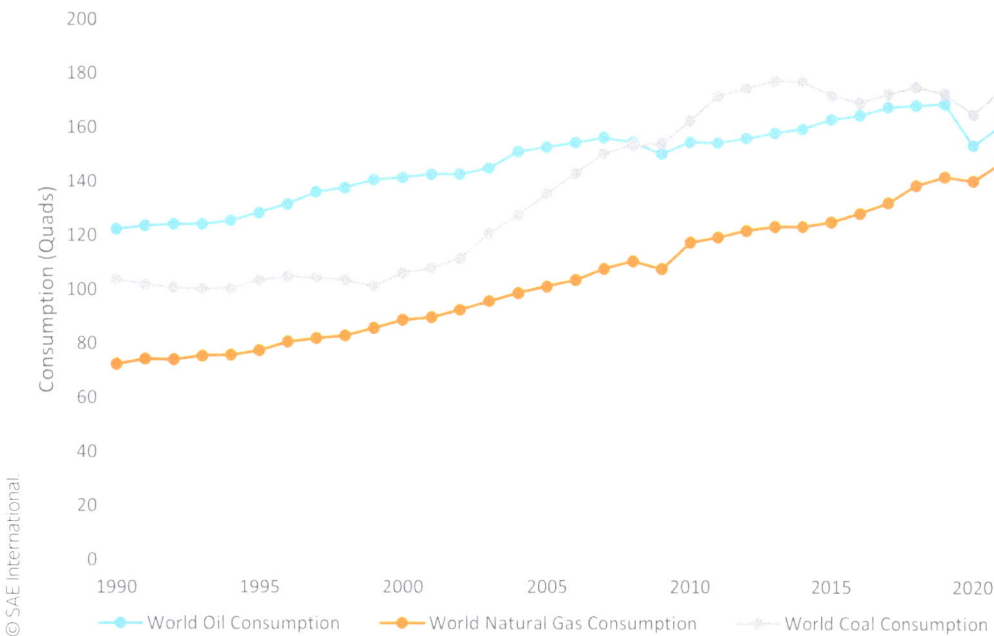

The data from Enerdata give petroleum consumption in millions of metric tons, natural gas in billions of cubic meters, and coal in millions of metric tons. To convert these numbers to quads and develop Figures 5.3, 5.4, and 5.5, average values for conversion factors were used. Using these average values, it was assumed a million metric tons of petroleum equals 0.0458 quad of energy. For natural gas, it was assumed a billion cubic meters equals 0.0353 quad of energy. For coal, a million metric tons equals 0.0222 quad of energy. Figures 5.3, 5.4, and 5.5 are based on these assumptions.

Energy usage in North America and Europe has been consistent over the past 30 years. There have been significant improvements in efficiency in electric power production, transportation, furnaces, and most of the other devices that we use. Our increase in energy needs has been offset by improvements in energy efficiency, so that overall annual energy usage has remained constant. For most of the rest of the world, energy use has increased and will continue to increase as they modernize and improve their lifestyle. The year 2009 was a bad year economically for most of the world, and most parts of the world showed a downward tick in energy consumption in 2009. 2020 was the COVID-19 year, and world energy consumption declined that year and has since recovered. If you look at individual countries, energy consumption declines when the country is facing economic difficulties, natural disasters, war, political strife, etc. Energy consumption is a good indicator of how well the standard of living in the country is improving or declining. Individual countries will experience difficult times, but they recover and their energy consumption and standard of living

increase. Since 1990, the trend in energy consumption and standard of living has been upward for all countries in the world. We tend to focus on the problems in the world. There are always problems to be solved. But overall the standard of living for people in the world is getting better each year, and that trend will likely continue.

Energy consumption is one indicator of quality of life. The United Nations and other international groups collect data for infant death mortality, average life expectancy, average level of education, number of people killed by violent crime, war, natural disasters, disease, famine, etc. If you look at the data for any of these indicators, the trends are all in the positive direction, indicating that the average quality of life in the world is improving. Individual countries will go through difficult times, and as I am writing this book I am sure the people in Ukraine and Palestine would not say their quality of life is improving. But if history is an indicator, they will recover and 10 years from now things will be much better for them.

The largest increase in energy usage has been in Asia, as shown in Figure 5.3. The data for Asia do not include the CIS, which is Russia and some of the former Soviet bloc countries, even though they are in Asia. The data for Asia includes China, India, Japan, and other countries in the Far East. About half of the world's population lives in Asia, and they have made great progress in modernizing and improving their lifestyles. If you compare life today for the average citizen in China or India to life in 1990, there has been a phenomenal improvement. The same is true for other countries in Asia. Increases in energy consumption in this part of the world will continue for at least a few more decades as they raise their quality of life to be comparable to the US and Europe.

The lines for Latin American, Middle East, and Africa on the chart appear to be flat in Figure 5.3, but this is because of scaling of the chart. In Figure 5.4 the y-axis was adjusted to show better how energy usage is changing in those parts of the world. Energy use in Latin America, the Middle East, and Africa has grown steadily in the last 30 years. Quality of life has improved steadily in the last 30 years. Individual countries have gone through political issues, war, and violence that caused a temporary decline in their economies and quality of life. It takes a few years to recover, but they recover and their energy consumption and quality of life improve.

The CIS line in the chart is Russia and some of the former Soviet Union countries. There were political issues in the 1980s and 1990s, which led to the breakup of the Soviet Union. The economy in that part of the world collapsed. The people saw their standard of living decline for several years. Bad governments and war can cause setbacks, but countries work through the problems and recover and their standard of living and energy consumption increase. Since the late 1990s, energy consumption has grown steadily in CIS. The Pacific line on the chart includes Australia, New Zealand, Indonesia, and the island countries in that part of the world. Energy consumption has increased steadily in that part of the world too.

Another way to look at world energy consumption is by the type of energy used. Figure 5.5 illustrates world energy consumption by the type of fossil fuel used. Petroleum was our largest source of energy until 2005 when coal became the largest source of energy in the world. Nuclear and renewable energy are much smaller sources in comparison and are not plotted in Figure 5.5. To reduce world carbon emissions, we will need to burn less fossil fuel, and in particular, we need to burn less coal. Europe and North America have significantly reduced the amount of coal used, which has led to a reduction in carbon emissions in Europe and North America. But there have been many coal plants built in Asia to produce the electricity they need. Building a coal plant is expensive and is normally considered to be at least a 50-year investment. Most of the coal plants built in the last 10 years will probably still be producing power in the year 2070. World carbon emissions will continue to increase each year in the near future.

An interesting comparison can be made between the United States and China, the two largest economies in the world. Compare the average carbon emissions per person, that is, divide total carbon emissions by the population. Using the Enerdata source, in 1990 the United States produced 4866 MMT carbon emissions with a population of 250 million. China produced 2257 MMT carbon emissions with a population of 1135 million. In 1990, the United States produced 9.8 times as much carbon emissions per person as China. Fifteen years passed, and in 2005 the United States produced 5803 MMT carbon emissions and the population was 296 million. China produced 5470 MMT carbon emissions with a population of 1304 million. In 2005, the United States produced 4.7 times as much carbon emissions per person than China. In 2022 the

United States produced 4735 MMT carbon emissions with a population of 333 million. China produced 10,504 MMT carbon emissions with a population of 1412 million. In 2022 the United States produced 1.9 times as much carbon emissions per person as China. In a 32-year period, the United States went from producing 9.8 times as much carbon emissions per person as China to a factor of only 1.9. Carbon emissions in China are increasing each year, while carbon emissions in the United States are declining each year. In 5 or 10 more years, it is likely that China will produce more carbon emissions per person than the United States. The results are summarized in Table 5.4.

Table 5.4 Comparison of United States and China carbon emissions.

	1990		2005		2022	
	United States	China	United States	China	United States	China
Carbon emissions (MMT)	4866	2257	5803	5470	4735	10,504
Population (Million)	250	1135	296	1304	333	1412
Carbon emissions per person (MMT/million)	19.5	1.99	19.6	4.19	14.2	7.4
Ratio (United States/China)	9.8		4.7		1.9	

© SAE International

Figure 5.6 shows world carbon emissions from 1990 to 2022 [3]. Carbon emissions are closely related to energy consumption because most of the energy we use comes from fossil fuels. The graph looks similar to Figure 5.3 except that the y-axis is carbon emissions rather than energy consumption. The COVID-19 year 2020 was a year when we did not use as much energy, and our carbon emissions were lower that year too. In 2022 the world emitted 33,811 MMT of carbon dioxide from burning fossil fuels [3].

Figure 5.6 World carbon emissions.

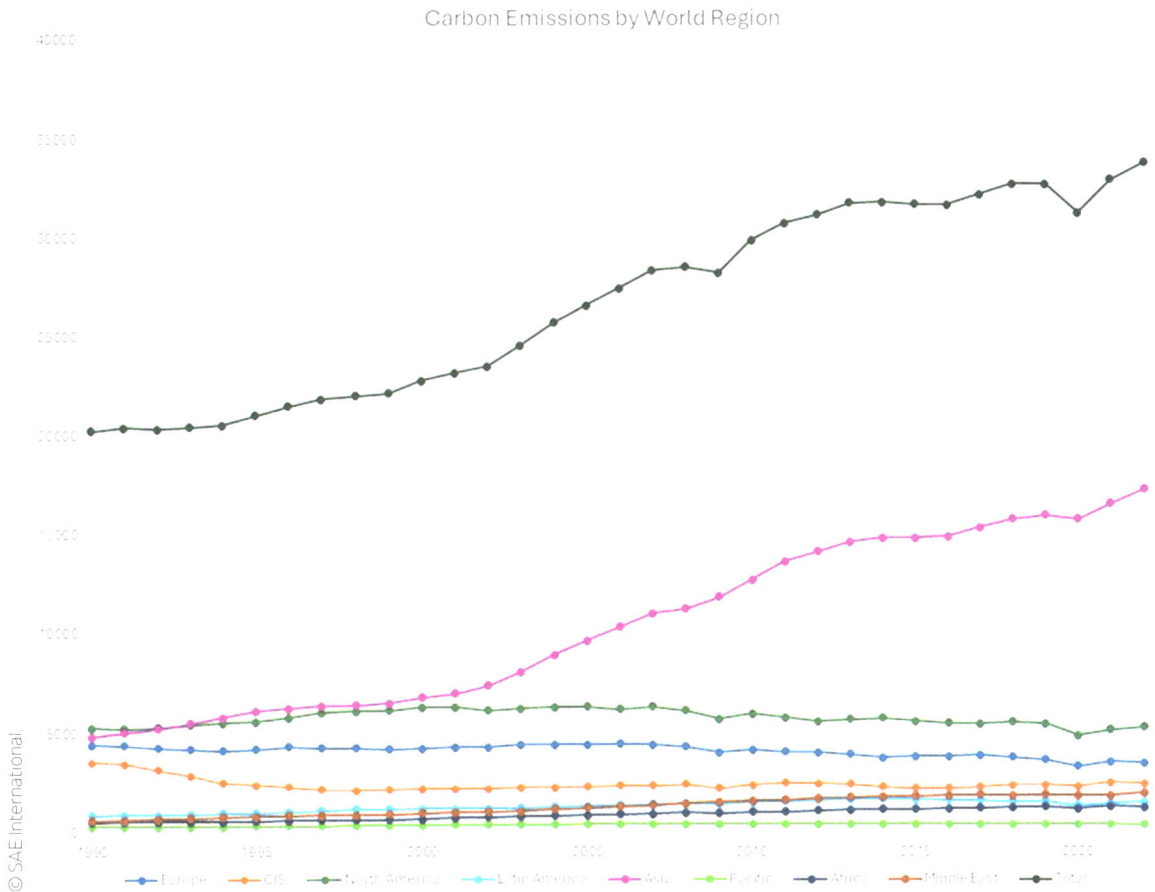

References

1.   https://flowcharts.llnl.gov/.

2.   https://www.epa.gov/system/files/
     documents/2024-04/us-ghg-inventory-2024-main-
     text_04-18-2024.pdf.

3.   https://www.enerdata.net/publications/world-
     energy-statistics-supply-and-demand.html.

When we burn fossil fuels, the combustion products are primarily water vapor and carbon dioxide, which go into the atmosphere. Carbon dioxide is a greenhouse gas, and it is linked to global warming. Fossil fuels also produce significant amounts of sulfur and trace impurities of mercury, lead, and other heavy metals. This is especially true for coal. There are things that can be done at the power plant to remove heavy metals like lead and mercury, but sulfur is difficult to remove. Most of the sulfur will go into the atmosphere as sulfur dioxide, which is the primary source of acid rain.

Electric vehicles are pollution free at first glance, but most of the electricity used to charge the vehicles is produced using fossil fuels. To be fair in evaluating electric vehicles, we should include the pollution and carbon emissions associated with producing the electricity the vehicles use. The approach used in this book will be to get an average value of carbon emissions and pollution generated per kW-h of electric energy. From there, we can estimate the amount of carbon dioxide and other pollutants that are generated indirectly from an electric vehicle. We will find that electric vehicles are low-emission vehicles but are not zero-emission vehicles. Hybrid-electric cars and trucks have emissions per mile traveled that are not much higher than electric vehicles.

Wind turbines, solar panels, and hydroelectric dams are considered pollution-free methods of generating electricity. But the production of wind turbines, solar panels, and hydroelectric dams requires that fossil fuels be used for mining, manufacturing, transportation, and maintenance. It is difficult to account for the fossil fuel used in production, and we will not try to account for it in this book. Distribution of fossil fuels to service stations and airports requires fossil fuels be consumed in transportation. It is difficult to account for the fossil fuel used in distribution of fossil fuel, and we will not try to account for it in this book.

In this unit, we will study the chemistry of combustion of fossil fuels. The easiest fuel to study is natural gas, so we begin with natural gas. Natural gas is composed mostly of methane (CH_4). We visualize methane in Figure 6.1. The carbon atom has four covalent bonds, and each of the hydrogen atoms has one covalent bond. On paper, we draw it in two dimensions as illustrated on the left side of Figure 6.1, but in reality, it is a three-dimensional molecule as illustrated on the right side of Figure 6.1. The four hydrogen atoms are arranged as the corners of a tetrahedron or pyramid around the carbon atom.

Figure 6.1 Molecular structure of methane.

© SAE International.

There are measurable amounts of ethane and propane in natural gas. Natural gas will be at least 95% methane, but it is not 100% methane. We can envision the ethane and propane molecules as shown in Figure 6.2. In reality, they are three-dimensional molecules that are twisted and not flat the way we draw them on paper, but the two-dimensional drawings are helpful.

Figure 6.2 Ethane and propane molecules.

© SAE International.

In addition to ethane and propane, there are smaller amounts of CO_2, N_2, H_2S, He, particulate material, and trace amounts of other materials in natural gas. We clean and purify the natural gas as best as we can, but it will never be 100% pure. Since natural gas is primarily methane, the combustion associated with methane is a good approximation for combustion of natural gas. The chemistry equation for complete combustion of methane is shown in Equation (6.1):

$$CH_4 + 2O_2 \rightarrow CO_2 + 2H_2O \tag{6.1}$$

The equation assumes that natural gas is all methane and that we achieve complete combustion. Air is 78.09% nitrogen, 20.95% oxygen, 0.93% argon, with trace amounts of other gasses. The percentages are by volume; the mass percentages are different. In an ideal gas,

each molecule takes up the same amount of volume regardless of its weight. For each 100 molecules in air, we could approximate the chemical composition of air with the following formula:

$$Air = 78N_2 + 21O_2 + Ar \tag{6.2}$$

Chemical reactions are based on probability, and there are many molecules involved in the chemical reaction. There will be trace amounts of

many different chemicals, but there is a strong driving force toward producing carbon dioxide and water. In a good combustion process, we will achieve near-complete combustion, but we can never achieve 100% complete combustion. There will be small amounts of carbon monoxide, and since the air used for combustion contains nitrogen, there will be small amounts of nitrous oxide created during combustion. The hydrogen sulfide impurity in the natural gas will create sulfur dioxide, and the particulate matter will not burn. There are trace amounts of other chemicals, but most of the combustion products are carbon dioxide and water.

For a good combustion process, we can estimate the pollution generated from burning natural gas. For each billion BTUs of heat energy we get from the natural gas, we will generate approximately [1]:

- 40 lb carbon monoxide (CO)
- 1 lb sulfur dioxide (SO_2)
- 92 lb nitrous oxide (NO_x)
- 7 lb particulate

We do not want any of these things in the air we breathe, but it is a part of the process of using natural gas. Carbon monoxide is poisonous. Sulfur dioxide causes acid rain and is mildly poisonous. Nitrous oxides cause smog and are not good for our health either. Breathing the fine particles in the particulate matter causes lung problems. Some of the particulate matter is carcinogenic. Natural gas is cleaner than other fuels that we burn, but it does generate some pollution.

Most of the chemical reaction of burning natural gas will be complete combustion of methane. Water and carbon dioxide are produced from combustion. Both water and carbon dioxide are greenhouse gasses in the sense that they absorb radiation, but the water does not stay in the atmosphere long. It falls as rain. Carbon dioxide

stays in the atmosphere longer. Plants gradually absorb carbon dioxide from the atmosphere using photosynthesis.

6.2.
A Theory for Global Warming

Global warming is a topic that is in the news and is controversial. When I first started offering the course in 2017, there were a lot of people who did not believe that global warming was real. They referred to it as a hoax. Students in the class were split, with most believing global warming was real, but a significant minority believed it was a hoax. Times have changed and the people who still believe that global warming is a hoax are a very small minority. My goal is to teach the science involved and how to do the calculations.

The earth has gone through warmer and cooler periods in its history. There have been times when the earth was much warmer and when it was much cooler than it is today. Geologists study the history of the earth, and the theory that scientists are using to explain global warming comes from geology. It is sometimes referred to as the "snowball earth" theory [2]. It is a good theory, but it is just a theory. It is not necessarily correct, and not all scientists agree with the theory. Other theories explaining how the earth goes through warmer and cooler periods involve the change in earth's rotational orbit known as Milankovitch cycles [3], changes in volcanic activity on the planet, changes in the intensity of the sun, and movement of the tectonic plates on the earth.

Most of the land area on earth is in the Northern Hemisphere. North America, Central America, Europe, and Asia are 100% in the Northern Hemisphere. Most of Africa and part of South

America are in the Northern Hemisphere. Australia and Antarctica are the only continents that are 100% in the Southern Hemisphere. In warmer periods, land areas can support a lot of plant life, and plants remove carbon dioxide from the atmosphere through photosynthesis. In colder periods, a large fraction of the land in North America, Europe, and Asia are covered with snow and ice and there is not as much plant life to remove carbon dioxide from the atmosphere. Carbon dioxide is soluble in water and the oceans will absorb carbon dioxide when the partial pressure of carbon dioxide in the atmosphere is higher, and release carbon dioxide when the partial pressure is lower. Equilibrium concentration in the ocean depends on the partial pressure of carbon dioxide in the atmosphere.

To illustrate the cycle, we will start with a cold period on earth, where a large fraction of the land masses is covered in ice. The ice reflects sunlight, reducing the solar energy absorbed by the earth. We will start with the carbon dioxide concentration in the atmosphere and oceans at a low level.

During the cold period, volcanic activity puts carbon dioxide into the atmosphere faster than plant life can remove the carbon dioxide. When a large fraction of the land area on earth is covered with ice, there is not enough plant life removing carbon dioxide from the atmosphere to keep up with volcanic activity putting carbon dioxide into the atmosphere. Volcanic activity is not just the eruptions of volcanoes. Yellowstone National Park adds a tremendous amount of carbon dioxide to the atmosphere each year from its hot springs and other thermal features. The volcanic activity in Hawaii is always adding carbon dioxide to the atmosphere. There are many places on the planet that have active volcanoes or other

volcanic activity adding carbon dioxide to the atmosphere.

As the carbon dioxide builds in the atmosphere, it absorbs some of the infrared radiation that the earth radiates into space. Carbon dioxide does not absorb much of the visible light coming from the sun, but it does absorb the infrared radiation coming from the earth radiating toward space. The reason that carbon dioxide absorbs infrared radiation and does not absorb visible light has to do with the physical chemistry of carbon dioxide. Wavelengths that the carbon dioxide molecule can absorb are in the infrared portion of the spectrum. The "greenhouse" effect of the carbon dioxide in the atmosphere helps the earth begin to warm. As the partial pressure of carbon dioxide increases in the atmosphere, the oceans absorb some of the carbon dioxide because it is soluble in water.

Melting the ice is a slow process because of the "heat of melting" phase change energy of the ice. It takes approximately 145 BTU to melt 1 lb of ice. The definition of a BTU is the energy to raise one pound of water 1°F. Converting the ice to water takes an additional 145 BTU/lb over and above the energy required to warm it to the melting temperature. It is the phase change energy of ice that makes it so great for cooling our sodas and for keeping things cool when we go camping. The earth will start to warm and melt the ice, but it will take many thousands of years to melt the ice from the large land masses in North America, Europe, and Asia. The carbon dioxide concentration in the atmosphere will continue to rise during this time period and will reach a relatively high value. The ocean will continue to absorb carbon dioxide too.

As the ice recedes, plant life will increase, and the plants will remove carbon dioxide from the atmosphere through the photosynthesis process. At some point, the plants will remove more carbon dioxide from the atmosphere than is added by volcanic activity, and carbon dioxide levels will begin to decline in the atmosphere and in the ocean.

As the carbon dioxide levels decline, the earth will begin to cool, but forming the ice at the poles is a slow process that will take many thousands of years. During that period, the plants will continue to thrive and remove carbon dioxide from the atmosphere and the carbon dioxide concentration in the atmosphere and ocean will decline to a relatively low value. For those who have studied control systems, the ice acts as a lag in the system causing an overshoot oscillation between lower and higher concentrations of carbon dioxide in the atmosphere. If there were no lag from the ice, the earth would move to an equilibrium concentration of carbon dioxide in the atmosphere and oceans, and the earth temperature would stabilize. But because of the ice, we have warmer and cooler periods on earth.

As the ice advances from the poles toward the tropics, the plant life is killed, and the plants are buried. Over a long period of time, the buried plants became the fossil fuel that we are using today. We talk about carbon sequestration today and envision methods of storing carbon dioxide somehow rather than releasing it into the atmosphere. The natural process has been for the plants to take carbon dioxide out of the atmosphere and then be buried in the ground, effectively storing carbon in the ground. As we burn fossil fuels, we are releasing the carbon dioxide back into the atmosphere. Most scientists are concerned that this will lead to very high levels of carbon dioxide in the atmosphere and oceans, which will lead to an exceptionally warm period on earth.

An exceptionally warm period on earth may lead to many environmental consequences. It will lead to melting of the ice in the polar regions,

which will cause the oceans to rise and damage our coasts. A warmer atmosphere will have more moisture and energy. Storms will be stronger and more frequent. The additional rain will help with drought in some parts of the world and will cause more flooding in other parts. Some parts of the world will benefit from a warmer atmosphere and some will be hurt.

The warming of the planet will not be evenly distributed. Models project a large amount of warming in the polar regions and a smaller amount of warming in the tropics. Some areas will benefit from global warming because it lengthens their growing season and allows them to grow more food. Winters will be less harsh in the high latitudes and polar regions. The temperature and carbon dioxide content of the ocean will rise, and we see that this is already killing coral in the ocean. Carbon dioxide dissolved in water forms a very mild carbonic acid, which causes a small change in the pH of the ocean. It is hard to assess the full impact of the increase of carbon dioxide in the ocean.

We have enough data to show clearly that the oceans are rising. Figure 6.3 is a plot of satellite data taken by NASA and is available at Ocean Level Data [4] https://climate.nasa.gov/vital-signs/sea-level/. The graph is based on satellite data starting in 1993. It shows the ocean has been rising on average 3.3 mm annually.

Figure 6.4 is based on tidal data [4], which has been recorded since 1880 and is available on the same website as Figure 6.3. Measuring ocean level to a tenth of a millimeter is difficult, and there is obviously noise in the data, but the trend is clearly upward. Since 1993 the ocean has risen about 90 mm, which is about 3.5 in. A rise of 3.5 in. in 25 years is substantial. Since 1880 the oceans have risen about 230 mm, which is about 9 in.

Figure 6.3 Satellite sea-level observations from NASA.

Source: climate.nasa.gov

Reprinted from climate.nasa.gov.

Figure 6.4 Sea-level ground data.

Reprinted from climate.nasa.gov

The purpose of presenting the sea-level data is to show that we have data clearly establishing that the oceans are rising. Part of the rise in ocean depth is due to the thermal expansion of the water in the oceans due to the temperature increase in the oceans. Part of the rise in ocean depth is due to the melting of ice in the polar regions on earth.

Figure 6.5 shows the rise in carbon dioxide in the atmosphere since 2005 [5]. On average, carbon dioxide has been increasing in the atmosphere at about 2.2 ppm each year since 2005. The data for this chart were collected by the National Oceanic and Atmospheric Administration (NOAA).

Figure 6.6 shows a much longer historical record of carbon dioxide in the atmosphere obtained from ice core samples [5]. As the earth has gone back and forth between warmer and cooler periods, the carbon dioxide has gone back and forth between lower and higher values. A typical cycle lasts about 100,000 years, as shown on the chart. The warmer periods on earth correspond to times where the

Figure 6.5 Atmospheric carbon dioxide since 1958.

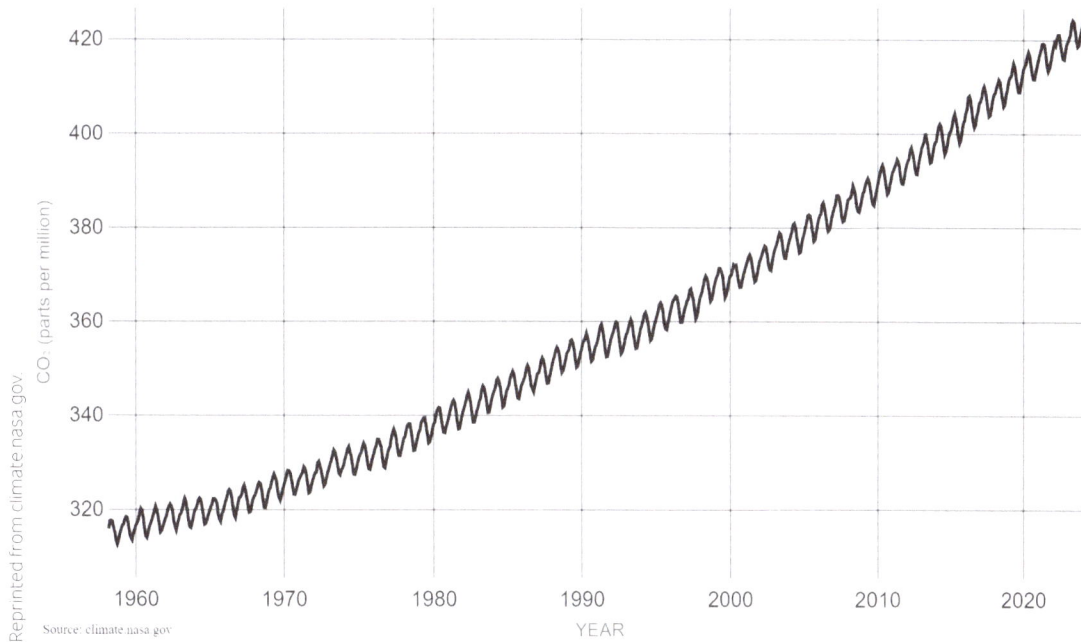

Reprinted from climate.nasa.gov

Source: climate.nasa.gov

Figure 6.6 Carbon dioxide measurements from ice core samples.

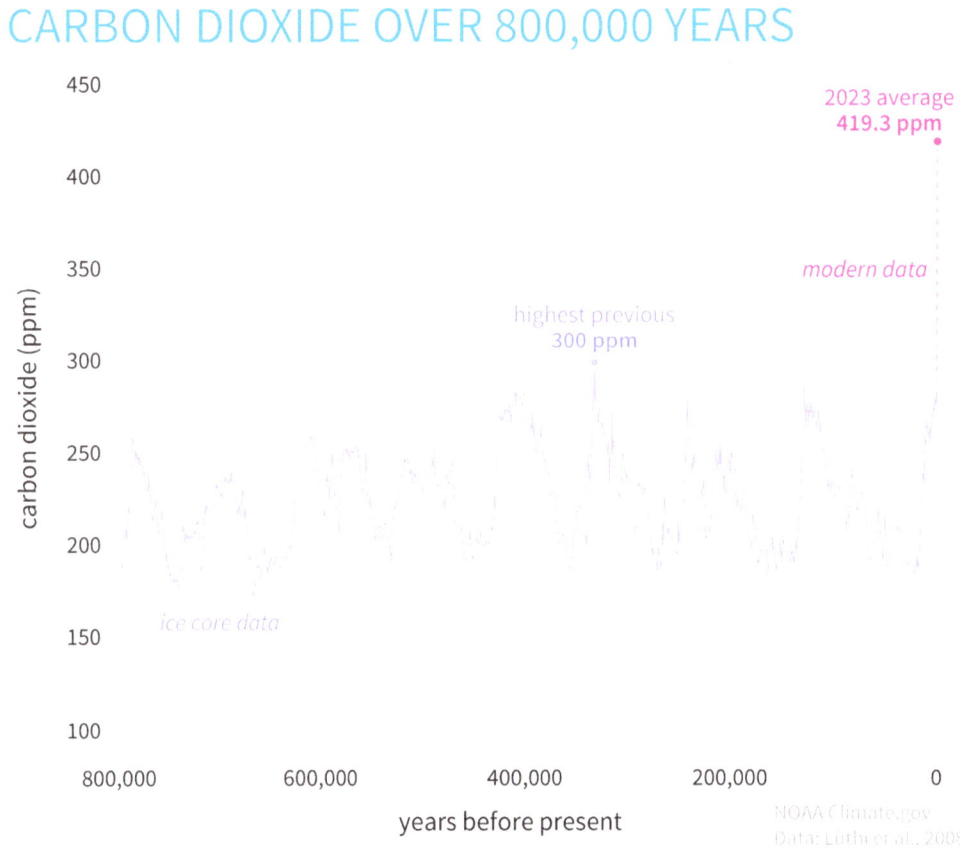

CARBON DIOXIDE OVER 800,000 YEARS

carbon dioxide was at relatively high values, and the cooler periods correspond to times when the carbon dioxide was at lower values. Since 1950 the earth has been at historic high values, much higher than any time in the last 400,000 years. Most scientists believe this historic high carbon dioxide reading in the atmosphere will lead to a very warm period in the history of the earth.

Figure 6.7 shows an average global land–ocean temperature index [5]. There is a lot of noise in these data because measuring the average earth temperature is a difficult thing to do, but the trend is upward. The average global temperature has risen approximately 1.2°C over the last 100 years.

Figure 6.7 Global land–ocean temperature index.

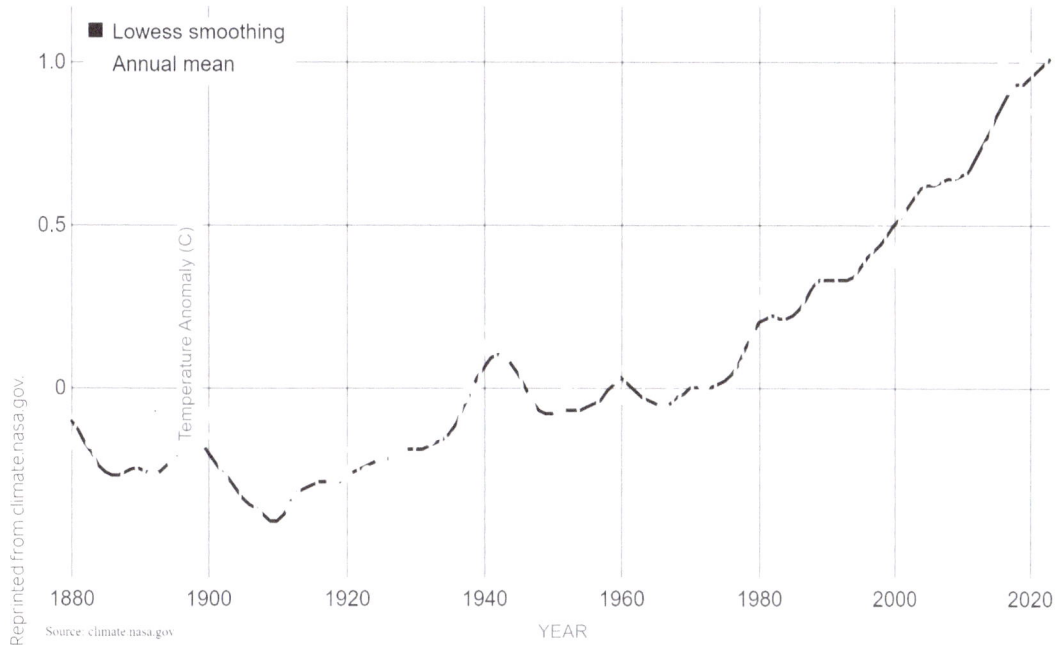

6.3.
Combustion of Natural Gas

In this chapter, we will study the combustion of fossil fuels with a focus on estimating the

pollution and carbon emissions for different types of fossil fuels. The first step is to analyze the combustion of natural gas in more detail as a prelude to analyzing more complicated fossil fuels. The basic equation for combustion of methane is:

$$CH_4 + 2O_2 \rightarrow CO_2 + 2H_2O \qquad (6.3)$$

The atomic weight of carbon is 12, meaning 1 mole (Avogadro's number) of carbon atoms will have a mass of 12 g. Hydrogen has an atomic weight of 1 and oxygen has an atomic weight of 16. In this book I am rounding the molecular weights to whole numbers. In reality the average

carbon atom has a molecular weight slightly higher than 12 but using the exact numbers does not make much difference in the final results. The molar weights of the compounds can be calculated as:

$$\text{Molar weight } CH_4 = 12 + 4(1) = 16 \text{ g / mole} \qquad (6.4)$$

$$\text{Molar weight } CO_2 = 12 + 2(16) = 44 \text{ g / mole} \qquad (6.5)$$

$$\text{Molar weight } H_2O = 2(1) + 16 = 18\,g\,/\,mole \qquad (6.6)$$

There is an important conclusion to this analysis. When we examine the molar weights and the chemical reaction, we conclude that burning 16 g of methane (CH_4) will produce 44 g of carbon dioxide (CO_2) and 36 g of water (H_2O). The reader should look at Equations (6.3) through (6.6) and make sure they understand. These are important conclusions.

Natural gas typically has 950–1050 BTU of energy per standard cubic foot. The standard cubic foot is defined at 60°F and 14.73 psi absolute pressure. We use the ideal gas law to compute the energy per mole for natural gas. The temperature T needs to be converted to Rankine, and we need the constant R for the ideal gas law equation.

$$T = 60 + 459.67 = 519.67°R \qquad (6.7)$$

$$R = \frac{0.023659\,lb\,ft^3}{in.^2\,mole°R} \qquad (6.8)$$

$$PV = nRT \qquad (6.9)$$

Substituting into Equation (6.9):

$$(14.73\,psi)(1\,ft^3) = n\left(\frac{0.023659\,lb\,ft^3}{in.^2\,mole°R}\right)(519.67°R) \qquad (6.10)$$

Solving, n = 1.198 moles per standard cubic foot of gas. Since natural gas has 950–1050 BTU per standard cubic foot and there are 1.198 moles in a standard cubic foot, the fuel has 793 to 876 BTU

of energy per mole. One mole of methane has a mass of 16 g, so dividing by 16 we get the energy density in BTU per gram for the fuel.

$$\text{Energy Density of Natural Gas} = 49.56\,\text{to}\,54.78\,BTU\,/\,g\,CH_4 \qquad (6.11)$$

We also know that 16 g of methane will produce 44 g of carbon dioxide and 36 g of water. By multiplying by (16/44) and (16/36), respectively, the following can be obtained:

- 18.02 to 19.92 BTU energy for each gram of CO_2 produced
- 22.03 to 24.35 BTU energy for each gram of H_2O produced

We can combine this information with the information on pollution discussed previously and make the following conclusions about the combustion of natural gas. For carbon dioxide assume on average we get 19 BTU energy for each gram of CO_2 produced, and for water assume 23 BTU energy per gram produced. In order to get a billion BTU of heat energy from the fuel, the following amounts of CO_2 and H_2O will be produced:

$$\frac{1 \times 10^9 \text{ BTU}}{\dfrac{19 \text{ BTU}}{\text{gram CO}_2}} = 5.263 \times 10^7 \text{ g CO}_2 = 116,000 \text{ lb CO}_2 \qquad \text{(6.12)}$$

$$\frac{1 \times 10^9 \text{ BTU}}{\dfrac{23 \text{ BTU}}{\text{gram H}_2\text{O}}} = 4.349 \times 10^7 \text{ g H}_2\text{O} = 95,600 \text{ lb H}_2\text{O} \qquad \text{(6.13)}$$

We burn fossil fuel for the energy. We might use the energy in a power plant to produce electricity, or we might use the energy to heat our homes. There are many industrial and commercial uses for natural gas. The best way to quantify pollution and carbon emissions of fossil fuel is in terms of a billion BTU of energy. We will compare the results for natural gas to the pollution and carbon emissions for gasoline, diesel fuel, and coal on the basis of a billion BTU of energy. This is how we will assess which fuels are "cleaner" or "dirtier." Cleaner fuels will create less pollution and carbon emissions in providing us with a billion BTU of energy. In summary, for one billion BTU of energy from natural gas we get [1]:

- 40 lb carbon monoxide (CO)
- 1 lb sulfur dioxide (SO_2)
- 92 lb nitrous oxide (NO_x)
- 7 lb particulate
- 116,000 lb carbon dioxide (CO_2)
- 95,600 lb water (H_2O)

The numbers above are good average values for natural gas. Natural gas is produced from many different regions of the country, and the sulfur dioxide and particulate pollution will vary significantly depending on the source. Assuming a good combustion process, the estimates for carbon monoxide, nitrous oxide, carbon dioxide, and water are independent of the source of the natural gas.

In examining the results above, it should be clear that most combustion products from burning natural gas are carbon dioxide and water. Most of the pollution generated comes from the other four products, though there are trace amounts of many different types of pollution. One billion BTUs is about 300 MW-h of heat energy. A power plant is about 33% efficient in converting the heat energy into electricity, so a billion BTU will produce about 100 MW-h of electric energy. This is a typical value for daily production in a natural gas power plant. Looking at the numbers above gives an estimate of the carbon dioxide, water, and pollution produced in a day at a typical natural gas power plant.

6.4.
Combustion of Diesel Fuel

Diesel fuel is a hydrocarbon fuel made from a blend of hydrocarbon molecules. We will study an average-sized hydrocarbon molecule for diesel fuel to get a representative chemical reaction. Diesel has an average of 23 hydrogen atoms for each 12 carbon atoms. It is a blend of sizes. In order to draw a sensible hydrocarbon molecule, there must be an even number of hydrogen atoms, so it is not possible to draw a sensible molecule for $C_{12}H_{23}$. In order to illustrate typical hydrocarbon molecules that may be included in the diesel fuel blend, consider the $C_{10}H_{20}$ molecule in **Figure 6.8**.

Figure 6.8 Typical $C_{10}H_{20}$ molecule.

$C_{10}H_{20}$

© SAE International.

Figure 6.9 Typical $C_{15}H_{28}$ molecule.

$C_{15}H_{28}$

© SAE International.

When drawing sensible molecules, each carbon atom needs to have four bonds and each hydrogen atom has one bond. Two carbon atoms can form a double bond as illustrated near the center of the $C_{10}H_{20}$ molecule. The double bond does not need to be in the middle as shown; it could be between any two carbon atoms. If there were two double bonds in the carbon chain, it would be a $C_{10}H_{18}$ molecule, and there may be $C_{10}H_{18}$ molecules in the blend that make up diesel fuel.

The carbon chains can be branched, in addition to having double bonds. To illustrate possible branching, consider the $C_{15}H_{28}$ molecule illustrated in **Figure 6.9**.

In all the hydrocarbon molecules, please recognize that the four carbon bonds are at the corners of a tetrahedron rather than in a plane as drawn in the figures. The molecules are more bent and twisted in a three-dimensional sense than can be illustrated in two dimensions. For complete combustion, all the molecules combine with oxygen to form carbon dioxide and water as the combustion products. There is a blend of sizes of molecules in the diesel fuel such that the average molecule has a chemical composition of $C_{12}H_{23}$, even though there are no molecules in the mix that have exactly that chemical composition. We can do the chemistry assuming a $C_{12}H_{23}$ molecule, and it gives the right answer even though the actual chemistry is more complex. Complete combustion yields:

$$4C_{12}H_{23} + 71O_2 \rightarrow 48CO_2 + 46H_2O \qquad (6.14)$$

We use air to provide the oxygen, which means there will also be nitrogen in the combustion chamber, and there will be nitrous oxides formed during combustion. Incomplete combustion results in small amounts of carbon monoxide being generated. Diesel fuel also contains small amounts of hydrogen sulfide, which will be converted to sulfur dioxide. The next step in the analysis is to compute the molecular weights of the components in the chemical reaction.

$$\text{Molecular weight of } 4C_{12}H_{23} = 4\left[(12)(12)+(23)(1)\right] = 668\,g \qquad (6.15)$$

$$\text{Molecular weight of } 48CO_2 = 48\left[12+(2)(16)\right] = 2112\,g \qquad (6.16)$$

$$\text{Molecular weight of } 46H_2O = 46\big[(1)(2)+16\big] = 828\,g \qquad (6.17)$$

The basic chemical analysis shows that 668 g of diesel fuel will produce 2112 g of carbon dioxide and 828 g of water. Experimental measurements show that diesel fuel has 138,700 BTU of energy per gallon and that the density of diesel fuel is 832 g/L. Recognizing there are 0.26417 gal in 1 L, it can be shown that the energy density of diesel fuel is:

$$\text{Energy Density Diesel Fuel} = \left(\frac{138,700\,\text{BTU}}{\text{gal}}\right)\left(\frac{0.26417\,\text{gal}}{\text{liter}}\right)\left(\frac{\text{liter}}{832\,g}\right) = 44.04\,\text{BTU}/g \qquad (6.18)$$

Since we get 2112 g carbon dioxide and 828 g water for each 668 g of diesel fuel burned, it follows that:

$$\left(\frac{44.04\,\text{BTU}}{g\,\text{diesel}}\right)\left(\frac{668\,g\,\text{diesel}}{2112\,g\,CO_2}\right) = \frac{13.93\,\text{BTU}}{g\,CO_2} \qquad (6.19)$$

$$\left(\frac{44.04\,\text{BTU}}{g\,\text{diesel}}\right)\left(\frac{668\,g\,\text{diesel}}{828\,g\,H_2O}\right) = \frac{35.5\,\text{BTU}}{g\,H_2O} \qquad (6.20)$$

Equation (6.20) is included to complete the analysis. The water vapor going into the atmosphere is not regarded as a problem. Carbon dioxide is what scientists believe is contributing to global warming. Equation (6.19) is the more important result. Looking back at the analysis for natural gas, it was shown that natural gas gives us 18.02–19.92 BTU of energy per g CO_2 generated, whereas diesel fuel only provides 13.93 BTU/g CO_2. Comparing the two fuels, the natural gas gives more energy for each gram of carbon dioxide generated. Therefore, we say that natural gas has a lower carbon footprint than diesel fuel. We get more energy out of the natural gas for the same amount of carbon dioxide generated. The next fuel to be studied is coal, and it will be shown that coal has a larger carbon footprint than diesel fuel, and a much larger carbon footprint than natural gas.

In addition to the carbon dioxide, diesel vehicles emit NO_X compounds, carbon monoxide, and particulates. The particulate or soot from diesel vehicles contains carcinogens and used to be much more significant than it is today. Improvements in diesel fuel allow diesel engines to emit far less particulate than they did 20 years ago. Sulfur has all but been eliminated from the diesel fuel we use today. In order to be able to compare to natural gas, I have used data collected in 2018 [6] and converted it into pounds per billion BTU of heat energy in the fuel. For each billion BTU energy in the fuel, the following emissions will be generated:

- 85 lb NO_X [compared to 92 lb for natural gas]
- 1480 lb CO [compared to 40 lb for natural gas]
- 0 lb SO_2 [compared to 1 lb for natural gas]
- 2.2 lb Particulate [compared to 7 lb for natural gas]
- 158,265 lb CO_2 [compared to 116,000 lb for natural gas]
- 62,050 lb H_2O [compared to 95,600 lb for natural gas]

These numbers are typical for diesel-powered cars and trucks, and for most ocean vehicles that burn diesel fuel (bunker fuel). In the past, ships at sea used a fuel that had a higher sulfur content and was less expensive. They used a cleaner fuel when in port and a dirtier fuel when at sea. Recent international agreements require ships to use a cleaner fuel even when at sea, though realistically it would be difficult to enforce the agreements on countries that wanted to allow ships to use the dirtier fuel. Today, most ships use a cleaner fuel even when at sea.

The chemistry associated with burning gasoline is assigned as one of the homework problems. Gasoline is a cleaner fuel in the sense that it produces essentially zero particulate and sulfur dioxide emissions. For comparison, NO_x emissions are approximately 127 lb/billion BTU and carbon monoxide emissions are approximately 1740 lb/billion BTU [7]. Gasoline engines emit high amounts of carbon monoxide compared to diesel engines. Carbon dioxide emissions for gasoline are slightly less than for diesel fuel, but there is not a lot of difference. In the grand scheme, diesel, gasoline, and jet fuel all have about the same carbon emissions.

Natural gas has significantly lower carbon emissions and coal has significantly higher carbon emissions than the petroleum fuels used by vehicles.

In the homework, you will calculate the carbon dioxide concentration in the atmosphere and compare it to the carbon emissions of the United States and of the world. You will find that the world's emissions of carbon dioxide from burning fossil fuels more than account for the increase in carbon dioxide in the atmosphere. Some of the carbon dioxide generated from burning fossil fuels is being removed by plants and absorbed by the oceans. You will also study the carbon dioxide emissions of cars in the homework and calculate the grams of CO_2 per mile traveled. In the literature, you will find the carbon emissions of cars given in grams of CO_2 generated per mi traveled (g CO_2/mi). If you know the mpg rating for the vehicle, it is a straightforward calculation. As an example, suppose a small, diesel-powered car gets 46 mpg. Diesel has 138,700 BTU/gal, and in Equation (6.19) we calculated that the fuel provides 13.93 BTU of energy for each gram of CO_2 generated. The calculation is as follows:

$$\left(\frac{138,700 \text{ BTU}}{\text{gal}}\right)\left(\frac{\text{gal}}{46 \text{ mi}}\right)\left(\frac{\text{g } CO_2}{13.93 \text{ BTU}}\right) = 216.5 \text{ g } CO_2 \text{ / mi} \tag{6.21}$$

Once you have worked through the chemistry for gasoline, you will be able to make the calculations for gasoline-powered cars too. For electric cars, we will use the information from the Department of Energy (Chapter 5) that says in the United States we generated 13.3 quads of electric energy in 2022. The Environmental Protection Agency (EPA) estimates that 1531.7 million metric tons (MMT) of carbon dioxide was generated in producing electricity in 2022. From these values you can calculate the number of grams of CO_2 generated for each kW-h of electric energy (393 g CO_2/kW-h). Then if an electric car can travel 100 mi on 40 kW-h of electric energy in the batteries, the carbon footprint is calculated as:

$$\left(\frac{393 \text{ g } CO_2}{\text{kW-h}}\right)\left(\frac{40 \text{ kW-h}}{100 \text{ mi}}\right) = 157.2 \text{ g } CO_2 \text{ / mi} \tag{6.22}$$

As you go through the calculations you will find that, for vehicles of similar size, the fully electric vehicles have the lowest carbon footprint. Hybrid-electric cars have a carbon footprint that is higher than electric vehicles, but not a lot higher. Regular gasoline-powered cars have a significantly higher carbon footprint than electric cars or hybrid-electric cars of similar sizes. Plug-in hybrid vehicles have carbon emissions that are between electric and hybrid vehicles. As long as we produce most of our electricity using fossil fuels, there is not a huge environmental benefit in driving electric vehicles instead of hybrid-electric vehicles or plug-in hybrid vehicles. If we can use renewable energy to generate our electricity, then electric vehicles are better for the environment.

6.5.
Combustion of Coal

Coal is complicated, because the composition depends on where it is mined. Coal also contains a significant amount of nonburnable material (mostly dirt and moisture). The softer coal contains more moisture than harder coal. Water shows up as hydrogen and oxygen in the spectrographic analysis of the coal. Coal can be divided into five basic categories:

1. Lignite is the softest coal that is used for generating power.
2. Sub-bituminous coal is a little harder and is the coal mined in the Powder River Basin in Wyoming and burned in many power plants.

3. Bituminous coal is a little harder than sub-bituminous. Eastern coal used in many power plants is bituminous.
4. Steam coal is a little harder than bituminous and softer than anthracite.
5. Anthracite coal is a little too hard to be ground up and used in power plants.

A spectrographic analysis can be performed on coal samples to determine the weight fractions of the different elements in the coal. Coal will have significant amounts of carbon, oxygen, hydrogen, and sulfur. There will be trace amounts of many other elements. For the purpose of combustion analysis, we will only consider carbon, oxygen, hydrogen, and sulfur. For generating electricity, we are primarily interested in sub-bituminous and bituminous coal, though significant amounts of lignite are used. Table 6.1 gives reasonable ranges of values for these types of coal [8].

The elemental percentages in Table 6.1 are in terms of mass. There is a lot of variation in coal depending on where it is mined and different authors may give values that are slightly different from what is given in the table, but these are good average values. Some coal has small amounts of nitrogen, which I have not included in the table. In order to explain the combustion of coal, we need reasonable values to work with, and the values in Table 6.1 will work well in helping the reader understand the combustion of coal.

Table 6.1 Typical chemical content and energy density of coal.

Coal Type	Carbon %	Hydrogen %	Oxygen %	Sulfur %	Energy (BTU/lb)
Lignite	60–75	6.0	33.5–16	0.5–3	7300–10,700
Sub-bituminous	75–82	5.9	18.1–11.1	1	10,400–11,700
Bituminous	82–88	5.8-5.6	11.2–5.4	1	11,700–12,800

© SAE International.

The oxygen in the coal is tied with hydrogen as water. The energy in the coal comes from burning the carbon and the hydrogen that is not tied to the oxygen as water. Burning the sulfur also contributes a small amount of energy. If there is nitrogen in the coal, it contributes almost nothing to the chemical reaction or the energy content of the coal.

The primary pollutants associated with burning coal are carbon monoxide, sulfur dioxide, nitrous oxides, and particulates. These are the same pollutants produced by burning other fossil fuels, but burning coal generates a lot more particulate than other fossil fuels. Coal is a solid fuel that is mined, and there is a significant amount of dirt, minerals, and moisture mixed with the coal. Companies have shaker operations to shake the dirt and minerals off and will wash the coal with water to remove more of the dirt and minerals. The coal is dried to remove most of the moisture. But it is inevitable that the coal delivered to power plants will contain a significant amount of dirt, minerals, and moisture. The dirt and minerals will be ground up with the coal and go through the combustion process in the furnace, but they will not burn. This is what generates the coal ash, or particulate, that comes from burning coal. As a general rule, you can estimate that about 5% of the coal delivered to the power plant will become coal ash (or particulate). This is primarily because approximately 5% of the coal delivered to the powerplant is dirt that could not be removed from the coal. If a large coal plant uses 20 train-car loads of coal per day, it will generate about one train-car load of coal ash per day.

Modern coal plants have the technology to remove almost all the particulate and prevent it from going up the smokestack. One hundred years ago, a large fraction of the particulate went up the smokestack and became air pollution. In addition to breathing the particulate, people who lived near the power plant would wake up every morning to a thin layer of black goo deposited on buildings, roads, and everything else. It was a very dirty environment. Modern coal plants emit very little particulate to the atmosphere. Coal plants are much cleaner than they were 100 years ago, but they generate a lot of coal ash, which must be landfilled. Coal ash contains trace amounts of heavy metals and carcinogens, and we need to be careful with disposal to prevent causing a health hazard.

We need a way to estimate the pollution from burning coal so we can compare it to the pollution from burning natural gas and other fossil fuels. The values for sulfur dioxide and particulate are going to vary a lot depending on where the coal is mined. The values in Table 6.2 are approximate. In the table, "brown coal" refers to lignite and sub-bituminous coal similar to what is mined in the Powder River basin. "hard coal" is bituminous and is similar to eastern coal. Please use Table 6.2 to estimate the pollution generated by burning coal. The numbers in the table are pounds of pollution for each billion BTU of heat energy in the coal [9].

Table 6.2 Typical pollution values for hard and brown coal.

Pollutant	Hard coal (lb)	Brown coal (lb)
SO$_2$ (lb/billion BTU)	1779	3166
NO$_X$ (lb/billion BTU)	679	426
CO (lb/billion BTU)	207	207
Particulate (lb/billion BTU)	2798	7569

Most of the particulate is removed at the power plant and is landfilled as coal ash. The other pollutants will go into the atmosphere.

If you compare to the values for diesel fuel and natural gas, the pollution from coal is higher in all cases. Burning gasoline produces more carbon monoxide than coal, but gasoline produces lower values for all other emissions. Particulate emission is much higher in coal than in other fuels, and this needs to be discussed.

With liquid and gas fuels, it is possible to put them in a large settling tank where most of the heavier impurities will settle to the bottom. It is an inexpensive way to reduce the contamination in the fuel. Gas and liquid fuels can also be run through a filter, further removing contamination. These simple inexpensive processes of settling and filtering remove nearly all the contamination and make gas and liquid fuels cleaner.

Coal is a solid material and it is much more difficult to remove contamination from a solid. There are mechanical processes to remove the dirt and minerals that inevitably come mixed with the coal. The industry does a good job of removing impurities. Only a few percent of the coal delivered is noncombustible material, but power plants use a lot of coal, so a few percent ends up being a lot of material. Coal plants remove almost all the particulate; very little is put into the atmosphere. Trace amounts of mercury and radon will be emitted as gasses because they cannot be removed. The particulate

matter becomes the coal ash that must be landfilled. Coal ash is largely dirt and minerals that have been ground into a fine powder and run through the furnace, with trace amounts of heavy metals and carcinogens. Breathing the dust is hazardous to our health, but properly landfilled coal ash does not present a large environmental problem. There have been power companies that did not properly landfill their coal ash, causing environmental problems. From a practical viewpoint, companies will occasionally be negligent in disposing of their coal ash and some of the coal ash will cause environmental problems.

Samples of coal are studied using a mass spectrometer to determine the mass concentration of the elements in the coal. Averaging several samples gives an average composition for the coal. The results of the analysis can be used to study the chemistry of the coal. As an example, assume that the analysis indicates the coal is 80% carbon, 5.9% hydrogen, 13.1% oxygen, and 1% sulfur. These percentages are by mass. The chemistry is done by number of atoms rather than by mass, so the first step in the analysis is to divide each mass percentage by the atomic mass of the element and add them up. As with the previous analysis, I have rounded the masses to 12 for carbon, 1 for hydrogen, 16 for oxygen, and 32 for sulfur.

$$\frac{80}{12} + \frac{5.9}{1} + \frac{13.1}{16} + \frac{1}{32} = 13.4167 \tag{6.23}$$

The atomic percentages are then obtained by dividing the appropriate term on the left side by

the sum on the right side of the equation. For example, the carbon percentage is:

$$Carbon\% = \frac{\frac{80}{12}}{13.4167} \times 100\% = 49.6894\% \tag{6.24}$$

The percentage by number of atoms in the coal is then:

- Carbon = 49.6894%
- Hydrogen = 43.9752%
- Oxygen = 6.1025%
- Sulfur = 0.2329%

The atoms are combined in the coal mostly as hydrocarbons, water, and hydrogen sulfide. The oxygen content indicates that there is significant moisture in the coal, and some of the hydrogen is tied to oxygen atoms as water. Some of the hydrogen is tied with carbon forming hydro-carbons. Some of the carbon exists as carbon graphite. The sulfur is probably tied with either hydrogen or oxygen as hydrogen sulfide or sulfur dioxide. The mass spectrometer analysis tells us the atomic mass percentages in the coal, but it cannot tell us how the atoms are tied together as molecules. When doing the chemistry of coal, we do not need to know exactly how the atoms are tied together. We will start with the percentage compositions above and write a balanced equation. This method of chemical analysis yields a good approximation to the burning of coal. The atomic components in the coal will be combined with oxygen in the air to form carbon dioxide, water, and sulfur dioxide. The left side of the chemical equation is written as:

$$(49.6894)C + (43.9752)H + (6.1025)O + (0.2329)S + (\underline{\hspace{1cm}})O_2 \rightarrow \qquad (6.25)$$

At this point in the analysis, we do not know how much oxygen will need to be added for complete combustion. All the carbon must become carbon dioxide. All the hydrogen must become water, and all the sulfur must become sulfur dioxide. The right side of the chemical equation is written as:

$$(49.6894)CO_2 + \left(\frac{43.9752}{2}\right)H_2O + (0.2329)SO_2 \qquad (6.26)$$

The oxygen required for combustion will be provided from air. There is a blank that goes with the oxygen in the air in Equation (6.25) that will be calculated. On the right side, all the carbon will form carbon dioxide, so the number with the carbon dioxide is 49.6894. The hydrogen will all form water, so the number with the water is half of 43.9752. All the sulfur will form sulfur dioxide, so the number with the sulfur dioxide is 0.2329. The oxygen on the left side from the air is then calculated so that the total oxygen on the left and right sides balance. Total oxygen on the right side is:

$$(49.6894)(2) + (21.9876) + (0.2329)(2) = 121.832 \qquad (6.27)$$

Subtract the 6.1025 oxygen already on the left side in the coal and then divide by two to get the O_2 that comes from air. The final balanced equation is:

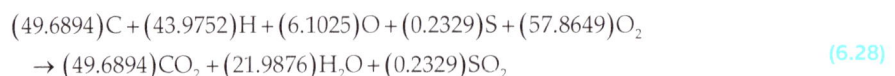

$$(49.6894)C + (43.9752)H + (6.1025)O + (0.2329)S + (57.8649)O_2$$
$$\rightarrow (49.6894)CO_2 + (21.9876)H_2O + (0.2329)SO_2 \qquad (6.28)$$

The total molar weight of the coal on the left side is:

$$(49.6894)(12)+(43.9752)(1)+(6.1025)(16)+(0.2329)(32)=733.341\,\text{g Coal} \qquad (6.29)$$

The molar weights for the carbon dioxide, water, and sulfur dioxide on the right side are:

$$(49.6894)\left[12+(2)(16)\right]=2186.33\,\text{g }CO_2 \qquad (6.30)$$

$$(21.9876)\left[(1)(2)+16\right]=395.777\,\text{g }H_2O \qquad (6.31)$$

$$(0.2329)\left[32+(2)(16)\right]=14.906\,\text{g }SO_2 \qquad (6.32)$$

The conclusion from the chemistry is that burning 733.341 g of coal will generate 2186.33 g CO_2, 395.777 g H_2O, and 14.906 g SO_2. This process is helpful in determining the amount of carbon dioxide and sulfur dioxide generated in burning coal.

A lab test must be performed to determine the heat energy that comes from burning the coal. For the example of coal, assume the heat energy is 11,200 BTU/lb. That is, burning 1 lb of the coal will generate 11,200 BTU of heat energy. Recognizing that 733.341 g coal generates 2186.33 g CO_2 and that there are 453.6 g/lb, the carbon footprint of the fuel can be calculated as:

$$\left(\frac{11,200\,\text{BTU}}{\text{lb Coal}}\right)\left(\frac{\text{lb}}{453.6\,\text{g}}\right)\left(\frac{733.341\,\text{g Coal}}{2186.33\,\text{g }CO_2}\right)=8.28\,\text{BTU / g }CO_2 \qquad (6.33)$$

The water and sulfur dioxide can be estimated in a similar manner. Water is not usually regarded as a problem as far as emissions. The estimate for sulfur dioxide is:

$$\left(\frac{11,200\,\text{BTU}}{\text{lb Coal}}\right)\left(\frac{\text{lb}}{453.6\,\text{g}}\right)\left(\frac{733.341\,\text{g Coal}}{14.906\,\text{g }SO_2}\right)=1214.8\,\text{BTU / g }SO_2 \qquad (6.34)$$

At this point, we have discussed natural gas, diesel fuel, and coal. With natural gas, we get 18.02 to 19.92 BTU of heat energy for each g CO_2 generated. The average value for natural gas is 18.97 BTU/g CO_2. Diesel fuel provides 13.93 BTU/g CO_2, and the coal example provides 8.28 BTU/g CO_2. It was discussed earlier that a typical value for a natural gas power plant was to use a billion BTU of energy in one day. Based on using a billion BTU of energy, a natural gas plant would produce 116,000 lb of CO_2. A diesel plant would produce 158,000 lb of CO_2, and a coal plant using the example coal would produce 266,000 lb of CO_2. Table 6.3 provides a summary of the fossil fuels studied and the emissions produced.

From Table 6.3 it should be clear that natural gas is the cleanest of the fossil fuels considered. Coal is the dirtiest of the fossil fuels. Gasoline was assigned as a homework problem. The carbon dioxide emissions from gasoline are similar to that of diesel fuel. Gasoline has essentially zero sulfur dioxide and particulate emissions. The carbon monoxide and nitrous oxide emissions are 1740 lb and 127 lb per billion BTU,

respectively. The carbon monoxide emissions for gasoline and diesel fuel are relatively high because they are used in internal combustion engines. If gasoline and diesel were used to fire the burner for a steam turbine, the carbon monoxide emissions would be much lower. Comparisons are based on a billion BTU, which is a typical amount of energy used in a medium-sized natural gas power plant on an average day. Large coal plants use several billion BTU of energy on a typical day, and a diesel-powered generating station would use much less than a billion BTU per day. Assuming a 33% thermal efficiency in the power plant, using a billion BTU would produce 97.7 MW-h of electric energy.

Table 6.3 Carbon dioxide and pollution emissions per billion BTU energy.

Emission (lb)	Natural gas	Diesel	Hard coal	Soft coal
CO_2	116,000	158,000	266,000	245,000
SO_2	1	Zero	1779	3166
NO_x	92	85	679	426
CO	40	1480	207	207
Particulate	7	2.2	2798	7569

© SAE International

6.5.1.
Carbon Emissions for Common Items

Most of the energy we use comes from fossil fuels. Virtually everything we purchase was partially manufactured and transported using fossil fuel. Everything we purchase has a "carbon footprint" because fossil fuels are used in manufacturing and transportation. There is a group of scientists who have tried to quantify how much carbon dioxide is generated from different products. The information is posted on the website http://www.co2list.org/. I included this information for the readers who are interested in how to calculate the carbon footprint of common everyday items. The website is maintained by a group of scientists who want to

reduce carbon footprint. It is a summary of many data sources and attempts to include everything in estimating the total carbon footprint.

The carbon footprint for natural gas and coal on the above website is higher than what we calculated in the previous section. Our calculations include only the carbon generated from burning the fuel. The website includes the carbon associated with mining and transporting the fuel to where it is burned in addition to the carbon associated with burning the fuel.

I have included Table 6.4 for curiosity. There is more information on the website for those who are interested [10] (Table 6.4).

Table 6.4 Carbon dioxide to produce common items.

Food

22 kilos CO_2 per kilo of red meat

5.9 kilos CO_2 per kilo of chicken, fish, or eggs

2.9 kilos CO_2 per kilo of cereal or carbohydrate

1.7 kilos CO_2 per kilo of fruit or vegetables

Miscellaneous

381 kilos CO_2 per square meter, to build houses

225 kilos CO_2 per square meter, to make solar cells

3.2 kilos CO_2 per liter of heating oil

2.4 kilos CO_2 per cubic meter of natural gas

0.8 kilos CO_2 per kW-h of electricity

0.5 kilos CO_2 per dollar spent in the United States, average

Transportation (These include making vehicles, roads, rails, airports, etc.)

0.47 kilos CO_2 per passenger-km on flights of 800 km

0.35 kilos CO_2 per passenger-km on flight of 1300 km

0.37 kilos CO_2 per passenger-km on cruise

0.34 kilos CO_2 per kilometer in a 12 kpl car (28.2 mpg)

0.28 kilos CO_2 per kilometer in a 17 kpl car (37.6 mpg)

0.11 kilos CO_2 per passenger-km in a train

0.13 kilos CO_2 per kilometer bicycling on bike lane

0.05 kilos CO_2 per kilometer bicycling without bike lane

0.20 kilos CO_2 per passenger-km in local bus

0.05 kilos CO_2 per passenger-km in long-distance bus

6.6.
Homework

Partial correct answers are included to give you a way to check your answers.

1. The mass of the earth's atmosphere is approximately 5.15×10^{18} kg. Assume that this corresponds to approximately 1.78×10^{20} moles of gas molecules. Our current estimate is that the carbon dioxide content of the atmosphere is about 420 ppm, which means that for every million gas molecules in the atmosphere, approximately 420 of them are carbon dioxide. The carbon dioxide content of the atmosphere

has been increasing about 2.2 ppm each year for the past 15 years.

For this homework problem make an estimate of the significance of the carbon footprint of the United States.

(a) Based on 420 ppm, calculate the number of moles of carbon dioxide in the atmosphere. Then calculate the total number of g CO_2 in the atmosphere.

(b) Based on an increase of 2.2 ppm annually, calculate the number of moles of CO_2 that are added to the atmosphere each year. Then calculate the total number of g of CO_2 added to the atmosphere.

(c) Based on data from the EPA the United States emitted 5053 MMT of CO_2 in 2022. How many g of CO_2 does the United States add to the atmosphere each year? (Answer: 5.2053×10^{15} g, or about 0.665 ppm in the atmosphere.)

(d) Compare the US CO_2 emissions each year to the 400-ppm total CO_2 in the atmosphere and the 2.2 ppm increase in CO_2 each year. Write a short paragraph about the comparison and explain whether you feel that US carbon emissions are significant or insignificant as far as building CO_2 in the atmosphere.

2. In 2022 the world CO_2 emissions were estimated as 33,811 MMT. This estimate is a little low because it only includes CO_2 emissions from the burning of fossil fuels. Most CO_2 emissions come from burning fossil fuels, so it is a low but reasonable estimate.

(a) With this estimate, how many grams of CO_2 does the world add to the atmosphere each year? (Answer: 33.811×10^{15} g, or about 4.32 ppm in the atmosphere each year.)

(b) Compare the world's CO_2 emissions each year to the 410-ppm total CO_2 in the atmosphere and the 2.2 ppm increase in CO_2 each year. Write a short paragraph about the comparison and explain whether you think the world's carbon emissions are significant or insignificant as far as building CO_2 in the atmosphere.

(c) If world carbon emissions are enough to add 4.32 ppm to the atmosphere each year, but the carbon content of the atmosphere only increases 2.2 ppm each year, where does the extra 2.12 ppm go?

3. Carbon footprint of electric vehicles. Electric vehicles are commonly regarded as having zero carbon footprint, but a significant portion of the electricity used in the cars will come from fossil fuels. In this problem you are to estimate the carbon footprint of four electric vehicles in grams of CO_2 per mile traveled (g CO_2/mi).

In the US in 2022, we produced 13.3 quads of electric energy according to the information from Lawrence Livermore National Laboratory. The EPA estimate is that we generated 1531.7 MMT of carbon dioxide in 2022 generating electricity. For this problem, assume that we will generate 1531.7 MMT carbon dioxide to generate 13.3 quads of electric energy. From this information, you should be able to calculate the grams of carbon dioxide generated per kW-h of electric power (g CO_2/kW-h). Then knowing the vehicle can travel on the battery capacity you can calculate the grams per mile CO_2 generated by the vehicle. (Answer: The United States generates 393 g of CO_2 for each kW-h of electric energy.) (*For those who are*

interested, in 2015 the United States generated 553 g CO_2/kW-h. The carbon emissions generating electricity have declined because of the increase in renewable energy and natural gas used to generate electricity and the decline in the use of coal.)

(a) The Chevy Bolt is a small electric car that has a typical cost of $30,000. The battery capacity is 65 kW-h and the range is 278 mi. If you pay an average of $0.14 per kW-h for electricity, calculate the cost of driving the car 15,000 mi. Calculate the carbon footprint in g CO_2 per mile traveled. (Answer: $491.00, 91.9 g CO_2/mi traveled.)

(b) The Tesla Model 3 base is a medium-sized electric car that has a typical cost of $40,000. The battery capacity is 60 kW-h and the range is 272 mi. If you pay an average of $0.14 per kW-h for electricity, calculate the cost of driving the car 15,000 mi. Calculate the carbon footprint in g CO_2 per mile traveled. (Answer: $463.00, 86.7 g CO_2/mi)

(c) The Kia EV6 is a small electric SUV that has a typical cost of $41,000. The battery capacity is 77.4 kW-h and the range is 310 mi. If you pay an average of $0.14 per kW-h for electricity, calculate the cost of driving the car 15,000 mi. Calculate the carbon footprint in g CO_2 per mile traveled. (Answer: $524.00, 98.1 g CO_2/mi)

(d) The Ford Lightning is an electric pickup truck that has a typical cost of $70,000. The battery capacity is 131 kW-h and the range is 300 mi. If you pay an average of $0.14 per kW-h for electricity, calculate the cost of driving the car 15,000 mi. Calculate the carbon footprint in g CO_2 per mile traveled. (Answer: $917.00, 171.67 g CO_2/mi)

4. Assume that the typical gasoline molecule has the formula C_8H_{18}.

  (a) Draw the molecule.

  (b) Work out the chemical reaction assuming complete combustion.

  (c) Calculate the number of BTUs of energy in the gasoline for each gram of CO_2 generated. Assume gasoline has 125,000 BTU/gal. Assume that the density of gasoline is 770 g/L. (Answer: 13.89 BTU/g CO_2)

  (d) Compare the carbon footprint of gasoline to natural gas, diesel, and coal.

5. Carbon footprint of hybrid-electric vehicles.

  (a) The Toyota Prius is a small hybrid-electric car that has a typical cost of $33,000. It is rated at 57 mpg on regular gas. Assume the gasoline has 125,000 BTU/gal. Assume gasoline cost of $3.00 per gallon and calculate the cost of driving 15,000 mi. Calculate the carbon footprint of the Toyota Prius in g CO_2 per mile. Use the results from problem 4 above in making your calculations. (Answer: $789.00, 157.9 g CO_2/mi)

  (b) The Toyota Camry is a medium-sized hybrid-electric car that has a typical cost of $33,000. It is rated at 52 mpg on regular gas. Assume the gasoline has 125,000 BTU/gal. Assume gasoline cost of $3.00 per gallon and calculate the cost of driving 15,000 mi. Calculate the carbon footprint of the Toyota Camry in g CO_2 per mile. Use the results from problem 4 above in making your calculations. (Answer: $865.00, 180.0 g CO_2/mi)

  (c) The Honda CRV is a hybrid SUV that has a typical cost of $38,000. It is rated at 38 mpg on regular gas. Assume the gasoline has 125,000 BTU/gal. Assume gasoline cost of $3.00 per gallon and calculate the cost of driving 15,000 mi. Calculate the carbon footprint of the Honda CRV in g CO_2 per mile. Use the results from problem 4 above in making your calculations. (Answer: $1184.00, 236.8 g CO_2/mi)

  (d) The Toyota Tundra is a hybrid pickup truck car that has a typical cost of $62,000. It is rated at 22 mpg on regular gas. Assume the gasoline has 125,000 BTU/gal. Assume gasoline cost of $3.00 per gallon and calculate the cost of driving 15,000 mi. Calculate the carbon footprint of the Toyota Tundra in g CO_2 per mile. Use the results from problem 4 above in making your calculations. (Answer: $2045.00, 409.1 g CO_2/mi)

6. Carbon footprint of gasoline vehicles.

  (a) The Honda Civic is a small gasoline car that has a typical cost of $27,000. It is rated at 35 mpg on regular gas. Assume the gasoline has 125,000 BTU/gal. Assume gasoline cost of $3.00 per gallon and calculate the cost of driving 15,000 mi. Calculate the carbon footprint of the Honda Civic in g CO_2 per mile. Use the results from problem 4 above in making your calculations. (Answer: $1286.00, 257.1 g CO_2/mi)

  (b) The Honda Accord is a medium-sized gasoline car that has a typical cost of $31,000. It is rated at 32 mpg on regular gas. Assume the gasoline has 125,000 BTU/gal. Assume gasoline cost of $3.00 per gallon and calculate the cost of driving 15,000 mi. Calculate the carbon footprint of the Honda Accord in g CO_2 per mile. Use the results from problem 4 above in making your calculations. (Answer: $1406.00, 281.2 g CO_2/mi)

(c) The Ford Escape is a small SUV that has a typical cost of $34,000. It is rated at 30 mpg on regular gas. Assume the gasoline has 125,000 BTU/gal. Assume gasoline cost of $3.00 per gallon and calculate the cost of driving 15,000 mi. Calculate the carbon footprint of the Ford Escape in g CO_2 per mile. Use the results from problem 4 above in making your calculations. (Answer: $1500.00, 300.0 g CO_2/mi)

(d) The Ford F 150 is a pickup truck that has a typical cost of $54,000. It is rated at 20 mpg on regular gas. Assume the gasoline has 125,000 BTU/gal. Assume gasoline cost of $3.00 per gallon and calculate the cost of driving 15,000 mi. Calculate the carbon footprint of the F 150 in g CO_2 per mile. Use the results from problem 4 above in making your calculations. (Answer: $2250.00, 450.0 g CO_2/mi)

7. Plug-in hybrid vehicles travel a significant distance on battery power before the gasoline engine is activated. Plug-in hybrid cars will typically allow the driver to use 80% of the battery capacity before activating the gasoline engine. People who have short daily commutes may be able to drive almost entirely on electric power. People who have longer commutes will drive using a mix of gasoline and electric power. When taking the vehicle on a long trip it runs almost entirely on gasoline power. At the time of this publication, the EPA method of determining carbon footprint is to assume 55% of the driving is on electric power and 45% on gasoline power. Hybrid vehicles are a relatively new innovation and the standards

could change, but for this problem assume that 55% of the driving is on electric power and 45% is on gasoline power.

(a) The Prius Prime is a small plug-in hybrid car that has a typical cost of $37,000. The usable battery capacity is 10.9 kW-h and the car can travel 40 mi on the battery before the gasoline engine is activated. Once the gasoline engine is activated, the car will get 48 mpg on regular gas. Assume the gasoline has 125,000 BTU/gal. Assume gasoline cost of $3.00 per gallon and electricity cost of $0.14/kW-h and calculate the cost of driving 15,000 mi. Calculate the carbon footprint of the Prius Prime in g CO_2 per mile. Use the results from problem 3 above for the electric range and from problem 4 above in making your calculations for the gasoline engine. (Answer: $737.00, 143.3 g CO_2/mi)

(b) The Volvo S60 Recharge is a medium-sized plug-in hybrid car that has a typical cost of $42,000. The usable battery capacity is 15.0 kW-h and the car can travel 41 mi on the battery before the gasoline engine is activated. Once the gasoline engine is activated the car will get 31 mpg on regular gas. Assume the gasoline has 125,000 BTU/gal. Assume gasoline cost of $3.00 per gallon and electricity cost of $0.14/kW-h and calculate the cost of driving 15,000 mi. Calculate the carbon footprint of the Volvo S60 in g CO_2 per mile. Use the results from problem 3 above for the electric range and from problem 4 above in making your calculations for the gasoline engine. (Answer: $1076.00, 209.7 g CO_2/mi)

(c) The RAV4 Prime is a small plug-in hybrid SUV that has a typical cost of $49,000. The usable battery capacity is 14.5 kW-h and the car can travel 36 mi on the battery before the gasoline engine is activated.

Once the gasoline engine is activated, the car will get 38 mpg on regular gas. Assume the gasoline has 125,000 BTU/gal. Assume gasoline cost of $3.00 per gallon and electricity cost of $0.14/kW-h and calculate the cost of driving 15,000 mi. Calculate the carbon footprint of the RAV4 Prime in g CO_2 per mile. Use the results from problem 3 above for the electric range and from problem 4 above in making your calculations for the gasoline engine. (Answer: $998.00, 193.6 g CO_2/mi)

(d) At the time of this publication, there is no plug-in hybrid pickup truck available in the USA. If/when such a truck becomes available assume it will cost $75,000. Assume the usable battery capacity is 17.5 kW-h and the truck can travel 40 mi on the battery before the gasoline engine is activated. Once the gasoline engine is activated, the truck will get 22 mpg on regular gas. Assume the gasoline has 125,000 BTU/gal. Assume gasoline cost of $3.00 per gallon and electricity cost of $0.14/kW-h and calculate the cost of driving 15,000 mi. Calculate the carbon footprint of the theoretical plug-in hybrid truck in g CO_2 per mile. Use the results from problem 3 above for the electric range and from problem 4 above in making your calculations for the gasoline engine. (Answer: $1426.00, 278.6 g CO_2/mi)

8. You have calculated the carbon footprint for four versions of a small car, medium-sized car, small SUV, and pickup. Consider which size/type of vehicle would be most appealing to you to purchase and drive. Please consider the initial cost of purchasing the vehicle, the fuel/electricity cost of driving the vehicle 15,000 mi per year, and the carbon emissions. Write a couple of paragraphs telling me which vehicle you think is the best purchase and why. There is no right or wrong answer to this question. Please tell me what you think is the best overall choice.

9. Suppose a power plant is generating electricity using coal. Assume that the coal being used is 80% C, 6% H, 13% O, and 1% S by weight. The energy density of the coal is 26,100 kJ/kg = 11,220 BTU/lb.

(a) Work out the chemical reaction for the coal.

(b) Calculate the number of BTUs of energy from the coal for each gram of carbon dioxide generated.

(c) Compare to the other fuels discussed in class.

10. Assume that we generate 225 kg of CO_2 in manufacturing 1 m² of solar panels, as is listed on the www.co2list.org web page. In Springfield, Missouri, we estimate that 1 m² of solar panel will produce 1000 kW-h of electric energy each year and that the solar panels will last for 20 years. Calculate the carbon footprint in grams per kW-h for the solar panels. (Answer: 11.25 g CO_2/kW-h.)

References

1. https://en.wikipedia.org/wiki/Natural_gas#Environmental_effects.

2. https://en.wikipedia.org/wiki/Snowball_Earth.

3. https://en.wikipedia.org/wiki/Milankovitch_cycles.

4. https://climate.nasa.gov/vital-signs/sea-level/.

5. https://climate.nasa.gov/vital-signs/carbon-dioxide/.

6. https://www.bts.gov/content/estimated-national-average-vehicle-emissions-rates-vehiclevehicle-type-using-gasoline-and.

7. https://www.bts.gov/content/estimated-national-average-vehicle-emissions-rates-vehiclevehicle-type-using-gasoline-and.

8. https://en.wikipedia.org/wiki/Coal_assay.

9. https://en.wikipedia.org/wiki/Coal.

10. http://www.co2list.org/.

If we drive electric cars, the electricity to power them will come from the electric power grid. It is important to have a basic understanding of how we generate and distribute electric power.

                illustrates hourly and seasonal variation in the demand for electric power in the mid-Atlantic region of the United States in 2009. In   , PJM stands for "Pennsylvania–Jersey–Maryland," the region of the country used in generating the figure. The region includes significant portions of the three states. The graph starts on Monday. Time zero on the graph is midnight separating Sunday and Monday [1].

For the mid-Atlantic region, and for most of the country, the highest use of electric energy is in the summer months because electricity is used for air-conditioning in most homes and many buildings. The lowest usage is in the spring and fall, and winter months are in between. Most people and businesses in the mid-Atlantic region have access to natural gas for heating, which makes the electric loads in winter less than in summer. In regions of the country where most of the customers use electricity for heat, the winter months will show more demand for electricity than the summer months. Electricity usage is larger during the week than on the weekend. The trend on this graph is typical for many parts of the country and the world. Demand for electricity is lowest in the middle of the night and highest in the afternoons and early evenings. Power companies must provide the electricity we need when we need it, and our need for electricity varies with the time of day, day of the week, and season.

Figure 7.1 illustrates that there is a daily cycle of electricity use. There is a large difference between how much electric power we use in the middle of the night and that used in the afternoons and evenings. There is also a large difference seasonally. In most parts of the United States, we use a lot of electric power in the summer to provide air-conditioning and in the winter to provide heat. We use lower amounts of electric power in the spring and fall when temperatures are moderate. The electric power industry must provide power when we need it, and they do a good job. We are seldom without electric power in the United States. Our electric power system is so reliable that we take it for granted, and it causes us hardship when there is a large weather event that knocks the power out. We are very fortunate. In many parts of the world, the electric power system is not reliable.

There is a considerable difference in the cost of electricity depending on where you live in the United States. According to the US Energy Information Administration (EIA) [2], in 2023 the cost per kW-h in New England is typically 27.5¢, Middle Atlantic 19.4¢, East North Central 16¢, West North Central 12¢, South Atlantic 14.2¢, East South Central 12.9¢, Arkansas–Louisiana–Oklahoma 11.3¢, Texas 14.6¢, Mountain 13.5¢, Oregan–Washington 11.8¢, California 29.1¢, Alaska 24.7¢, Hawaii 41.6¢. The overall average cost in the United States in 2023 was 15.7¢ per kW-h. For examples and homework problems, I have used electricity costs typical for the middle of the country. Please recognize that if you live in New England, California, or Hawaii the cost of electricity is much higher than in the middle of the country.

Figure 7.1 Average hourly load.

7.1.1.
Why Electricity?

A question that sometimes arises is "Why do we use so much electricity?" Electricity is not a naturally occurring source of energy. We use primary sources of energy to produce electricity, and it would be more efficient to use the primary sources of energy directly. The answer is that electricity can be transmitted over large distances and distributed to homes and businesses with very little energy loss. That is what makes electricity valuable. We can take any primary source of energy, convert it into electricity, and use the grid to distribute the energy to customers. When we use electricity, we use a blend of the primary energy sources that were used to make the electricity: wind, solar, natural gas, coal, etc. Electricity is only as "clean and green" as the sources used to create the electricity.

Power companies must always provide exactly the amount of electric power that people want to use. If demand exceeds supply, we will have a blackout. If supply exceeds demand, the voltage will rise and begin destroying the power system. Balancing the power is largely automated, but there are people in control rooms monitoring the power system, making sure that supply and demand are always balanced. Power plants go online and offline as needed to keep the power balanced. By using weather forecasts and historical records of the electricity demand in the past, they can plan for how much power needs to be produced during the day. The overall goal is to produce the required electric power from the least expensive sources available, but there are many other considerations. In practice, balancing the grid is complex, and the details are beyond the scope of this book.

There is protection circuitry that will shut the system down and protect the power grid if supply and demand get out of balance. The protection circuitry prevents destruction of the power grid when there is a major weather event or other events that stress the grid. The grid system will shut down and protect itself if the power gets out of balance. Ideally, we would have a way to store large quantities of electric energy so that we could store excess electric energy when the demand is low and provide the energy later at a time when demand is high, but we do not have a good way to store this energy. Electricity must be generated as it is needed to keep the system in balance. There is a great deal of research being done to develop ways of storing large quantities of energy, which will be discussed later. If we are successful it will greatly change how we balance the power grid. The following discusses the current methods used to balance the electric power grid.

7.1.2.
Base Load Power

Base load power is power that must be generated all the time. Base load power is approximately the minimum power in Figure 7.1 that must be generated each day in the middle of the night. For base load power we want to use power plants that have a low fuel cost per kilowatt-hour (kW-h) to generate the electricity, since they will be running all the time. Base load power plants tend to have low fuel costs and high capital costs.

7.1.3.
Peak Load Power

Peak load power is the power necessary to handle the higher loads during the day. Peak load power plants are brought online during the day and turned off at night. Fuel cost is less important for the peak load power plants because they are not operating all the time. Capital cost and operating cost are more important for the peak load power plants.

Electric power is bought and sold all the time to keep supply and demand in balance. Power companies can produce power from their power plants, or they can buy power from another power company. Power generated in the middle of the night is less valuable than power generated during the day. Overall, base load power is less valuable per kW-h than peak power. In most of the country, power companies are required by law to charge individual households the same cost per kW-h regardless of when they use the power, but businesses and industries may pay a different price for the electricity depending on when they want to use it. In 2002, Texas became more deregulated than the rest of the country, and household customers can get real-time market prices in Texas, paying a different price for electricity depending on when they use it. The buying and selling of electricity in Texas is significantly different from the rest of the country because of deregulation.

7.1.4.
Sources of Electricity

Wind and solar energy have a zero fuel cost. All the costs for wind and solar power are from the capital costs of the equipment and the maintenance costs. We cannot control how much power we get from wind and solar energy, which can cause problems in balancing the grid. We can use weather forecasts to estimate how much wind and solar power we will get at different times of the day, but it is not possible to control the wind or the sun or clouds. We currently use other power plants, mostly natural gas power plants, to keep the grid in balance. The way we currently use wind and solar power is to get all the energy we can from the wind and sun and use other power plants to balance the load. Wind and solar can be used for a portion of the base load power but are not reliable enough to provide all the base load power. Battery systems are being developed to absorb energy from wind and solar sources when there is an excess and to put energy into the grid when it is needed. This will help keep the grid balanced. At the time of this writing, we are many years away from being able to produce enough battery capacity to balance the wind and solar energy. But in the future, large battery systems can help make wind and solar energy more practical.

Starting in 1992 wind power was subsidized in the United States by the federal government at a rate of $0.023 per kW-h. The "Production Tax Credit" subsidy continued until 2016 at which time it was gradually decreased each year to $0.015 per kW-h in 2021. Wind power put in place prior to 2022 earns the subsidy for each kW-h of energy generated. Starting in 2022 the subsidy was eliminated. Today it is economically viable to install wind power without the subsidy.

There is a federal tax credit for homeowners installing solar panels on their house or property. The subsidy varies from 22 to 30% of the cost of purchasing the solar panels, depending on which year the panels were installed, for homeowners who meet the eligibility requirements. State subsidies vary from state to state. Utilities that install solar farms can receive a production tax credit of 2.75¢/kW-h. Solar subsidies are set to expire in 2035, but wind power subsidies were extended several times by Congress, so the solar subsidies may be extended too. The purpose of subsidies is to encourage development and cost reduction to make solar power economically viable.

The cost and subsidies associated with the wind and solar power that has been installed are such that owners want to produce all the energy they can generate. Production tax credits are only given for electricity that is produced. The subsidy is based on the number of kW-h of energy produced and shutting the wind turbine or solar panels down means losing the subsidy, which is not a good financial decision. We do not want to turn them off. Local power companies have indicated that the cost of wind and solar has declined to the point that they plan to install more wind and solar power even if the subsidies end. As wind and solar become less expensive, and as the subsidies end, it will make sense to use wind and solar to help balance the power grid. Development of large battery systems will help make wind and solar power more economically viable. We can install a large excess of wind and solar generation capacity and put some of the excess in batteries, or turn off what we do not need. It is similar to what we do with natural gas peaking plants today. It does not make sense financially to do that with the wind and solar installed to date. But if the cost of electricity generated from wind and solar continues to decline, and if the subsidies end, it will make sense to do that in the future.

Nuclear power has a very high capital cost and a low fuel cost. Nuclear power is used as base load power. Ideally, we run the nuclear power plant at full capacity all the time to minimize the overall cost of the electricity generated. We can control the power generated in the nuclear power plant and reduce or increase power when necessary to balance the grid.

Coal power has a relatively high capital cost and a low fuel cost. Coal provides base load power. We reduce the power from the coal plants at night, but do not shut them down. It takes quite a while to get a coal plant started once it has been shut down. The capital cost of a coal plant is much lower than for a nuclear power plant. The fuel cost for a coal plant is higher than for a nuclear plant.

Natural gas power plants are the primary way we handle the peak loads that happen during the day. They have a lower capital cost and higher fuel cost than coal power plants. Natural gas power plants can be shut down and restarted each day. With fracking, the cost of natural gas has declined in recent years and the overall cost of producing electricity from natural gas is comparable to coal, or in many places less expensive than coal. In the current market, building a natural gas plant makes more sense economically than building a coal plant.

Hydropower is generally used to assist with peak load power. Hydro is low-cost electricity to produce, and it generates essentially no pollution or carbon dioxide. Environmentalists complain that building the lake destroys natural habitat, but overall hydropower is the cleanest and greenest energy source. There are a limited number of places where it makes sense to build a lake and dam. We do not have unlimited hydropower available. On most days there is a limited amount of water that can be drawn from the lake. We must maintain a suitable lake level. Power companies use hydropower when they feel they can get the most value from it, which is generally during peak hours.

Oil power plants and diesel generators provide a very small percentage of the electric power we use but are important in generating the peak power during the summer months, when usage is high or for emergencies. These power plants have a low capital cost and a very high fuel cost. Some of these power plants may only be used a few times each year, so the capital cost is most of the cost of operating them.

Fuel prices are variable. I will summarize fuel prices today as I write this book, but the numbers I give will become inaccurate with time. A good source for current fuel prices is the EIA [3]: https://www.eia.gov/state/seds/data. php?incfile=/state/seds/sep_prices/total/ pr_tot_US.html&sid=US.

Producing electricity from fossil fuels requires heat energy (BTU) to fire the burner, generate the steam, and power the steam turbine. The heat energy can come from any energy source. It can be used to compare the fuel cost for the commonly used fuels.

- Coal: The average cost in the United States for steam coal delivered to the power plant in 2022 was $54.46 per short ton, up from $31.41 in 2020, and the highest cost ever historically. Data for 2022 were the most recently published by EIA. There is considerable variation in the cost of coal depending on location, and the BTU content varies depending on where the coal was mined. The average for coal in the United States is about 19,270,000 BTU per short ton. Thus, the average cost for a million BTU is $54.46/19.27 = $2.83 per million BTU.

- Natural gas: Power plants purchase natural gas for less than what homeowners pay. The average cost for power plants in the United States in 2023 was $3.49 for 1000 ft^3. On average natural gas has 1030 BTU/ft^3, so the cost of the natural gas for power plants is $3.59 for a million BTU. The average residential cost for natural gas in 2023 was $15.23 per thousand cubic feet.

- Propane: For wholesale pricing in 2022, the average was $1.48 per gallon. It requires 10.93 gal to make one million BTU, so the cost for a million BTU is $16.17. Residential propane in 2018 was $2.99 per gallon, which is $32.68 per million BTU.

- Gasoline: The US average cost in 2023 was $3.43 per gallon. There are 125,000 BTU/gal, so this translates to a cost of $27.44 for a million BTU.

- Diesel: The US average cost in 2023 was $4.25 per gallon. There are 138,700 BTU/gal, so this translates to a cost of $30.84 for a million BTU.

- Nuclear: People working in the power industry say that nuclear fuel is the lowest cost for the amount of energy, and I feel confident the statement is true. However, reliable data are unavailable. I believe the cost of a million BTU of energy from nuclear fuel will be less than the cost of coal.

- Wind and sun: There is no fuel cost associated with wind and solar energy. The wind and sun are free.

Historically, the fuel cost for coal has been significantly less than for natural gas, but the capital cost of building the power plant is much less for natural gas than for a coal plant. The operational and maintenance cost for a natural gas plant is less than for a coal plant. In the United States and Western Europe, coal plants are shutting down, as the cost of the electricity generated by the coal plants is expensive compared to natural gas and wind power. Coal has higher carbon emissions than other fuels, which is part of the reason the coal plants are being shut down. Worldwide, especially in Asia, the number of coal plants has increased significantly in the past 10 years.

In the United States, natural gas is relatively inexpensive today. Historically, when all costs are included, electricity produced in coal plants has been less expensive than electricity produced in natural gas plants. Today natural gas and wind power are both less expensive than coal, and we see the coal plants being closed in the United States. Fracking technology produces a considerable amount of natural gas, and the abundance of natural gas has caused the price to decline. It is difficult to predict the future. If we reduce the amount of fracking, the cost of natural gas will probably increase, and, at some point coal could become less expensive than natural gas. Building a power plant is a 50+ year investment and power companies need to make their decisions on what they think will be best over the next 50+ years.

There is a perpetual myth that propane is as economical as natural gas in heating your house. It is a myth. Natural gas is less expensive and has always been less expensive than heating with propane. Propane is a good economical choice if natural gas is not available.

7.2.
Nuclear Power

Nuclear power is used to generate 22.1% of the electricity used in the United States. Nuclear energy is used to heat water and convert it to steam, and then the steam drives a steam turbine, which drives a generator to make the electric power. The process is similar to coal and natural gas power plants.

The radiation from nuclear fuel is dangerous and hazardous to our health. A large part of the cost of building a nuclear power plant is related to containing and controlling the radiation from the nuclear fuel. Special materials must be used for components in the nuclear plant because not all materials can withstand the nuclear radiation.

There is an essentially unlimited amount of nuclear fuel that could be used in power plants. The fuel cost for a nuclear power plant is low compared to other fuel sources. The capital cost of building a nuclear power plant is very high. Getting permits and approvals to build a nuclear power plant is difficult. There are always public protests against building a nuclear power plant, and the protests lead to delays and cost overruns. Nuclear accidents, such as Pennsylvania's Three Mile Island in 1979, Ukraine's Chernobyl in 1986, and Japan's Fukushima in 2011, have had a large impact on turning public opinion against nuclear power. In the early years of the Obama presidency, plans were made to build new nuclear power plants in the United States, but the Fukushima accident caused many of those plans to be delayed or canceled. We completed the Watts Bar Unit 2 in Tennessee that began operation in 2016. That was the first new nuclear power plant in the United States to begin operation in 20 years. (The Watts Bar Unit 1 plant in Tennessee began operation in 1996.) In 2023 the Vogtle Unit 3 came online in Georgia, and the Vogtle Unit 4 is scheduled to come online in 2024. These two units were built at the same facility as Vogtle Units 1 and 2, which came online in 1987 and 1989, respectively. As of 2024 there are no new nuclear power plants planned or under construction in the United States.

In my opinion, the future of nuclear power currently looks bleak. In the United States and Europe, nuclear plants are closing faster than they are being built. The high cost of building nuclear plants makes it difficult to find investors. The low cost of electricity produced by natural gas and wind power makes the economics of nuclear power questionable. On the positive side, we have an almost unlimited amount of nuclear energy and nuclear power does not cause air pollution or global warming. Properly contained nuclear power is a very clean energy source. Disposing of nuclear waste is a difficult issue. There is no site in the United States approved for nuclear waste disposal. In the United States, all the nuclear waste is stored on-site at the nuclear power plants. The biggest challenge for nuclear power is to convince the public that nuclear power is safe. It seems that every time we get to a point where the public is ready to embrace nuclear power there is another accident that turns the public against it. There are many good things about nuclear power, but it cannot be successful without public support.

7.3.
Hydroelectric Power

Hydroelectric power is an old and successful technology. Water wheels were used for hundreds of years to provide power for grinding grain, sawing lumber, etc. Modern water turbines are used to capture energy from the flowing water and produce electric power. Hydroelectric power is one of the greenest sources of energy. There are carbon emissions associated with building the dam and lake, but very little carbon or pollution emissions in operating the hydroelectric power plant. Building a lake destroys habitat and covers private property. There are always protests and public concerns when building the lake. The primary justification for acquiring the land to build the lake is usually flood control. Lakes are used to absorb large rainfall events and protect property from flood damage. It would be difficult to justify acquisition of the property for other reasons. Once the lake is built it provides a good public water supply, recreational opportunities, and hydroelectric power. There are many uses for the lake, but the primary justification of the project is usually flood control.

Hydropower is low in cost. If there were more suitable places to build hydroelectric dams, we would build more of them. The energy associated with hydroelectric power is gravitational energy, mass × gravity × height, as taught in physics classes. By creating a lake behind the dam, we have a large mass of water at a height above the water turbine. The illustration in Figure 7.2 is helpful in illustrating the energy available from the lake.

Figure 7.2 Schematic of hydropower.

Usable water in lake

h

Turbine

© SAE International

If V is the usable volume of water in the lake and ρ is the density of the water, then the mass of usable water in the lake is ρV. The volume V will be the surface area of the lake multiplied by the depth of water that can be removed and run through the water turbines. The water turbines will be about 90% efficient in converting the energy into electricity. Putting this together the following equation can be developed.

$$\text{Hydro}\left(\text{Gravitational}\right)\text{Energy} = \rho Vgh = \gamma Vh \tag{7.1}$$

where γ is the weight density of the water. The area of a lake is usually measured in acres or square miles. Drawing one foot of water from a lake is a large volume and mass of water and represents a large amount of energy. In most cases we do not want to raise or lower the lake level significantly on a day-to-day basis. Water is used as it flows into the lake from tributaries, maintaining an approximately steady lake level. The amount of water that can be drawn from the lake varies with rainfall and evaporation, which to some extent varies according to time of year. There are times when the water must be removed from the lake faster than it can flow through the water turbines. The floodgates are opened, and energy is dumped down the spillways. There will be extremely dry times when we will not be able to draw any water from the lake. The power company will try to maximize the value of the electricity produced, using the hydro energy to produce electricity when it is most valuable.

Example 7.1

Table Rock Lake in Missouri covers approximately 43,100 acres [3]. How many MW-h of electric energy could be produced by lowering the lake level 6 in.? Assume that the weight density of water is 62.4 lb/ft³, that the water turbine is 90% efficient, and that the height of the water above the turbine is 185 ft.

An acre is 43,500 ft². The volume of water in 6 in depth can be calculated as:

$$\text{Volume} = \left(43,100 \text{ acre}\right)\left(\frac{43,500 \text{ ft}^2}{\text{acre}}\right)\left(\frac{6}{12}\text{ ft}\right) = 9.3872 \times 10^8 \text{ ft}^3 \tag{7.2}$$

The gravitational energy is the weight of the volume of water multiplied by the 185 ft height above the water turbines. One MW-h of electric energy is 2.655×10^9 ft-lb. Since the water turbines will be about 90% efficient in converting the gravitational energy to electricity, the electric energy is calculated as:

$$\text{Electric Energy} = \left(0.90\right)\left(9.3872 \times 10^8 \text{ ft}^3\right)\left(\frac{62.4 \text{ lb}}{\text{ft}^3}\right)\left(185 \text{ ft}\right)\left(\frac{\text{MW-h}}{2.655 \times 10^9 \text{ ft-lb}}\right) = 3673 \text{ MW-h} \tag{7.3}$$

To put this in perspective, the power station at Table Rock Lake has four 50-MW generators, a total of 200 MW capacity. The energy in 6 in. of water in the lake would be enough to power the generators for 18.4 h. On most days there is less than 6 in. of usable water in the lake, and the generators will operate for shorter periods of time. Also please understand that we do not reduce the level of the lake 6 in. when drawing the water out. There is water constantly flowing into the lake and we need to balance maintaining a steady water level with drawing water out of the lake to make electric power.

The dam and lake serve more purposes than just generating electricity. Justification for building the dam and lake will usually be based on flood control, which is the most important reason for building the dam and lake. The lake will provide a source of recreation that many people will enjoy. The lake provides a reliable source of fresh water. If we divide the cost of the lake and dam by all the uses and benefits, the cost of generating electricity from hydropower is very low. Even if we charge the whole cost of building the lake and dam to generating electricity, the cost is 3–5¢/kW-h. Electricity generated using hydropower is inexpensive compared to other methods of generating electricity.

There are negative aspects of hydropower too. When building a hydroelectric power plant, we must build a dam and lake. The property for the lake must be acquired. Some owners will sell willingly, and some property will need to be acquired by eminent domain. Environmentalists will be concerned that the lake destroys natural habitat. Building the dam will reduce the flow of water in the river below the dam, which will have negative impacts on the people who live below the dam. The people below the dam may be concerned about the failure of the dam. More damage and death are caused by dam failure than any other electric power plants, except perhaps nuclear plants. There will need to be public hearings and political activity to obtain the land and permits to build the dam and lake.

7.4.
Wind Power

Wind power was a vital part of transportation for hundreds of years. Wind powered the ships. Europeans colonized the world using wind power. Developing powerful sailing naval ships allowed European nations to become dominant world powers. Sailing vessels averaged about 5 mph on the oceans. The ships traveled faster when the wind was blowing, and slower when it was not blowing.

As we developed the steam engine during the 1800s, sailing ships were replaced by steamships. The steam engines were more expensive than sails. Steam engines were noisy and dangerous and required fuel and lots of maintenance. Boilers exploded and ships burned. There were lots of problems with the early steam engines, but with all their faults, they were an improvement over the sailing ships. Steam power was faster and more reliable.

The wind is free and there is always a temptation to use a free energy source. Wind turbines have been used for hundreds of years to pump water. One of the fundamental problems in using wind is that it is not possible to control the wind. You must be opportunistic and use the wind when it is available. Pumping water from a well into a tank or pond is a good use of wind power because you do not need to be pumping all the time. Windmills have been used for many years for pumping water. In the Netherlands, wind power was used to drain the fields by pumping water from low areas into the ocean. Wind turbines are not a new idea, but they are important in generating electricity.

The electric grid must be balanced. We must generate exactly the amount of power that we are using. Starting in 1992, wind power was subsidized in the United States by the federal government at a rate of $0.023 per kW-h. The "Production Tax Credit" subsidy continued until 2016 at which time it was gradually decreased each year to $0.015 per kW-h in 2021. Wind power put in place prior to 2022 earns the subsidy for each kW-h of energy generated. Starting in 2022, the subsidy was eliminated. The owners of wind turbines would like for them to produce as much electricity as possible because the subsidy they receive from the government is proportional to the amount of electric energy produced. That approach has worked fine as long as wind power is a small percentage of the electricity produced. As wind power becomes a larger percentage of the electricity generated, we will need to be able to turn the wind turbines down at night to reduce the amount of energy they generate. It is technically possible to turn the wind turbines down or off now. The turbines are all designed so that we can twist the blades and cause the rotation to slow down or stop, but this makes them less profitable for the owners. The owners will lose the subsidy for each kW-h the wind turbine could have produced, but we must balance the electric grid. As wind power becomes a larger percentage of the electricity we use, it will be necessary to reduce power from the wind turbines at night.

We use a peak amount of electric power during the day and into the early evenings. We must have a way to handle these peak loads. If we have excess wind power available from wind turbines that are shut off, then we can bring them online to generate the power we need. This requires a large capital investment in wind turbines that are turned off most of the time. Most of the cost associated with wind power is capital cost. If the wind turbines are off most of the time, the number of kW-h produced each year will be much less than if they are on all the time. This makes the cost of electricity much more expensive because the annual capital depreciation cost and other annual costs are divided by a smaller number of kW-h of electricity. It becomes a matter of cost. If wind turbines become inexpensive enough, we can afford to have excess wind power and use it to balance the grid and handle peak loading.

There are other ways to balance the grid. Solar power is available during the day, and in the future, it may be possible to handle a large part of the daytime peak loads using solar power. In the near future we will continue to use natural gas power plants to handle most of the peak loads, along with hydropower and a small amount of oil-fired power plants and diesel generators. Coal and nuclear will be used as baseline power in the near future but are currently more expensive than using wind or natural gas. With the current costs, we will continue to see less electricity generated using coal and nuclear because they are the more expensive solution. However, it is hard to predict the future. If natural gas and wind become more expensive in the future, we may see a resurgence of coal and nuclear power.

In an ideal situation, we would store some of the excess electric energy generated at night and use it during the day when we need more electricity. We have been experimenting with this for more than 100 years. It is not a new idea. The problem has always been that we do not have a good economical way to store large quantities of electric energy. One of the most successful experiments has been using excess electric power to pump water into a lake, and then using water turbines to extract the energy from the lake when we need it. In Missouri, the Taum Sauk project was built in the early 1960s using this idea. In the middle of the night, electricity produced at the power plant has almost no value, so the electricity is used to pump water into a lake. During the day when the electricity has more value, the water in the lake is run through water turbines and used to make electricity. There is a large loss of energy due to pumping and turbine efficiencies and evaporation from the lake. Lots of energy can be stored in a lake, but with the pumping losses and evaporation, we will get less than half of the energy back as electricity. It is not very efficient, but it is cost-effective and it is a way we can store large quantities of energy.

Another approach to storing excess electric energy generated at night is to store it in batteries. The batteries can then be used to handle the peak loads during the day. In Unit 1, we studied electric cars and learned that the batteries for electric cars must be lightweight for the amount of energy they store. Weight is not as critical for utilities. Utility batteries will be stationary, so the main design considerations will be cost and reliability. Storing enough energy to power a city will require a huge amount of battery capacity. Utilities are experimenting with pilot projects using batteries to store small amounts of electric power. Utility batteries are not practical today, but they could become practical in the future.

The energy associated with wind is kinetic energy. There is a stream of air moving through the wind turbine blades. The cross-sectional area of the stream of wind is πR^2, where R is the turbine radius. Assume there is a mass of air of thickness dx moving toward the wind turbine at speed V. Figure 7.3 illustrates the differential element of air moving toward the wind turbine.

Figure 7.3 Differential element in airstream.

© SAE International

Recognizing that kinetic energy is ½mV², we can write the mass and energy of the differential mass of air as shown below:

$$\text{Differential Mass} = \rho\pi R^2 dx \tag{7.4}$$

$$\text{Differential Energy} = \frac{1}{2}\left(\rho\pi R^2 dx\right)V^2 \tag{7.5}$$

The power of the wind approaching the wind turbine is the derivative of energy with respect to time. Dividing Equation (7.5) by dt and

recognizing that dx/dt is the velocity V of the wind, it follows:

$$\text{Power} = \frac{1}{2}\rho\pi R^2 V^3 \tag{7.6}$$

Equation (7.6) is the power in the wind stream approaching the turbine blades. The turbine will not be able to absorb all the energy in the wind, because absorbing all the energy in the wind would require that the turbine stop the wind, that is, bring the velocity of the wind to zero as it hits the turbine. This is not physically possible, because the wind must flow through the turbine blades. The turbine blades will slow the speed of the wind as it flows through but not stop it. Equation (7.6) will greatly overestimate the amount of power the wind turbine can produce. We will need a better model for doing the analysis of wind turbines.

Wind turbines are similar to propellers except they absorb power from the moving fluid rather than adding power to the moving fluid. The model we will use was originally developed by William Froude. For the turbine to slow the speed of the wind stream, the conservation of mass law in fluid mechanics requires that the radius of the incoming airstream must be smaller than the existing airstream behind the turbine. Figure 7.4 is similar to the figure for developing the propeller model.

Figure 7.4 Wind turbine model.

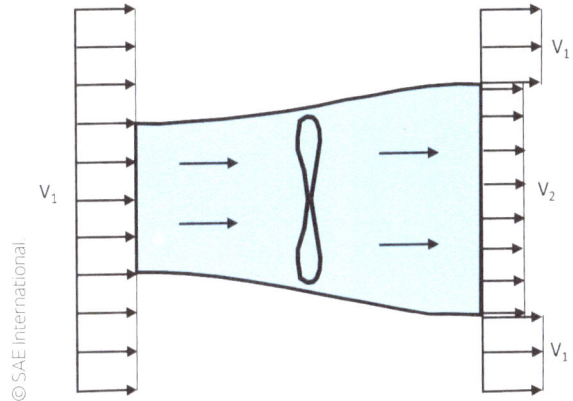

© SAE International

In the analysis, V_1 is the speed of the wind and V_2 is the speed of the wind after it has been slowed by the wind turbine. If R is the radius of the turbine blades, then the radius of the airstream approaching the turbine must be smaller than R and the radius of the airstream behind the turbine must be greater than R. Conservation of mass requires that $A_1 V_1 = A_2 V_2$, where A_1 is the area of the stream of wind approaching the turbine and A_2 is the area of the stream of wind behind the turbine. Since $V_1 > V_2$ it must be true that $A_2 > A_1$. Just as with the propeller derivation, Froude assumed the velocity V of the wind at the turbine is the average of V_1 and V_2.

$$V = \frac{V_1 + V_2}{2} \tag{7.7}$$

With this assumption, the conservation of momentum formula can be used to find the force acting on the turbine:

$$\text{Force} = \frac{1}{2}\rho\pi R^2\left(V_1^2 - V_2^2\right) \tag{7.8}$$

The power that the wind turbine absorbs from the stream of wind is the product of the force and the velocity V of the air at the turbine.

$$\text{Power} = \frac{1}{2}\rho\pi R^2\left(V_1^2 - V_2^2\right)V = \frac{1}{2}\rho\pi R^2\left(V_1^2 - V_2^2\right)\left(\frac{V_1 + V_2}{2}\right) \tag{7.9}$$

The theoretical efficiency of the turbine e_{turb} is given by Equation (7.10):

$$e_{turb} = \frac{1}{2}\left[1 - \frac{V_2^2}{V_1^2}\right]\left(1 + \frac{V_2}{V_1}\right)$$

(7.10)

The above formulas assume an optimum design for the turbine blades. The maximum theoretical efficiency is 59.3%. That is, the maximum amount of power that can be absorbed by the wind turbine is 59.3% of the energy in the wind stream approaching the wind turbine, due to the nature of how it works. In reality, the efficiency will be 40–50% for most wind turbines.

Equation (7.10) can be used to prove the 59.3% maximum possible efficiency. Substitute $x = V_2/V_1$ so that the efficiency is only a function of x. Take the derivative with respect to x and set it equal to zero to find the x values where the efficiency is at minimums and maximums. The solution will yield $x = 1/3$ for the maximum. Back substitute into the efficiency equation and the value is 59.3%.

The goal for the information on wind turbines is to give the reader a way to estimate the energy and power that is possible based on the specifications of the wind turbine. Companies sometimes give optimistic estimates of the energy that can be recovered from the wind turbine. The reader should be able to perform analysis to estimate what is possible for a well-designed wind turbine.

Example 7.2

A utility-scale wind turbine has a diameter of 150 m and the installed cost is $3.6 million assuming that the utility buys several of the wind turbines for a wind farm. The rated wind speed is 19 mph. Assume that the density of air is 1.22 kg/m³.

1.   For a first approximation, assume the ideal turbine efficiency and use the analysis to estimate power output. How much power will the turbine produce in a 19 mph wind?

     Solution: The wind speed V_1 = 19 mph, which is equal to 8.492 m/s. The radius R = 75 m. Assuming the ideal efficiency of the turbine means that V_2/V_1 = 1/3, so V_2 = 8.492/3 = 2.831 m/s. Plugging everything into the power equation [Equation (7.9)]:

$$\text{Power} = \frac{1}{2}(1.22)\pi(75^2)(8.492^2 - 2.831^2)\left(\frac{8.492 + 2.831}{2}\right) = 3.911 \text{ MW}$$

(7.11)

     The analysis shows that a 150-m-diameter turbine under the most ideal conditions can produce 3.911 MW 19-mph wind. This assumes a perfectly designed turbine and that the generator and inverter are 100% efficient, which is not possible or realistic, but it illustrates the idealized assumption.

2. As a more realistic set of assumptions, assume that the turbine is 48% efficient and that the generator and inverter are 90% efficient. Under these assumptions, how much power will the turbine produce in a 19-mph wind?

Solution: When the turbine efficiency is given, start with the efficiency formula [Equation (7.10)] and solve for the V_2/V_1 ratio. You will probably need to use the root finder in a calculator, and it is convenient to substitute $x = V_2/V_1$ into the formula using the calculator. The equation is:

$$0.48 = \frac{1}{2}\left[1 - x^2\right]\left(1 + x\right) \tag{7.12}$$

Solving this equation yields x = 0.6454, and since we know V_1 = 8.492 m/s, it follows that V_2 = (0.6454) (8.492) = 5.481 m/s. In analyzing wind turbines, it is common to start with the wind speed (V_1) and the efficiency of the turbine. When this is the case, start with the efficiency formula [Equation (7.10)] and solve for x and the velocity of the air behind the wind turbine (V_2). The turbine power is then calculated using the power formula:

$$\text{Turbine Power} = \frac{1}{2}\left(1.22\right)\pi\left(75^2\right)\left(8.492^2 - 5.481^2\right)\left(\frac{8.492 + 5.481}{2}\right) = 3.168 \; MW \tag{7.13}$$

This is a realistic estimate of the mechanical power the wind turbine would produce in a 19-mph wind. If the generator and inverter are 90% efficient in converting the turbine power into electric power, it follows that the electric power generated is:

$$\text{Electric Power} = \left(0.90\right)\left(3.168\right) = 2.85 \; MW \tag{7.14}$$

3. Assume that the wind turbine operates in a 19-mph wind all the time generating 2.85 MW of power and estimate the amount of electric energy generated in a year.

$$\text{Annual Energy} = \left(2.85 \; MW\right)\left(24 \; h \, / \, day\right)\left(365 \; day \, / \, year\right) = 24,966 \; MW\text{-}h \tag{7.15}$$

Assume that the average value of the electricity generated is $50.00 per MW-h (including the federal wind subsidy) and estimate the annual value of the electricity generated.

$$\text{Annual Value} = \left(24,966\right)\left(\$50.00\right) = \$1.248 \; \text{million} \tag{7.16}$$

To invest $3.7 million and have an annual return of $1.248 million is a sound investment. The federal wind subsidy started out at 2.3¢/kW-h, or $23.00 per MW-h. The subsidy was phased out in 2022. With the example presented here, wind power installed today without the federal subsidy would be worth only $27.00 per MW-h instead of $50 per MW-h because there is no federal subsidy today. The annual return would be $0.674 million, which is an 18% annual return on the $3.7 million invested.

Example 7.3

The wind turbine has a rotor diameter of 100 m and an efficiency of 44% in a 12 m/s wind. Assume the density of air is 1.22 kg/m³. Cost is $2,000,000 installed. The goal for this example is to evaluate the power output and the value of the electricity produced in a variable wind speed.

Solution: We are given the efficiency of the propeller, so use the efficiency formula to get the V_2/V_1 ratio and the value of V_2.

$$0.44 = \frac{1}{2}\left[1 - x^2\right](1 + x) \tag{7.17}$$

The solution is $x = V_2/V_1 = 0.693$ and $V_2 = (0.693)(12 \text{ m/s}) = 8.316$ m/s. The thrust force on the turbine is calculated using Equation (7.8):

$$\text{Thrust Force} = \frac{1}{2}(1.22)\pi\left(50^2\right)\left(12^2 - 8.316^2\right) = 358,753 \text{ N} \tag{7.18}$$

This thrust force creates a large bending moment at the base of the wind turbine. The geotechnical work for the foundation and the structural work for the tower will need to be substantial to carry the large bending moment. The power is calculated using Equation (7.9) and multiplying by the 90% efficiency.

$$\text{Power} = \frac{1}{2}(1.22)\pi\left(50^2\right)\left(12^2 - 8.316^2\right)\left(\frac{12 + 8.316}{2}\right)(0.90) = 3.28 \text{ MW} \tag{7.19}$$

A 12 m/s wind is equal to a 26.85 mph wind. Getting a typical sustained wind that high is uncommon in most of the country. The average wind speed will depend on where the wind turbine is located.

To help illustrate how location impacts the power output of the wind turbine, weather data were collected from weather.com for one day in Springfield, Missouri, and is presented in Table 7.1. Analyzing data for a day is not adequate to decide whether or not to buy a wind turbine, but it illustrates the principle. In practice the reader would want to extend the analysis to a year or more of data in determining the economics of purchasing a wind turbine. Table 7.1 is the spreadsheet that was used to analyze the wind turbine for the weather data gathered. The wind speed data were put into the V_1 column in the spreadsheet, and the analysis was done using the assumptions above.

In Table 7.1 it was assumed that the wholesale value of the electricity was $0.05 per kW-h. There is a subsidy on wind power of $0.023 per kW-h, so this assumption is that the actual wholesale value is $0.027 per kW-h. The actual value will depend on the time of day and time of year, but this is a reasonable assumption. For a $2 million investment, this wind turbine would return approximately $150,000 per year in Springfield, Missouri. This is not a good return on the investment. Springfield, like many other places in the country, is not a good place for wind power. A second set of data were collected for Sioux Falls, South Dakota, and are presented in Table 7.2.

Table 7.1 Wind turbine analysis for Springfield, Missouri.

Time	V1 (mph)	V1 (m/s)	V2 (m/s)	V (m/s)	Power (W)	Energy (kW-h)	Value		
Prop. dia.	100	m			V2/V1	0.6927	0.44		
Efficiency	0.44								
Air density	1.22	kg/m³							
Value	$0.05	/kW-h							
Generator	0.90							Total (kW-h)	8437.87
Midnight	8	3.5756	2.4770	3.0263	86,765	86.76	$4.34	Total ($)	$421.89
1:00	8	3.5756	2.4770	3.0263	86,765	86.76	$4.34	Monthly ($)	$12,657
2:00	6	2.6817	1.8577	2.2697	36,604	36.60	$1.83	Yearly ($)	$153,991
3:00	4	1.7878	1.2385	1.5131	10,846	10.85	$0.54		
4:00	2	0.8939	0.6192	0.7566	1356	1.36	$0.07		
5:00	2	0.8939	0.6192	0.7566	1356	1.36	$0.07		
6:00	2	0.8939	0.6192	0.7566	1356	1.36	$0.07		
7:00	6	2.6817	1.8577	2.2697	36,604	36.60	$1.83		
8:00	8	3.5756	2.4770	3.0263	86,765	86.76	$4.34		
9:00	10	4.4694	3.0962	3.7828	169,462	169.46	$8.47		
10:00	10	4.4694	3.0962	3.7828	169,462	169.46	$8.47		
11:00	10	4.4694	3.0962	3.7828	169,462	169.46	$8.47		
Noon	12	5.3633	3.7154	4.5394	292,831	292.83	$14.64		
1:00	12	5.3633	3.7154	4.5394	292,831	292.83	$14.64		
2:00	14	6.2572	4.3347	5.2959	465,005	465.00	$23.25		
3:00	16	7.1511	4.9539	6.0525	694,118	694.12	$34.71		
4:00	18	8.0450	5.5731	6.8091	988,304	988.30	$49.42		
5:00	18	8.0450	5.5731	6.8091	988,304	988.30	$49.42		
6:00	20	8.9389	6.1924	7.5656	1,355,698	1355.70	$67.78		
7:00	18	8.0450	5.5731	6.8091	988,304	988.30	$49.42		
8:00	14	6.2572	4.3347	5.2959	465,005	465.00	$23.25		
9:00	14	6.2572	4.3347	5.2959	465,005	465.00	$23.25		
10:00	12	5.3633	3.7154	4.5394	292,831	292.83	$14.64		
11:00	12	5.3633	3.7154	4.5394	292,831	292.83	$14.64		

This was probably an exceptionally windy day in Sioux Falls. But if this were a typical day the return on investment would be very good. Investing $2,000,000 and getting an annual return of $1,431,633 is an excellent investment. Wind power would make sense in Sioux Falls. To do the analysis correctly, the reader should collect wind data for a year or multiple years at a particular location and estimate the return for a typical year. Wind data are available on many weather sites for many locations in the country. It is tedious to obtain the data and run the analysis, but it is a straightforward process.

Table 7.2 Wind turbine analysis for Sioux Falls, South Dakota.

Prop. dia.	100	m			V2/V1	0.6927	0.44			
Efficiency	0.44									
Air density	1.22	kg/m³								
Value	$0.05	/kW-h								
Generator	0.90									
Time	V1 (mph)	V1 (m/s)	V2 (m/s)	V (m/s)	Power (W)	Energy (kW-h)	Value	Total (kW-h)	78,446	kW-h
Midnight	33	14.7492	10.2174	12.4833	6,089,966	6089.97	$304.50	Total ($)	$3922	
1:00	33	14.7492	10.2174	12.4833	6,089,966	6089.97	$304.50	Monthly ($)	$17,668	
2:00	32	14.3022	9.9078	12.1050	5,552,940	5552.94	$277.65	Yearly ($)	$1,431,633	
3:00	30	13.4083	9.2886	11.3485	4,575,482	4575.48	$228.77			
4:00	27	12.0675	8.3597	10.2136	3,335,526	3335.53	$166.78			
5:00	22	9.8328	6.8116	8.3222	1,804,434	1804.43	$90.22			
6:00	14	6.2572	4.3347	5.2959	465,005	465.00	$23.25			
7:00	14	6.2572	4.3347	5.2959	465,005	465.00	$23.25			
8:00	12	5.3633	3.7154	4.5394	292,831	292.83	$14.64			
9:00	16	7.1511	4.9539	6.0525	694,118	694.12	$34.71			
10:00	18	8.0450	5.5731	6.8091	988,304	988.30	$49.42			
11:00	23	10.2797	7.1212	8.7005	2,061,848	2061.85	$103.09			
Noon	24	10.7267	7.4309	9.0788	2,342,647	2342.65	$117.13			
1:00	26	11.6206	8.0501	9.8353	2,978,469	2978.47	$148.92			
2:00	26	11.6206	8.0501	9.8353	2,978,469	2978.47	$148.92			
3:00	30	13.4083	9.2886	11.3485	4,575,482	4575.48	$228.77			
4:00	32	14.3022	9.9078	12.1050	5,552,940	5552.94	$277.65			
5:00	33	14.7492	10.2174	12.4833	6,089,966	6089.97	$304.50			
6:00	32	14.3022	9.9078	12.1050	5,552,940	5552.94	$277.65			
7:00	30	13.4083	9.2886	11.3485	4,575,482	4575.48	$228.77			
8:00	28	12.5144	8.6693	10.5919	3,720,036	3720.04	$186.00			
9:00	26	11.6206	8.0501	9.8353	2,978,469	2978.47	$148.92			
10:00	24	10.7267	7.4309	9.0788	2,342,647	2342.65	$117.13			
11:00	24	10.7267	7.4309	9.0788	2,342,647	2342.65	$117.13			

A problem with wind energy that is difficult to model is the fact that the best places for wind energy are areas that are low in population. When wind turbines are located far from population centers, transmission lines must be constructed to get the energy to where it will be used, and building transmission lines is expensive. The cost of building the transmission lines needs to be part of the capital cost. Wind turbines can be located offshore in the ocean, which will be close to population centers.

7.5.
Solar Power

Solar power varies with the time of day and the time of year. To analyze solar power, there is a standard model used to estimate the intensity of the sunlight on a clear day. The model assumes the earth is spherical and rotates around the sun in a circular idealized orbit. With these assumptions, on a clear day, the intensity of the sunlight and sun angle depend on the latitude, longitude, and time of day. Cloudiness is a major factor in solar intensity and is difficult to predict from weather forecasts. It is possible to add a cloudiness factor to the model, but from a practical viewpoint, it is difficult to get an accurate value for the cloudiness factor. Historical data on cloudiness, if available, are helpful but are difficult to get for most locations.

This model neglects many secondary effects. The earth is closest to the sun in early January and furthest from the sun in early July, and this has a secondary impact on the solar intensity. There is less moisture in the air in the winter and early spring, and more moisture in the summer. Moisture in the air absorbs a small amount of solar radiation and reduces solar intensity. The sun emits higher and lower levels of radiation; it is not a constant energy source. In the early morning and late evening, the sunlight must pass through a longer distance of the atmosphere because of the low sun angle, which reduces the intensity of the sunlight. The earth is not exactly spherical, and the rotation around the sun is slightly elliptical rather than circular. All these things are minor secondary effects that are not accounted for in the model. The secondary effects are all small compared to the effect of the position of the sun relative to the earth and the clouds. The model developed in this section is a good approximation for modeling solar energy.

7.5.1.
Angle of Declination

On the summer solstice (approximately June 22), the sun is directly over the Tropic of Cancer at 23.44° north latitude, and on the winter solstice (approximately December 21) the sun is directly over the Tropic of Capricorn at 23.44° south latitude. For this model we will assume a sinusoidal variation of the declination angle (θ_D) with a small correction factor according to Equation (7.20) [4]:

$$\theta_D = -\arcsin\left[0.39779\cos\left(0.98565°\left(N+10\right)+1.914°\sin\left(0.98565(N-2)\right)\right)\right] \tag{7.20}$$

Equation (7.20) is an old formula, and the angles are given in degrees. It works for finding the declination angle on a calculator, if the calculator is set in degree mode. When programming in Excel, it is convenient to have a formula that uses radians for angle measurement because Excel uses radians. Recognizing 0.98565° = 0.017203 rad and 1.914° = 0.033406 rad, the formula can be modified for Excel as:

$$\theta_D = -\arcsin\left[0.39779\cos\left(0.017203(N+10)+0.033406\sin\left(0.017203(N-2)\right)\right)\right] \tag{7.21}$$

For Equations (7.20) and (7.21), N is the day of the year such that January 1 is N = 1. At solar noon, when the sun is the highest in the sky, the angle between a normal to the surface of the earth and the sun is equal to the latitude location of the place on earth minus the declination angle. Example 7.4 illustrates how to calculate the angle of the sun at solar noon.

Example 7.4

Consider a city located at a latitude of 37.2° north. Calculate the angle between a normal to the surface of the earth and the sun at solar noon on March 26, 2024. If a solar array were placed parallel to the ground at this location, this would be the angle between the solar array and the sun at solar noon. The intensity of the sunlight striking the solar array is proportional to the cosine of this angle. To do the analysis, we find the day of the year for March 26, 2024, and substitute into Equation (7.20) to find the declination angle.

$$N = 31 + 29 + 26 = 86 \qquad (7.22)$$

$$\theta_D = 2.62° \qquad (7.23)$$

$$\text{Sun Angle} = 37.2° - 2.62° = 34.58° \qquad (7.24)$$

On March 26, 2024, at a latitude of 37.2°, the angle between a normal to the earth surface and the sun at solar noon is 34.58°. The intensity of the sunlight striking the solar array is proportional to the cosine of this angle, which is 0.823, or 82.3% of the direct sunlight. On average, direct sunlight has an intensity of approximately 950 W/m². If the solar array were tilted so it pointed directly at the sun, the intensity of sunlight striking the solar array would be 950 W/m². If the solar array is parallel to the ground, the intensity of the sunlight will be 82.3% of the 950 W/m², which is 782 W/m². The solar cells used for terrestrial applications are typically about 18% efficient in converting sunlight into electricity.

Using Excel, it is possible to plot how the declination angle and sun angle vary throughout the year. Continuing from the example above, the plot was developed assuming that the place on earth where the solar array is located is at latitude 37.2° and that the sun angle is plotted for solar noon (Figure 7.5). The declination angle varies from −23.44° at the winter solstice (December 21) to +23.44° at the summer solstice (June 22). Variation through the year is approximately sinusoidal. The high noon sun angle for 37.2° latitude at the winter solstice, measured as the angle between a normal to the surface of the earth and the sun, is 37.2° + 23.44° = 60.64°. Another way to look at this is that at the winter solstice, the sun will only rise 90° − 60.64° = 29.36° above the horizon. At the summer solstice, the high noon sun angle is 37.2° − 23.44° = 13.76°. The sun will be 90° − 13.76° = 76.24° above the horizon at solar noon.

Figures 7.6 and 7.7 were developed to help the reader understand how the earth rotates and progresses around the sun. The earth is tilted 23.44° and spinning on an axis as it moves around the sun. The axis is always tilted in the same direction. For the summer solstice (June 21) the earth is tilted toward the sun as illustrated in Figure 7.6 with the Northern Hemisphere tilted toward the sun. From the earth it appears as if the sun is directly over the Tropic of Cancer.

Figure 7.5 Declination and sun angle for 37.2° latitude.

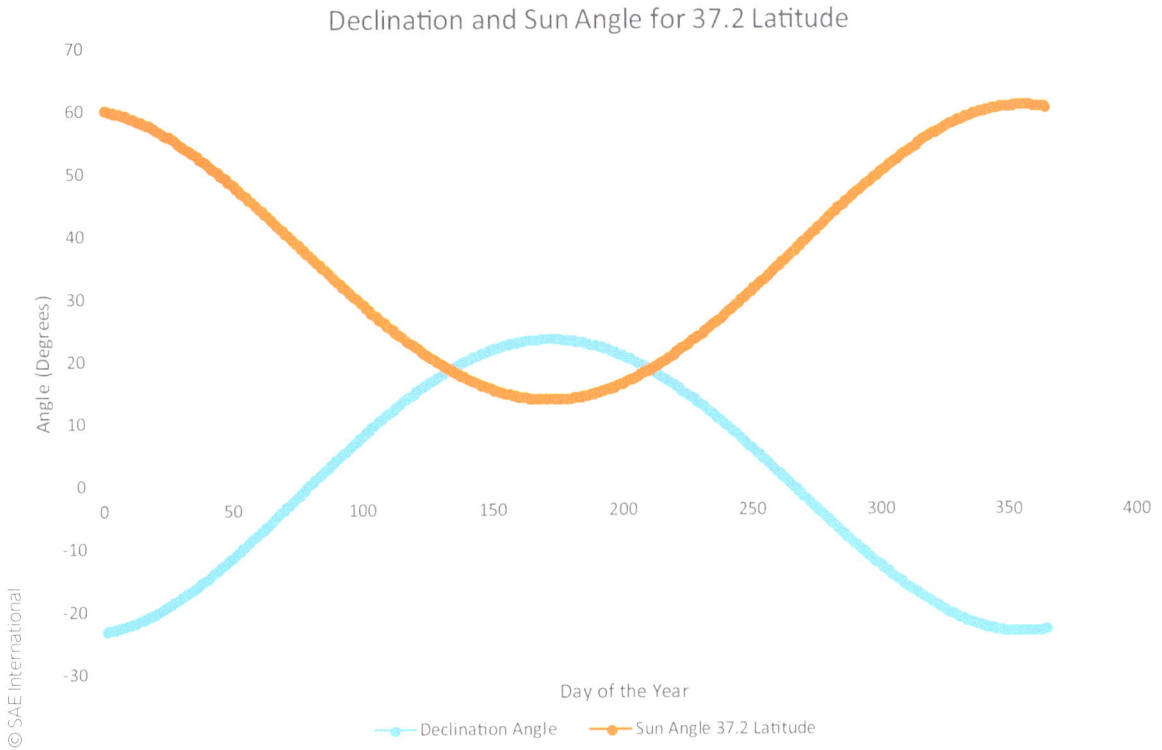

Declination and Sun Angle for 37.2 Latitude

Figure 7.6 Summer solstice.

The earth is spinning around the North Pole axis such that if your right thumb is pointed in the direction of the North Pole your fingers will cup in the direction the earth is rotating. If the point in question is located at 37.2° latitude, then solar noon occurs as the earth rotates around its axis so that the point in question is toward the sun. At that point in time, the angle between the sunlight and a normal to the surface of the earth will be 37.2° − 23.44° = 13.76°. As the earth spins on its axis, regions near the South Pole will never see the sun. Regions near the North Pole will have 24-h daylight.

Figure 7.7 illustrates the winter solstice (December 21) when the earth has rotated around the sun, so it is now on the right side of the sun as illustrated in the figure. At the winter solstice it appears from earth that the sun is directly over the Tropic of Capricorn. If the point in question is located at 37.2° latitude, then solar noon occurs as the earth rotates around its axis so that the point in question is toward the sun. At that point in time the angle between the sunlight and a normal to the surface of the earth will be 37.2° + 23.44° = 60.64°. As the earth spins on its axis, regions near the North Pole will never see the sun. Regions near the South Pole will have 24-h daylight.

Figure 7.7 Winter solstice.

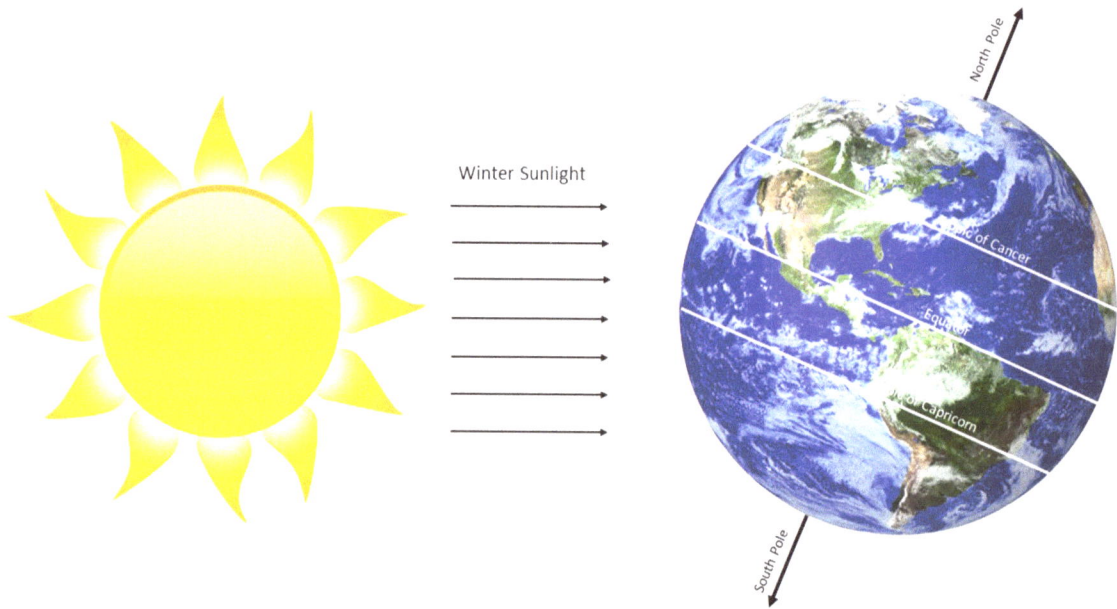

Two times each year, September 21 and March 21, the earth is in equinox position. It is hard to draw a good diagram for the equinox position. Imagine that the earth is located above the page directly over the sun. It will appear from earth that the sun is directly over the equator during the equinox. Northern and Southern Hemispheres will get the same amount of daylight at the equinox.

7.5.2.
Hour Angle

The declination angle is used to find the sun angle at solar noon, but as the earth spins on its axis, the angle between the solar array and the sun varies with time of day. The hour angle accounts for the movement of the sun during the day. The hour angle depends on the longitude location. The earth rotates 15° per hour, which makes it appear from earth that the sun moves 15° per hour around the earth. If we use Greenwich Mean Time (GMT) as the time, the longitudinal location of the sun is at −15*(GMT-12). The negative is because the sun moves in the negative longitude direction, and the −12 is because the longitudinal location of the sun is at Greenwich, England, at noon GMT. Missouri is in the central time zone, which is 6 h behind GMT. Eastern time is 5 h behind GMT, and Pacific time is 8 h behind GMT.

Example 7.5

Find the longitude location of the sun at 10:00 am (regular time) in Missouri.

Solution: The sun is at −15 × (10 + 6 − 12) = −60° longitude. Please notice that if we are on daylight savings time the central time zone is only 5 h behind GMT and the longitude location of the sun would be −45° longitude.

Example 7.6

Springfield, Missouri, is at −93.29156° longitude. What is the hour angle between the sun and Springfield at 10:00 am standard time?

Solution: To solve this problem, take the longitude position of Springfield and subtract the longitude position of the sun. At 10:00 am the longitude angle is −93.29156° − (−60°) = −33.29156°. The negative sign indicates that the sun will appear to be east of Springfield in the sky. The hour angle for Springfield, Missouri, at 10:00 am standard time is −33.29156°. If it were 10:00 am daylight savings time the hour angle would be −48.29156°.

7.5.3.
Hour Angle Definition

The hour angle is defined as the longitude position of the point on earth minus the longitude position of the sun (city longitude – sun longitude position). A negative hour angle means the sun will appear to be east of the city. An hour angle of zero is solar noon for the city. A positive hour angle means the sun will appear to be west of the city.

The power that will be produced by a solar array is proportional to the cosine of the angle between the solar array and the sunlight. For the derivation, we will assume that the solar array is parallel to the surface of the earth. Stationary arrays are usually tilted toward the south to capture more solar energy. Once the formula is developed for solar arrays parallel to the earth surface, it is relatively easy to incorporate the angle of tilt.

In the derivation, the sun is located by the declination angle θ_D and the hour angle θ_H. The position of the solar array on the earth is located from its latitude φ_{Lt} and longitude φ_{Lo}. Define a coordinate system so that the z-axis is the North Pole axis. The x-axis comes from the center of the earth through the equator directly south of where the solar array is located. The hour angle is the east or west location of the sun relative to where the solar array is located. $\mathbf{n}_E$ is a unit normal vector perpendicular to the surface of the earth where the solar array is located. **Figure 7.8** illustrates the coordinate system used.

Figure 7.8 Sun–earth position and vectors.

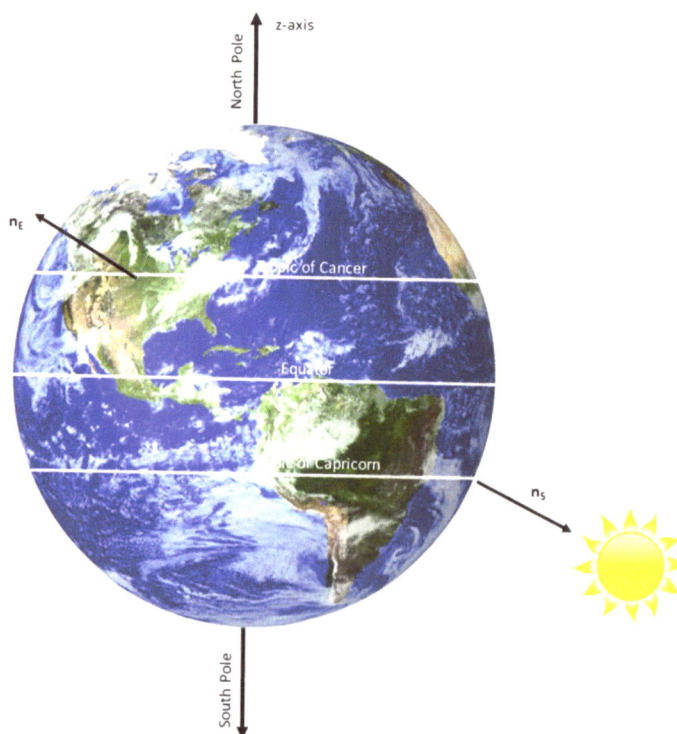

With the coordinate system defined, the unit normal to the surface of the earth $\mathbf{n}_E$ is defined as:

$$\mathbf{n}_E = \cos(\varphi_{Lt})\mathbf{i} + 0\,\mathbf{j} + \sin(\varphi_{Lt})\mathbf{k} \tag{7.25}$$

The unit normal vector for the sun depends on the declination angle θ_D and the hour angle θ_H. The z-component of the normal vector of the sun is the sine of the declination angle. The sun normal is then projected into the x–y plane and the projection is divided into x and y components. The unit normal vector of the sun $\mathbf{n}_S$ is then defined as:

$$\mathbf{n}_S = \cos(\theta_D)\cos(\theta_H)\mathbf{i} + \cos(\theta_D)\sin(\theta_H)\mathbf{j} + \sin(\theta_D)\mathbf{k} \tag{7.26}$$

The dot product of $\mathbf{n}_E \cdot \mathbf{n}_S$ is the cosine of the angle between the sunlight and the normal to the surface of the earth. Since the power produced by the solar array is proportional to the cosine of the angle, the parameter needed is the cosine of the angle. Defining θ as the angle between the sunlight and the solar array, it follows:

$$\cos(\theta) = \cos(\theta_D)\cos(\theta_H)\cos(\varphi_{Lt}) + \sin(\theta_D)\sin(\varphi_{Lt}) \tag{7.27}$$

Equation (7.27) is the factor that will be used to multiply the solar intensity for a solar array that is mounted parallel to the ground. That is, if the solar intensity is 950 W/m², the intensity of the sunlight striking the solar array will be (950 W/m²) cos (θ). The process is to use the day of the year (N) and Equation (7.20) to find the declination angle θ_D. The time of day is used to find the longitudinal location of the sun. The hour angle θ_H is the longitudinal location of the city minus the longitudinal location of the sun. Equation (7.27) is then used to find cos (θ). If the solar array is parallel to the ground, the intensity of the sunlight striking the solar array is proportional to cos (θ) in Equation (7.27). The model developed is useful for modeling the solar array power received by a solar-powered car, assuming the solar array on the car is parallel to the ground.

The next step is to develop a spreadsheet model for the cosine factor for a particular location on earth (latitude and longitude) and the time of day. This model gives a good approximation of solar intensity during the day for a solar array that is parallel to the ground on a clear sunny day. The average solar constant at sea level is approximately 950 W/m². The constant varies slightly with time of year and cycles of the sun. Solar intensity increases with altitude to a maximum of approximately 1400 W/m² above the atmosphere.

Example 7.7

Assume the declination angle (θ_D) was calculated to be 20° on the day in question. Assume the solar array is in Springfield, Missouri, at latitude 37.2° and longitude −93.2916°. Develop a spreadsheet model for the solar factor cos (θ) of Equation (7.27). Plot the solar factor as a function of time of day.

Solution: The first column in the spreadsheet is time of day, GMT, with a time of 0 representing noon GMT. The second column is the longitude location of the solar array. The third column is the sun location. The longitudinal location of the sun moves 15° per hour as shown in the spreadsheet. Column 4 is the hour angle θ_D, which is column 2 minus column 3. Columns 5 and 6 are the declination angle and latitude location of the solar array, which are constant for this problem. The cosine factor column is calculated using Equation (7.27). Springfield is in the central time zone, which is 6 h behind the time in Greenwich. On standard time, noon GMT would be 6:00 am in Springfield, or on daylight savings time, it would be 7:00 am. The column for Springfield time is included so the reader can relate the solar factor to local time. The "Solar factor" column is equal to the "Cosine factor" column except when the cosine factor is negative. A negative value in Equation (7.27) indicates the sun has set. A logic statement was added for the "Solar factor" column to set it equal to zero when the cosine factor is negative. Table 7.3 is the spreadsheet.

Table 7.3 Spreadsheet model for solar intensity.

GMT with noon = 0	Longitude	Sun angle	Hour angle	Declination angle	Latitude	Cosine factor	Springfield time	Solar factor
0	−93.2916	0	−93.2916	20	37.2	0.163808	6	0.163808
1	−93.2916	−15	−78.2916	20	37.2	0.358677	7	0.358677
2	−93.2916	−30	−63.2916	20	37.2	0.543195	8	0.543195
3	−93.2916	−45	−48.2916	20	37.2	0.704787	9	0.704787
4	−93.2916	−60	−33.2916	20	37.2	0.832442	10	0.832442
5	−93.2916	−75	−18.2916	20	37.2	0.917458	11	0.917458
6	−93.2916	−90	−3.2916	20	37.2	0.954044	12	0.954044
7	−93.2916	−105	11.7084	20	37.2	0.939705	1	0.939705
8	−93.2916	−120	26.7084	20	37.2	0.875418	2	0.875418
9	−93.2916	−135	41.7084	20	37.2	0.765566	3	0.765566
10	−93.2916	−150	56.7084	20	37.2	0.617633	4	0.617633
11	−93.2916	−165	71.7084	20	37.2	0.441702	5	0.441702
12	−93.2916	−180	86.7084	20	37.2	0.249762	6	0.249762
13	−93.2916	−195	101.7084	20	37.2	0.054893	7	0.054893
14	−93.2916	−210	116.7084	20	37.2	−0.12963	8	0
15	−93.2916	−225	131.7084	20	37.2	−0.29122	9	0
16	−93.2916	−240	146.7084	20	37.2	−0.41887	10	0
17	−93.2916	−255	161.7084	20	37.2	−0.50389	11	0
18	−93.2916	−270	176.7084	20	37.2	−0.54047	12	0
19	−93.2916	−285	191.7084	20	37.2	−0.52613	1	0
20	−93.2916	−300	206.7084	20	37.2	−0.46185	2	0
21	−93.2916	−315	221.7084	20	37.2	−0.352	3	0
22	−93.2916	−330	236.7084	20	37.2	−0.20406	4	0
23	−93.2916	−345	251.7084	20	37.2	−0.02813	5	0
24	−93.2916	−360	266.7084	20	37.2	0.163808	6	0.163808

A declination angle of 20° corresponds to May 20 or July 24, plus or minus a day depending on leap year. Once the spreadsheet is developed, it can be modified to work for any city where the solar array is located and for any day of the year. A plot of the solar factor is shown in Figure 7.9.

Figure 7.8 illustrates how the power produced by a solar array parallel to the ground would vary with the time of day. The maximum power output would occur approximately at GMT 6:00 pm, which will be noon standard time in Springfield, or 1:00 pm daylight savings time. The output is lower at other times of the day. The spreadsheet model developed in Table 7.3 is a good starting point for modeling solar arrays that are parallel to the ground.

Figure 7.9 Plot of solar factor for Springfield, MO, May 20.

7.5.4.
Process for Tilted Arrays
Stationary solar arrays are usually tilted to the south to allow them to collect more solar energy.

This changes the problem because the normal to the earth is no longer normal to the solar array. If the array is tilted to the south by an amount θ_T, the normal to the solar array $\mathbf{n}_A$ is given by the following:

$$\mathbf{n}_A = \cos(\phi_{Lt} - \theta_T)\mathbf{i} + 0\,\mathbf{j} + \sin(\phi_{Lt} - \theta_T)\mathbf{k} \tag{7.28}$$

The angle for the sun $\mathbf{n}_S$ does not change from what it was for the array parallel to the ground. Taking the dot product $\mathbf{n}_A \cdot \mathbf{n}_S$ is the cosine factor for a tilted solar array in Equation (7.29). Unfortunately, it is not quite this simple. The simple dot product approach assumes the solar array is high enough above the surface of the earth that the earth is not blocking the sunlight from reaching the solar array. If the sun can reach the solar array the cosine factor can be obtained from Equation (7.29).

$$\cos(\theta) = \cos(\theta_D)\cos(\theta_H)\cos(\varphi_{Lt} - \theta_T) + \sin(\theta_D)\sin(\varphi_{Lt} - \theta_T) \qquad (7.29)$$

This can be incorporated into the solar array spreadsheet that was developed for the array parallel to the ground. First, obtain the solar factor for when the solar array is parallel to the ground as was done in the previous spreadsheet. If the cosine factor for the array parallel to the ground is positive, then the sun has risen above the horizon and is able to shine on the solar array. A logic statement is added so that Equation (7.29) is used to find the tilted array solar factor when the flat solar factor is positive. It is a little confusing, but this is the best way to model tilted solar arrays.

For summer months in the Northern Hemisphere, the sun rises in the northeast and sets in the northwest. In early morning and late afternoon, the tilted solar array produces less power than the flat solar array. In summer months, there is not much difference overall in how much energy is collected during the day whether the solar array is flat or tilted to the south. In spring, fall, and winter, the tilted solar array will collect significantly more solar energy than a flat solar array. The optimum angle of tilt depends on time day and time of year. Mechanical systems have been used to adjust the tilt of the solar array during the day, and/or track the sun and keep the array angled at the sun during the day. These mechanical systems allow the solar array to collect more energy, but the mechanical complexity makes the system more expensive, less reliable, and prone to being damaged by the weather. Most solar panels are mounted in a fixed position and tilted to the south at an angle to increase the amount of electricity they produce during the year. The ideal angle of tilt depends on the latitude location of the solar array. In home construction, the most

common roof used is the 4:12 pitch, which is an angle of 18.4°. If the solar panels are mounted directly on the south-facing roof, they will be tilted 18.4° to the south, which is a good angle of tilt for most places in the United States. It is not the ideal angle of tilt, but when you consider the cost of mounting the cells on the roof and the fact that they are less likely to be damaged by the weather if mounted flat on the roof, this is a good design.

Equation (7.29) was developed assuming the city is in the Northern Hemisphere so that φ_{Lt} is a positive latitude. Tilting the solar array to the south makes θ_T a positive angle. The normal situation in the Northern Hemisphere makes φ_{Lt} and θ_T positive angles. Equation (7.29) works in the Southern Hemisphere where φ_{Lt} is a negative latitude. In the Southern Hemisphere, the solar array would be tilted to the north, making θ_T a negative angle. The normal situation in the Southern Hemisphere makes φ_{Lt} and θ_T negative angles, but Equation (7.29) still yields the correct answer if the negative values are used.

A column can be added to the spreadsheet model in Table 7.4 titled "Tilt solar factor" to total the solar factors for the day. The tilt angle can be adjusted to maximize the total of the solar factors, and the reader can discover the optimum tilt of the array to maximize the energy collected during the day. For a declination angle of 20° and a latitude of 37.2°, it turns out that the optimum tilt is 0°. That is, the maximum energy would be collected by putting the solar array parallel to the ground. Tilting the solar array will increase the peak power in the middle of the day but will decrease the power in the morning and evening, and overall the maximum energy will be collected with the array parallel to the ground.

Table 7.4 Spreadsheet model for tilted solar array.

| Declination | 0 | degrees | | Latitude | 37.2 | degrees | | |
| | | | | Tilt angle | 37.2 | degrees | | |

GMT with noon = 0	Longitude	Sun angle	Hour angle	Declination angle	Latitude	Cosine factor	Flat solar factor	Tilt solar factor
0	−93.2916	0	−93.2916	0	37.2	0.163808	0	0
1	−93.2916	−15	−78.2916	0	37.2	0.358677	0.16164	0.202931
2	−93.2916	−30	−63.2916	0	37.2	0.543195	0.358	0.44945
3	−93.2916	−45	−48.2916	0	37.2	0.704787	0.529963	0.66534
4	−93.2916	−60	−33.2916	0	37.2	0.832442	0.66581	0.835888
5	−93.2916	−75	−18.2916	0	37.2	0.917458	0.756282	0.949472
6	−93.2916	−90	−3.2916	0	37.2	0.954044	0.795216	0.99835
7	−93.2916	−105	11.7084	0	37.2	0.939705	0.779957	0.979193
8	−93.2916	−120	26.7084	0	37.2	0.875418	0.711545	0.893306
9	−93.2916	−135	41.7084	0	37.2	0.765566	0.594642	0.746541
10	−93.2916	−150	56.7084	0	37.2	0.617633	0.437215	0.5489
11	−93.2916	−165	71.7084	0	37.2	0.441702	0.249994	0.313853
12	−93.2916	−180	86.7084	0	37.2	0.249762	0.045735	0.057418
13	−93.2916	−195	101.7084	0	37.2	0.054893	0	0
14	−93.2916	−210	116.7084	0	37.2	−0.12963	0	0
15	−93.2916	−225	131.7084	0	37.2	−0.29122	0	0
16	−93.2916	−240	146.7084	0	37.2	−0.41887	0	0
17	−93.2916	−255	161.7084	0	37.2	−0.50389	0	0
18	−93.2916	−270	176.7084	0	37.2	−0.54047	0	0
19	−93.2916	−285	191.7084	0	37.2	−0.52613	0	0
20	−93.2916	−300	206.7084	0	37.2	−0.46185	0	0
21	−93.2916	−315	221.7084	0	37.2	−0.352	0	0
22	−93.2916	−330	236.7084	0	37.2	−0.20406	0	0
23	−93.2916	−345	251.7084	0	37.2	−0.02813	0	0
24	−93.2916	−360	266.7084	0	37.2	0.163808	0	0
						Total	6.085999	7.640641

If the declination angle is 0° (March 21 and September 21) and the latitude is 37.2°, then the solar array will collect more energy if it is tilted to the south. The optimum tilt for this case is 37.2°, which is the latitude location. If the declination angle is zero, the ideal tilt for the solar array is equal to the latitude location of the solar array. The spreadsheet for this case is shown in Table 7.4. Tilting the array to 37.2° makes the total solar factor for the day 7.64, whereas keeping the array parallel to the ground gives a total solar factor of 6.89. Tilting the solar array will increase the energy recovered by 25.5%.

If the declination angle is near −20° (December and early January), the maximum energy for the day will be collected by tilting the solar array 63.5°. If the declination angle is near +20° (June and early July), it is best to have the array parallel to the ground. To summarize for a latitude of 37.2°, the array should be parallel to the ground in summer, tilted to the south 37.2° in spring and fall when the sun is at the equinox, and tilted 63.5° to the south in the winter. This model is developed assuming the solar array is tilted to the south. Tilting the array in other directions or having a mechanism to allow the array to track the sun will yield different results.

7.6.
Solar Array Power Equation

The model developed in this section is based on research done by Courtney Green in 2006, supervised by Doug Carroll [5]. There is extensive solar data available from the National Renewable Energy Lab (NREL). The data were collected at five different locations around the country (Elizabeth City State University in North Carolina, Central Oklahoma, Bluefield State College in Virginia, and two locations in Colorado). A model was developed with parameters, and the parameters were adjusted to fit the experimental data. The best-fit parameters are shown below the equation. The model gives the solar intensity of the sunlight P if the normal to the solar array is aimed directly at the sun at midday.

$$P = \left[P_\infty + P_V \cos\left(\frac{360(N-3)}{365} \right) \right] \left(\frac{P_0}{P_\infty} \right)^{\frac{t-h}{t(\cos(\theta))^m}} \tag{7.30}$$

In using Equation (7.30), the input parameters are N (day of the year), where January 1 is 1, and h = height above sea level in meters. The declination angle θ_D is calculated using Equation (7.20), and the hour angle θ_H is calculated using the method in Section 7.5. The other parameters in the model were adjusted to achieve the best fit to the data available from NREL. The best fit parameters are listed below:

- P_∞ = 1400 W/m²
- P_0 = 1087.445 W/m²
- P_v = 116.4844 W/m²
- t = 10,304.022 m
- m = 0.65
- $\cos(\theta) = \cos(\theta_D) \cos(\theta_H) \cos(\varphi_{Lt}) + \sin(\theta_D) \sin(\varphi_{Lt})$

In developing the model, the parameters had a physical meaning, but their values were adjusted to make the model fit the data. P_∞ represents the solar intensity in space above the atmosphere. P_0 represents the solar intensity at sea level. P_v represents the variability of solar intensity with the time of year since the earth is closest to the sun in January and furthest from the sun in July. The parameter t represents what the thickness of the atmosphere would be if the gasses in the atmosphere were compressed to sea-level density. The parameter m is empirical and helped us get a better fit to the data. Using these parameters gives a good overall fit between the model and the available data.

To get the power from the solar array, multiply the power from Equation (7.30) by the cosine of the angle between the solar array and the sunlight. For solar arrays parallel to the ground, use Equation (7.27), and for solar arrays tilted to the south, use Equation (7.29).

Example 7.8

Assume a sunny day April 27, 2017, and calculate the power output for a 20 m² solar array tilted 18.4° to the south. The solar array is 18% efficient in converting sunlight into electricity. Assume latitude ϕ_{Lt} = 30.32° and the hour angle θ_H = 23.5° for the time in question. Assume the solar array is located at an altitude of 1400 ft = 426.7 m above sea level.

Solution: Equation (7.30) is complex, and I prefer to evaluate it a piece at a time and then put the pieces together. First find the day of the year for April 27, 2023, and the first part of the equation:

$$N = 31 + 28 + 31 + 27 = 117 \tag{7.31}$$

$$P_\infty + P_V \cos\left(\frac{360(N-3)}{365}\right) = 1355.5 \text{ W / m}^2 \tag{7.32}$$

When taking the cosine in Equation (7.32), the angle is in degrees (not radians). Equation (7.20) is used to calculate the declination angle. θ_D = 13.8296° on April 27, 2017. The latitude is 30.32° and the hour angle is 23.5°. The cos (θ) term in Equation (7.30) is calculated as:

$$\cos(\theta) = \cos(13.8296)\cos(23.5)\cos(30.32) + \sin(13.8296)\sin(30.32) = 0.889347 \tag{7.33}$$

The altitude correction in Equation (7.30) is computed as:

$$\left(\frac{P_0}{P_\infty}\right)^{t\left(\left(\cos(\theta)\right)^m\right)^{t-h}} = \left(\frac{1087.445}{1400}\right)^{10,304.022\left(0.889347^{0.65}\right)^{\frac{10,304.022-426.7}{}}} = 0.770003 \tag{7.34}$$

Equations (7.32) and (7.34) are combined to find the solar power striking a solar array with the normal to the solar array pointed directly at the sun as:

$$P = \left(1355.5 \text{ W / m}^2\right)(0.770003) = 1044 \text{ W / m}^2 \tag{7.35}$$

The model predicts that at a latitude of 30.32° and altitude of 1400 ft above sea level on April 27, 2017, the solar intensity striking a solar array aimed directly at the sun will be 1044 W/m². The solar array in question is tilted 18.4° to the south, so it is not aligned with the sun. Equation (7.31) is used to account for the misalignment with the sun:

$$\cos(\theta) = \cos(13.8296)\cos(23.5)\cos(30.32 - 18.4) + \sin(13.8296)\sin(30.32 - 18.4) = 0.920646 \tag{7.36}$$

Remember that the tilted solar array factor can only be positive if the flat solar array factor calculated in this case in Equation (7.33) is positive. The logic statement discussed before is still required for the tilted solar array equation. In this case it is positive, so the 0.920646 factor calculated in Equation (7.36) is valid. The solar array has an area of 20 m² and the solar cells are 18% efficient in converting the sunlight into electricity. The solar array power is computed as:

$$\text{Array Power} = \left(1044 \text{ W} / \text{m}^2\right)\left(0.920646\right)\left(20 \text{ m}^2\right)\left(0.18\right) = 3460 \text{ W} \qquad (7.37)$$

This is the solar array power when the hour angle is 23.5°, which would be approximately 1:30 pm standard time. The exact time depends on where the solar array is in the time zone.

Homework

1. Assume that the United States generates 12.9 quad of electricity each year and that we also generate 1584 MMT of carbon dioxide each year producing electric energy from the various energy sources used.

   (a) For comparison assume that all the electricity was generated using coal in power plants that are 33% efficient in converting the energy in the coal into electric energy. Calculate the metric tons of carbon dioxide that would be generated each year. Assume that the coal provides 9.8 BTU of heat energy for each gram of carbon dioxide generated. Compare the carbon dioxide that would be generated using coal to produce all the electricity to the carbon dioxide that is generated using the mix of energy sources actually used. (Answer: 3989 MMT, 2.52 times as much carbon dioxide)

   (b) Calculate the amount of sulfur dioxide, nitrous oxide, and particulate generated from burning the coal. Assume that for each billion BTU of energy from burning the coal, we generate 2500 lb of sulfur dioxide, 560 lb of nitrous oxides, and 5150 lb of particulate matter. Assuming that a typical train car can hold 100 tons (2000 lb/ton) of particulate matter (coal ash), how many train car loads does this represent? (Answer: 201.3 billion pounds of particulate, or a little over one million train car loads. Nearly all the particulate will be recovered in the power plant. It will not be put into the atmosphere. The other gasses will be put into the atmosphere)

2. Assume that the United States generates 12.9 quad of electricity each year, and that we also generate 1584 MMT of carbon dioxide each year. Assume that all the electricity was generated using natural gas in power plants that are 33% efficient in converting the energy in the natural gas into electric energy.

   (a) Calculate the metric tons of carbon dioxide that would be generated each year. Assume that natural gas provides 19.1 BTUs of energy for each gram of carbon dioxide generated. Compare this number to coal and to the 1584 MMT generated with the mix of energy used.

   (b) Calculate the amount of sulfur dioxide, nitrous oxides, and particulate generated from burning the natural gas. Assume that for each billion BTU of energy from burning the natural gas, we generate 1 lb of sulfur dioxide, 92 lb of nitrous oxides, and 7 lb of particulate matter. Compare these values to the emissions generated by burning coal.

3. Lake Mead covers approximately 150,000 acres of land. The water turbines in the Hoover Dam are located 590 ft below the surface of the lake. Calculate the electric energy that is produced by lowering the lake level one inch. Assume that the weight density of water is 62.4 lb/ft^3 and the water turbines are 90% efficient in converting the gravitational energy of the water into electric energy. Also calculate the value of the electricity assuming it is worth $0.12 per kW-h. (Answer: 6795 MW-h, at $0.12 per kW-h = $815,400.00 of electricity.)

4. A commonly used wind turbine is made by General Electric and is rated to provide 1.5 MW in a 27-mph wind. The diameter of the wind turbine blade is 232 ft. Assume that the density of air is 1.225 kg/m^3. Based on this information, calculate the efficiency of the wind turbine. (Answer: 35.5%)

5. The wind turbine in problem 4 is to be installed in Springfield, Missouri. According to weather data, the wind speed over the next 48 h is forecast according to Table 7.5 (speeds in mph).

   (a) Develop a spreadsheet and estimate the number of kW-h of electric energy that will be generated over the 48 h period.

   (b) Assuming that the electricity is worth $0.07 per kW-h, extrapolate and estimate the value of the electricity generated in one month and in one year.

   (c) If the wind turbines cost $2.5M each, how long would it take to recover your

investment in Springfield? Assume the wind data in Table 7.5 are wind speed in mph each hour for 48 h. If the wind data in Table 7.5 are typical for Springfield, do you feel that Springfield is a good location for wind power?

Table 7.5 Wind speed in mph over a 48 h period in Springfield, Missouri.

14	7	14	9	11
14	7	15	9	10
13	7	16	9	12
12	8	17	10	12
9	8	17	9	12
10	8	17	9	14
10	8	15	9	14
9	9	12	9	15
9	10	9	10	
7	13	9	10	

6. Rework problem 5 with the data from Sioux Falls, South Dakota in Table 7.6. Copy your spreadsheet to a second sheet and then modify it with the data in Table 7.6.

Table 7.6 Wind speed in mph for Sioux Falls, South Dakota over a 48 h period.

16	10	14	16	13
15	10	15	15	13
14	9	16	15	13
13	10	17	14	13
11	10	17	15	14
12	10	17	15	15
12	10	16	14	15
11	11	16	14	14
11	12	16	14	
10	13	17	14	

7. Calculate the declination angle for Rolla, MO, on July 14, 2023. Develop a spreadsheet and calculate the cosine factor for the solar intensity as a function of time of day for objects parallel to the ground. Plot the cosine factor vs time of day. (Answer: Declination = 21.688°)

8. Assume that a 10 m² solar array is placed on the roof of a house in Rolla, MO, facing south, and tilted 18° to better capture the solar energy. If the solar cells are 18% efficient in converting the sunlight to electricity, how much electric energy would be collected on a sunny day April 23? Use the solar array power equation to estimate the intensity of the sunlight. The hour angle and the cosine factor will vary with the time of day. Use an Excel spreadsheet to integrate the solar power during the day and get the total solar energy collected. (Answer: Declination = 12.5262°. At solar noon, the angle between the solar panel and the sun is 7.4253°. If you assume 950 W/m² average solar intensity the answer is approximately 13.56 kW-h of energy on a sunny day. If you use the more complex power equation the answer is approximately 14.16 kW-h on a sunny day. I used Rolla, MO, at latitude 37.9485°, longitude −91.7715°, and altitude 335 m.)

References

1. https://www.e-education.psu.edu/ebf200/node/151.

2. https://www.eia.gov/electricity/monthly/epm_table_grapher.php?t=epmt_5_6_a.

3. https://www.eia.gov/state/seds/data.php?incfile=/state/seds/sep_prices/total/pr_tot_US.html&sid=US.

4. https://en.wikipedia.org/wiki/Table_Rock_Lake.

5. https://en.wikipedia.org/wiki/Position_of_the_Sun.

6. Green, C., "Solar Power Model," in *Proceedings of Opportunities for Undergraduate Research*, University of Missouri-Rolla, 2006.

Climatologists have been saying that green-house gases (GHGs) increase the temperature of the earth and lead to global warming. There is concern that the carbon dioxide emitted when burning fossil fuels will lead to global warming. The first step to understanding how GHGs impact the temperature of the earth is to look at a radiative heat transfer of the earth.

The earth receives heat from the sun as electro-magnetic radiation. Most of the energy from the sun is in the visible range because of the temper-ature of the sun. Objects emit radiation in a range of energy that is related to their temper-ature, with hotter objects emitting higher energy radiation and cooler objects emitting lower energy radiation. The temperature of the sun causes it to emit radiation mostly in the visible range, with smaller amounts of radiation in higher and lower energy ranges.

The flame from a campfire is very hot and emits radiation in the visible range. This allows us to see the flame. Once the fire burns down, there are still glowing red and orange coals that are hot enough to emit radiation in the lower energy frequencies of the visible range. Within the visible range, red and orange are lower energy frequencies, and blue and violet are higher energy frequencies. Once the fire burns down completely, it does not emit visible light anymore, but it is still hot. The fire has cooled to the point that it is emitting radiation in the infrared range. We cannot see infrared radiation, but we can feel the heat.

The surface of the earth emits radiation in the infrared range. The earth receives heat energy from the sun mostly in the visible range and emits energy back into space as infrared energy.

For energy balance, the earth must receive the same amount of energy from the sun that it emits as infrared radiation. If the earth were to receive more energy from the sun than it emits, the earth would become warmer. Conversely, if the earth were to receive less energy from the sun than it emits, it would become cooler. We see this happen every day. The earth will tend to warm during the day when it receives lots of solar energy and will tend to cool at night. The average solar radiation [1] striking the top of the atmosphere is 1362 W/m². Solar radiation is not constant because the radiation emitted by the sun is not constant. The orbit of the earth around the sun is almost circular, but is slightly elliptical. The earth is closer to the sun in January and further from the sun in July, which impacts the intensity of the solar radiation. But if we average through the year, the average solar intensity is 1362 W/m².

Sunlight strikes the earth in profile so that if R is the radius of the earth (plus atmosphere), the total energy striking earth is (1362 W/m²) πR². The earth will emit infrared radiation in all directions from the surface of the earth, and the surface of the earth (plus atmosphere) is a sphere of surface area 4πR². To convert the 1362 W/m² striking a circular profile area to an average radiation over the spherical surface of the earth, we need to ratio the areas. The average solar radiation striking the top of the atmosphere is:

$$\text{Average Solar Radiation} = \left(1362 \, \text{W} / \text{m}^2\right)\left(\frac{\pi \text{R}^2}{4\pi \text{R}^2}\right) = 340.5 \, \text{W} / \text{m}^2 \tag{8.1}$$

If 100% of the solar energy striking the top of the atmosphere were absorbed by the earth and atmosphere, then the earth would receive 340.5 W/m² average solar radiation over its surface all the time. In reality, some parts of the earth receive more solar radiation than others and this causes some parts of the earth to be warmer than others. Assuming 100% absorption of the energy from the sun, the average solar radiation would be 340.5 W/m². But not all radiation from the sun is absorbed by the earth.

Some of the energy from the sun is reflected into space by the clouds and the surface of the earth. The reason we can see the earth from space is because of the reflected sunlight. If all sunlight was absorbed by the earth and atmosphere it would not be possible to see the earth from space. To develop the radiative heat transfer model, we need to know how much solar energy is absorbed by the earth and atmosphere and how much is reflected into space. Through experimental work, atmospheric scientists have found that on average the earth absorbs 239 W/m² of radiation [2].

$$\text{Average Solar Radiation Absorbed} = 239 \, \text{W} / \text{m}^2 \left(\text{Experimental Value}\right) \tag{8.2}$$

This means that 340.5 – 239 = 101.5 W/m² is being reflected into space. In percentages, 70.2% of the sunlight is absorbed by the earth and atmosphere and 29.8% is reflected into space.

The model developed averages the energy balance over the surface of the earth. It is the simplest model that can be developed to quantify how GHGs cause the average temperature of the earth to increase. It is a good starting point for people who wish to study the climate science of global warming.

GHGs do not absorb much visible light (sunlight). The reason has to do with the physical chemistry of gases. Gases absorb radiation in a range of frequencies, depending on their molecular structure. Carbon dioxide and methane are not effective at absorbing visible light, but are effective at absorbing frequencies in the infrared range that earth is emitting. A more precise way of saying this is that GHGs will absorb a small amount of sunlight, but are more effective at absorbing the IR radiation from the surface of the earth. The energy balance of the earth must be maintained. As GHGs increase, a higher percentage of the IR radiation from the earth will be absorbed by the atmosphere. This will cause the earth to get warmer. As the earth gets warmer, it emits more IR radiation, because warmer objects emit more radiation. The earth will warm until the IR radiation emitted through the atmosphere into space is equal to the radiation received from the sun. The energy balance is achieved by increasing the average temperature of the earth. This is a qualitative explanation of global warming. The goal of this chapter is to develop a quantitative model for global warming.

The next step in understanding the radiative heat transfer of the earth is to understand how the earth emits infrared radiation back into space. The average temperature of the earth is constant. Scientists have averaged temperature readings over the surface of the earth and measured the average earth temperature as 288°K, which is 59°F. That is, if you average the temperature over the entire earth, day and night and throughout the year, the average temperature is 288°K.

The Stefan–Boltzmann relationship says that a body emits radiation according to its temperature raised to the fourth power. The Stefan–Boltzmann constant in its metric form is 5.67×10^{-8} W/m^2 °K^4. Since the average earth temperature is 288°K, the infrared radiation from earth is calculated as:

$$\text{Earth Surface Radiation} = \left(\frac{5.67 \times 10^{-8} \, \text{W}}{\text{m}^2 °\text{K}^4} \right) \left(288°\text{K} \right)^4 = 390 \, \text{W} / \text{m}^2 \tag{8.3}$$

Some of the infrared radiation from earth is absorbed by GHGs. For the average earth temperature to be constant, the energy absorbed by the earth and atmosphere from the sun (239 W/m^2) must equal the infrared energy leaving the earth from the top of the atmosphere. The earth's surface is emitting 390 W/m^2 of infrared energy, so the difference must be absorbed by the GHGs in the atmosphere. The energy balance yields:

$$\text{GHG Absorption} = 390 \, \text{W} / \text{m}^2 - 239 \, \text{W} / \text{m}^2 = 151 \, \text{W} / \text{m}^2 \tag{8.4}$$

In percentages, 61.3% of the infrared radiation from the surface of the earth will find its way through the atmosphere and into space. In total 38.7% of the infrared radiation will be absorbed by the GHGs in the atmosphere. Figure 8.1 is helpful in illustrating the energy balance of the earth.

Figure 8.1 Radiative energy balance for earth.

340.5 W/m² sunlight striking the earth and atmosphere.

101.5 W/m² sunlight reflected by earth and atmosphere into space.

390 W/m² infrared emitted by earth surface upward

151 W/m² infrared absorbed by atmosphere and/or reflected back to and absorbed by earth.

A net of 239 W/m² sunlight absorbed by the earth and atmosphere.

A net of 239 W/m² infrared emitted by the earth and atmosphere into space.

© SAE International.

The most common GHGs are water vapor, carbon dioxide, methane, halocarbons [chloro-fluorocarbons (CFCs)], NOx, and ozone. Water vapor is the strongest GHG as far as absorbing infrared radiation. The amount of water vapor in the atmosphere depends on the average temperature of the atmosphere. Warmer air will hold more moisture than cooler air. The moisture in the air is somewhat self-regulating because an excess of moisture in the air will cause rain, which removes moisture from the atmosphere. Carbon dioxide is the second most important GHG, and we are concerned about the increase in carbon dioxide in the atmosphere causing the earth to warm. Methane is more effective at absorbing infrared radiation than carbon dioxide, but the concentration of methane in the atmosphere is very small, so overall methane absorbs less infrared radiation than carbon dioxide. Some scientists are concerned that thawing of the polar regions will release massive quantities of methane into the atmosphere, which would greatly increase the amount of energy absorbed by GHGs. Halocarbons, NOx, and ozone all contribute to GHG absorption, but to a lesser extent than water vapor, carbon dioxide, and methane.

Example 8.1

Assume that an increase in GHGs in the atmosphere allows the atmosphere to absorb 40% of the infrared radiation from earth rather than the 38.7% it is absorbing now. How will this affect the average temperature of the earth?

Solution: As the temperature of the earth increases, the earth will emit more infrared radiation due to the temperature increase. If GHGs absorb 40% of the infrared radiation, then 60% will leave the atmosphere and go into space. The earth will still absorb 239 W/m² from the sun, and in order for the earth to stabilize to a new temperature the radiation leaving the top of the atmosphere going into space must equal the 239 W/m². The radiation from the surface of the earth is calculated as:

$$\text{Earth Surface Radiation} = \frac{239 \text{ W} / \text{m}^2}{0.60} = 398.3 \text{ W} / \text{m}^2 \tag{8.5}$$

The Stephan–Boltzmann relationship can then be used to determine the average earth temperature required to emit 398.3 W/m² radiation from the surface of the earth.

$$\left(\frac{5.67 \times 10^{-8} \text{ W}}{\text{m}^2 \text{K}^4} \right) (T)^4 = 398.3 \text{ W} / \text{m}^2 \tag{8.6}$$

Solving Equation (8.6), the earth's average temperature is 289.5°K. The solution shows that increasing the absorptivity of the atmosphere from 38.7 to 40% will increase the average temperature of the earth from 288 to 289.5°K, an increase of 1.5°K (or 1.5°C). Another way of looking at it is to say increasing the absorption of the atmosphere from 151 to 156 W/m² will cause the 1.5°C temperature increase. An average temperature increase of 1.5°C will significantly change our environment. The purpose of developing this model is to show that a relatively small increase in the absorptivity of the atmosphere will cause significant climate change.

The radiation model presented in this chapter is a great oversimplification. In reality, the temperature on earth varies significantly from place to place. The model uses the average earth temperature and assumes the temperature is constant over the surface of the earth. The earth receives sunlight only during daytime hours, and there is a significant variation in solar intensity on the surface of the earth. The model uses the average solar radiation and assumes the radiation is spread evenly over the surface of the earth. It is easy to dismiss the model developed here as a great oversimplification of what is actually happening. The advantage of the model developed in this chapter is that it is relatively simple, and most scientists and engineers can do the calculations and make a good estimate of how much GHGs will impact the average earth temperature. Climate scientists have worked through much more detailed models than those presented here. The results of this simple model are similar to the results of the more complex models. Though not perfect, the simple model developed in this unit can be used to estimate how GHGs affect the average temperature of the earth.

8.2.
Carbon Dioxide Portion of GHG Absorption

Carbon dioxide is the second most important GHG in the atmosphere. Water vapor accounts for more of the 151 W/m² infrared radiation absorbed by the atmosphere than all the other GHGs combined. It is difficult to get an accurate measurement of the amount of energy absorbed by individual GHGs because they are mixed in the atmosphere. We can do experiments with individual GHGs and measure the amount of infrared radiation that will be absorbed, but we cannot simply add the contributions of the different GHGs. An infrared photon absorbed by water vapor will not also be absorbed by carbon dioxide. It is absorbed by one or the other but not both. If we test the individual GHGs and add the results, we are double-counting some of the absorption and it leads to a high estimate of the infrared energy absorbed.

Recognizing the limitations discussed in the paragraph above, it is possible to estimate the amount of energy absorbed by carbon dioxide if it were the only GHG in the atmosphere.

This model yields a high estimate of the energy absorbed by carbon dioxide, but it is a place to start the discussion. Stull et al. measured the infrared energy transmitted through carbon dioxide as a function of the carbon dioxide concentration [3]. The concentration of carbon dioxide was measured in mole/m², and the device measured the fraction of infrared radiation transmitted through the experimental chamber. The results are presented in Table 8.1.

Table 8.1 Results from Stull et al.

Concentration of carbon dioxide (mole/m²)	Fraction of energy transmitted through
40.9	0.5917
81.8	0.5328
204.5	0.4535
409	0.3810

© SAE International

We are more interested in the fraction of energy absorbed, which is one minus the fraction transmitted. The absorption increases as the carbon dioxide concentration increases. The data follow a logarithmic function. A good fit to the data is:

$$\text{Absorption Fraction} = (0.09083)\ln(\text{Concentration}) + 0.0681 \tag{8.7}$$

The logarithm in Equation (8.7) is the natural log (not the base 10 log). Figure 8.2 illustrates how absorption varies with the concentration of carbon dioxide. The orange line is Equation (8.7), and the blue dots are the data points.

Figure 8.2 Plot of absorption vs carbon dioxide concentration.

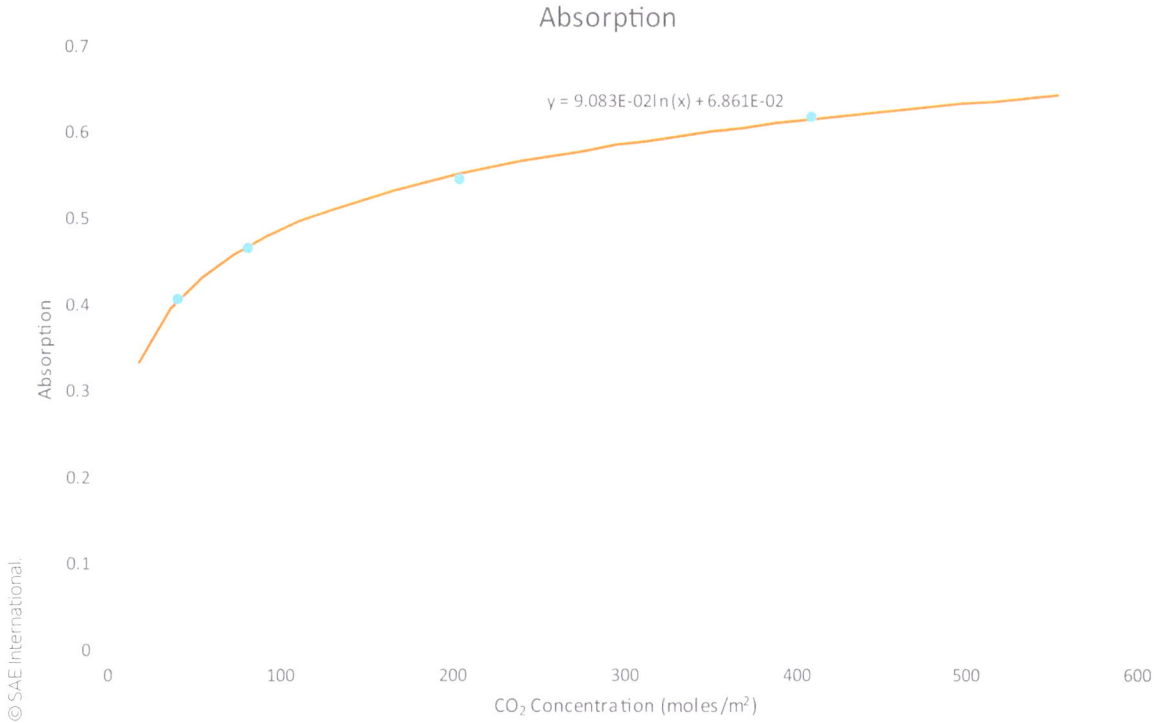

Carbon dioxide can absorb a range of frequencies of infrared radiation with wavelengths in the 13–17 μm range. Approximately 48.9% of the infrared radiation from earth falls in this range. When we look at the amount of infrared radiation absorbed, we need to multiply the equation by 0.489 because most of the infrared radiation emitted from the surface of the earth cannot be absorbed by carbon dioxide.

The carbon dioxide molecules will absorb the infrared photons and then reemit photons in a random direction. Approximately half of the reemitted photons will travel up and escape the atmosphere, and approximately half will travel downward and be absorbed. Equation (8.7) must also be multiplied by a factor of ½ to account for this.

The next step is to calculate the concentration of carbon dioxide on earth in moles per square meter surface area of the earth. There are 1.78×10^{20} moles of gas in the atmosphere. The radius of the earth is R = 6,370,000 m. Dividing the moles of gas in the atmosphere by the surface area of the earth yields:

$$\frac{1.78 \times 10^{20} \text{ mole}}{4\pi (6,370,000 \text{ m})^2} = 349,085 \text{ mole} / \text{m}^2 \tag{8.8}$$

This is the total number of gas molecules per square meter of surface area on the earth.

We keep track of the carbon dioxide in parts per million (ppm). The atmosphere currently has a little over 400 ppm carbon dioxide, which means that of every million molecules in the atmosphere, a little over 400 are carbon dioxide molecules. If we define C_{ppm} as the parts per million carbon dioxide in the atmosphere, then the CO_2 concentration in moles per square meter earth surface is calculated as:

$$CO_2 \text{ Concentration} = (0.349)C_{ppm} \text{ mole} / m^2 \qquad (8.9)$$

Putting all of this together, the energy absorbed is calculated as:

$$\text{Energy Absorbed by } CO_2 = \frac{1}{2}(0.489)\left(390 \text{ W}/m^2\right)\left[(0.09083)\log\left((0.349)C_{ppm}\right)+0.0681\right] \qquad (8.10)$$

As an example calculation, if there are 400 ppm CO_2 in the atmosphere, the CO_2 concentration is calculated from Equation (8.9) as (0.349) (400) = 139.2 mole/m². This is within the range of experimental values measured by Stull et al., so it is reasonable to use their results. What this means is the infrared radiation from 1 m² of surface area on the earth will travel upward through the atmosphere where there are 139.2 moles of CO_2 that might absorb the infrared radiation. The earth at 288°K emits 390 W/m² infrared radiation. At 400 ppm carbon dioxide, the energy absorbed by the carbon dioxide in the atmosphere can be calculated using Equation (8.10) as 49.3 W/m².

According to the model developed, the carbon dioxide will absorb 49.3 W/m² of the total 151 W/m² absorbed by the atmosphere. As a percentage, carbon dioxide absorbs 32.6% of the energy that is absorbed by the atmosphere. As stated above, the model yields a high estimate because some of the infrared radiation that could be absorbed by carbon dioxide will be absorbed by water vapor and other GHGs. The carbon dioxide will actually absorb less than the 49.3 W/m² calculated.

Water and other GHGs absorb infrared radiation in different bands of frequencies or wavelengths. There is overlap in the frequency bands so that more than one GHG can absorb some frequencies of infrared radiation. These things make it difficult to determine exactly how much energy is absorbed by the carbon dioxide in the atmosphere, and how much is absorbed by other GHGs.

As the CO_2 in the atmosphere is near 400 ppm, it makes sense to take the derivative of Equation (8.10) and evaluate it at 400 ppm. This gives the slope of the curve at 400 ppm and allows us to estimate how small changes in C_{ppm} impact the average temperature of the earth. The derivative of Equation (8.10) evaluated at C_{ppm} = 400 ppm is 0.0217. This means that increasing the CO_2 in the atmosphere by 1 ppm will increase the infrared radiation absorbed by 0.0217 W/m². This information can be used to estimate how increasing the carbon dioxide in the atmosphere will impact global warming.

Example 8.2

Assume that at 400 ppm CO_2 in the atmosphere the earth is at an average temperature of 288°K, that the surface of the earth is emitting 390 W/m², and that 151 W/m² is being absorbed by the atmosphere. Assume that increasing the CO_2 in the atmosphere by 1 ppm will increase the energy absorbed by the atmosphere by 0.0217 W/m². Estimate the amount of global warming caused by increasing the CO_2 to (a) 450 ppm and (b) 500 ppm.

Solution: An increase to 450 ppm will increase the energy absorbed by (50) (0.0217) = 1.085 W/m². The atmosphere will absorb 151 + 1.085 = 152.085 W/m². To balance the energy, the earth surface must radiate 390 + 1.085 = 391.085 W/m². The Stefan–Boltzmann equation is then used to calculate the temperature of the surface of the earth:

$$\left(\frac{5.67 \times 10^{-8} \text{ W}}{m^2 K^4} \right)(T)^4 = 391.085 \text{ W} / m^2 \tag{8.11}$$

Solving Equation (8.11) yields T = 288.185°K. From this model, increasing the CO_2 in the atmosphere to 450 ppm would cause a 0.185°C increase in temperature. Increasing to 500 ppm would cause the atmosphere to absorb 151 + 2.17 = 153.17 W/m². The increase in global temperature would be 0.385°C.

8.3.
Keeling Curve

Dr. Ralph Keeling began taking measurements of carbon dioxide in the atmosphere in 1958 at the Mauna Loa Observatory in Hawaii [4]. The data are plotted in Figure 8.3.

The average level of CO_2 in the atmosphere has increased each year. The fluctuations during the year are because there are more land area and more plants in the Northern Hemisphere than in the Southern Hemisphere. During summer in the Northern Hemisphere, the plants will remove carbon dioxide from the atmosphere. As plants die in the winter, the carbon dioxide in the atmosphere increases. This cycle and the upward trend have been going on since we began taking measurements in 1958.

If you take data close to the earth's surface where there are many plants around, the plants will have a large impact on carbon dioxide concentration during the day. That is, there will be significant daily fluctuations. Mauna Loa was chosen because there are few plants around the site, and it is high enough to measure a more consistent concentration in the atmosphere. There are many other sites on earth where carbon dioxide concentration is measured today, and all sites are showing the concentration in the atmosphere is increasing.

The slope of the Keeling Curve increased significantly in the mid-1990s. During the 1960s, carbon dioxide was increasing at a rate of about 1 ppm each year. In the most recent 20 years, carbon dioxide has been increasing at a rate of a little over 2 ppm each year. Worldwide, we burn enough fossil fuel annually to amount to 4.5 ppm in the atmosphere. Approximately half of the carbon dioxide generated from burning fossil fuels goes into the atmosphere and about half is absorbed by the oceans.

Figure 8.3 Keeling Curve.

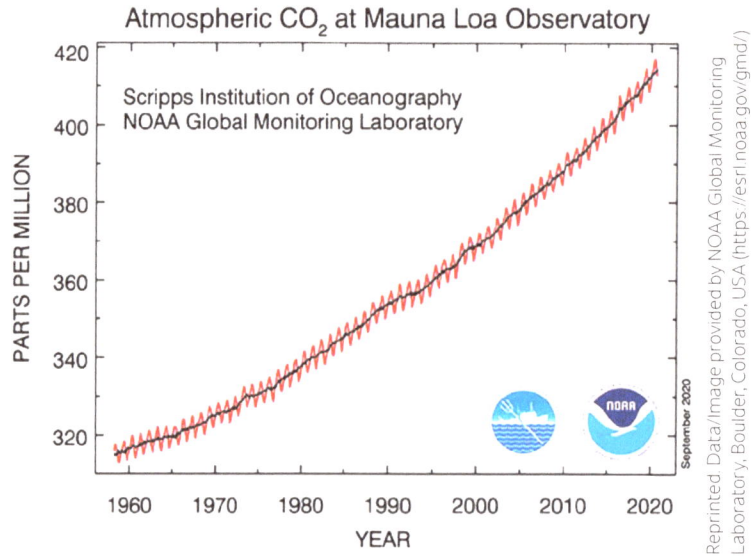

Atmospheric CO_2 at Mauna Loa Observatory

Some projections assume the carbon dioxide concentration will continue to increase at 2 ppm, or that it accelerates its increase as we continue to burn an increasing amount of fossil fuel. It is hard to predict the future. Plants remove a significant amount of carbon dioxide in the atmosphere. The Keeling Curve shows significant seasonal variation in carbon dioxide content because most of the land masses are in the Northern Hemisphere. Carbon dioxide content decreases when it is summer in the Northern Hemisphere because plants are removing lots of carbon dioxide from the atmosphere. Carbon dioxide increases when it is winter in the Northern Hemisphere because there are fewer plants on earth removing carbon dioxide from the atmosphere. A warmer earth will allow more plants to grow in Canada and Siberia, which are huge land masses. A large increase in plant life on earth will reduce the amount of carbon dioxide in the atmosphere.

There may come a point where the plants take more carbon dioxide out of the atmosphere than humans put into the atmosphere. On the other side of the argument people say that parts of the world will become so hot that plants will die, which would tend to increase carbon dioxide content in the atmosphere. The Keeling Curve has shown a steady increase in carbon dioxide, but that steady increase may or may not continue. In the near term, it is likely that the increase in carbon dioxide will continue, but projecting out 100 years is a projection of unknown accuracy.

The Keeling Curve can be combined with the models above to estimate the increase in earth temperature due to rising levels of carbon dioxide in the atmosphere. In the year 2000, there were 368 ppm carbon dioxide and the level has been increasing at 2.2 ppm annually since then. If we extrapolate this into the future, the carbon dioxide in the atmosphere (C_{ppm}) is:

$$C_{ppm} = (year - 2000)(2.2) + 368 \qquad (8.12)$$

Once this is known, Equation (8.10) is modified to include that the emissivity of the earth increases as the atmosphere absorbs more infrared radiation. That is, as the atmosphere absorbs more infrared radiation, the average infrared radiation from the surface of the earth

will increase to values above 390 W/m² to balance the energy emitted by and received by the earth. Letting "Emissivity" be the average emissivity of the surface of the earth, Equation (8.13) is the modified version of the equation.

$$\text{Energy Absorbed by CO}_2 = \frac{1}{2}(0.489)(\text{Emissivity})\left[(0.09083)\log\left((0.349)C_{ppm}\right)+0.0681\right] \qquad (8.13)$$

Equation (8.13) is used to get the energy absorbed by CO_2 in the atmosphere. A spreadsheet can be developed to calculate the average earth temperature as a function of the CO_2 in the atmosphere. Table 8.2 is the first few lines of the spreadsheet.

A plot of the results is shown in Figure 8.4. This model shows a temperature increase of 0.86°K over 100 years or about 1.5°F. The model is an oversimplification of what actually happens,

because it assumes the earth is of constant temperature and that the radiation it receives and emits is constant over the surface of the earth. It is easy to find faults with the model. But this model includes the main effects of carbon dioxide absorbing infrared radiation and how it increases the temperature of the surface of the earth. Climate scientists have developed much more sophisticated models, but the results are similar to the results of this model.

Table 8.2 Extrapolating Keeling Curve to find average earth temperature.

Year	C_{ppm}	Moles/m²	Absorption fraction	Absorbed by CO_2 (%)	Other GHGs (W/m²)	Total absorbed (W/m²)	Emitted (W/m²)	Temperature (°K)	Temperature difference (°K)
2000	368	128.4	0.5091	48.55	102.45	151	390	287.985	0
2001	370.2	129.2	0.5097	48.60	102.45	151.05	390.05	287.995	0.010
2002	372.4	130.0	0.5102	48.66	102.45	151.11	390.11	288.006	0.020
2003	374.6	130.7	0.5113	48.71	102.45	151.17	390.17	288.016	0.031
2004	376.8	131.5	0.5113	48.77	102.45	151.23	390.23	288.027	0.042
2005	379	132.3	0.5118	48.83	102.45	151.28	390.28	288.038	0.052
2006	381.2	133.0	0.5123	48.89	102.45	151.34	390.34	288.048	0.063
2007	383.4	133.8	0.5128	48.94	102.45	151.40	390.40	288.059	0.073

Figure 8.4 Global temperature increase.

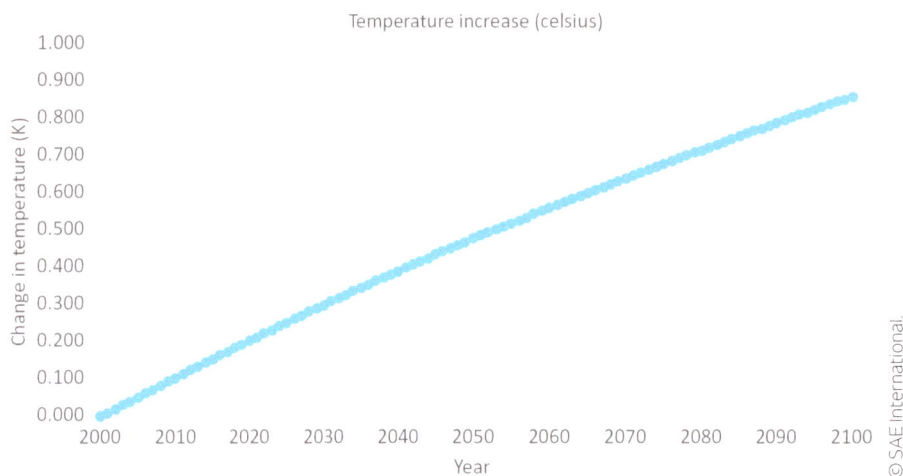

Temperature increase (celsius)

Homework

1. Assume that on average the earth absorbs 239 W/m^2 of radiative energy from the sun. (This is the number reported by climate scientists who study this topic.) Assume that the earth surface emits radiation according to the Stefan–Boltzmann relationship: Emission = $(5.67 \times 10^{-8})\ T^4\ W/m^2$, where T is the average earth temperature in Kelvin.

   (a) If the average earth temperature is 60°F, calculate the energy being emitted by the surface of the earth in W/m^2 and the energy being captured by GHGs in the atmosphere.

   (b) Use the number calculated in (1) above for the amount of energy captured by GHGs and increase this number by 3% to simulate an increase in the amount of GHGs in the atmosphere. Calculate the temperature increase of the earth. (Answer: 0.845°C or 1.52°F. If the atmosphere were to absorb 3% more infrared radiation than it currently does,

the average earth temperature would increase by 1.52°F.)

2. Bonus Problem: This bonus problem is for students who want to deepen their understanding of how GHGs affect the average temperature of the earth. Read the textbook to help you with this problem. It is not required for the class. This problem goes into the physical chemistry of how carbon dioxide absorbs infrared radiation and is more difficult than the previous problem.

   (a) An approximation to the Keeling Curve is that the carbon dioxide concentration in the atmosphere in parts per million varies according to (year – 2000) (2.2) + 368. In the year 2000, the carbon concentration was 368 ppm, and it has increased approximately linearly at 2.2 ppm each year since then. Assume that the carbon dioxide concentration will continue to increase in the future according to the equation.

   (b) To estimate the amount of infrared radiation absorbed by carbon dioxide, we need to calculate the moles of carbon dioxide in the atmosphere per square meter of surface area on the earth. Go through the details of this calculation in your homework. You should conclude that if C_{ppm} is the concentration of carbon dioxide in the atmosphere in ppm, the number of moles of carbon dioxide per square meter of earth surface is $(0.349)C_{ppm}$.

(c) Assume that the average temperature of the earth in the year 2000 was 288°K. Use the Stefan–Boltzmann equation to relate the average temperature of the earth to the average infrared radiation from the surface of the earth. In the model you are developing, the average temperature of the earth will increase slightly each year.

(d) Using the experimental work by Stull et al. on the absorptivity of carbon dioxide, assume that the absorption factor varies according to Equation (8.7). Assume that 48.9% of the infrared radiation from earth is in the range of frequencies that can be absorbed by carbon dioxide and assume that half of the radiation absorbed by the carbon dioxide is reemitted into space while the other half is reemitted back to earth and absorbed.

(e) Develop a graph showing that the earth temperature would increase by 0.86°K over 100 years according to the model. The model you have developed uses average earth temperatures and average radiation and is not as good as the more detailed models developed by climate scientists. But your model includes the important scientific principles explaining how carbon dioxide concentration in the atmosphere impacts the temperature of the earth. The results of the model you develop are similar to the results of the more complex models.

References

1. https://en.wikipedia.org/wiki/Solar_irradiance.

2. https://www.ess.uci.edu/~yu/class/ess200a/lecture.2.global.pdf.

3. Stull, V.R., Wyatt, P.J., and Plass, G.N., "The infrared transmittance of carbon dioxide," *Applied Optics* 3, no. 2 (1964): 250.

4. https://en.wikipedia.org/wiki/Keeling_Curve.

Before reading through this unit, I think it is helpful to watch the NOVA movie "Search for the Super Battery." It is a little outdated, but there is good information in the movie about batteries. As we move toward having electric and hybrid-electric cars, the battery system is going to be very important.

The ideal battery would be able to store and provide energy at any rate. That is, we would be able to charge and discharge the battery as slowly or as quickly as desired. The battery would be 100% efficient when charging and discharging. None of the electric energy would be converted to heat. The battery would be small and lightweight.

Real batteries are not like the ideal battery. We must be careful how fast we charge and discharge batteries to prevent damaging them. It takes hours to fully charge a battery, whereas it takes only a few minutes to fill a car with gas. Long road trips in electric cars will require waiting for the batteries to recharge during the trip. For most people, gasoline cars and hybrid-electric cars are better options for long road trips.

Real batteries are heavy and take up considerable space in a vehicle. Batteries have good efficiency, but some of the energy is always converted to heat when charging and discharging a battery. A good battery system is about 80% efficient in typical use, meaning that about 80% of the electric energy used to charge the battery system will be recovered when the battery system is discharged. The efficiency is rate sensitive.

Charging and/or discharging the battery system quickly forces it to operate at a lower efficiency.

Batteries can be classified into two types: primary batteries and secondary batteries. Primary batteries are small batteries that are used around the house. They are discharged once and then thrown away. Primary batteries are not useful for electric vehicles and will not be discussed. Secondary batteries are rechargeable and are the type of batteries that are useful for electric vehicles. There are many types of secondary batteries: lithium-ion, lead-acid, nickel–cadmium, and nickel–metal hydride are the most common. Other secondary battery types have specialized uses. Common battery terminology is as follows:

- Energy density: The amount of energy that can be stored in a battery divided by its mass (W-h/kg).
- Amp-hour capacity: The amount of charge the battery will hold in amp-h (amp-h = 3600 C).
- Watt-hour capacity: The amount of energy the battery will hold in W-h. The watt-hour capacity of a battery depends on the rate the battery is discharged. Discharging the battery at a high rate will cause the energy to be drawn out at a lower voltage, which reduces the amount of energy that can be drawn out of the battery.
- C rating: This is the charge or discharge rate for the battery that was used to measure the amp-hour and watt-hour capacity. A rate of C/3 means the battery was charged or discharged completely in 3 h. C/8 means it was charged or discharged in 8 h, etc.

Figure 9.1 shows a comparison of different energy sources. The x-axis in Figure 9.1 is the energy density associated with the mass of the energy source (MJ/kg). The y-axis is the energy density associated with the volume of the energy source (MJ/L). Ideally, an energy source would provide a lot of energy for its mass and the volume it occupies. The perfect energy source would be in the top right-hand corner of the chart. No such perfect energy source exists. Hydrogen gas has the highest energy content for its mass, but it takes up a lot of volume on the vehicle. Burning aluminum metal yields the highest amount of energy for the volume it occupies. Metals such as aluminum, silicon, iron, zinc, and magnesium are not practical fuels for most applications. The energy to make these elements as a fuel would exceed the energy recovered when they are burned.

Figure 9.1 Energy density of common energy sources.

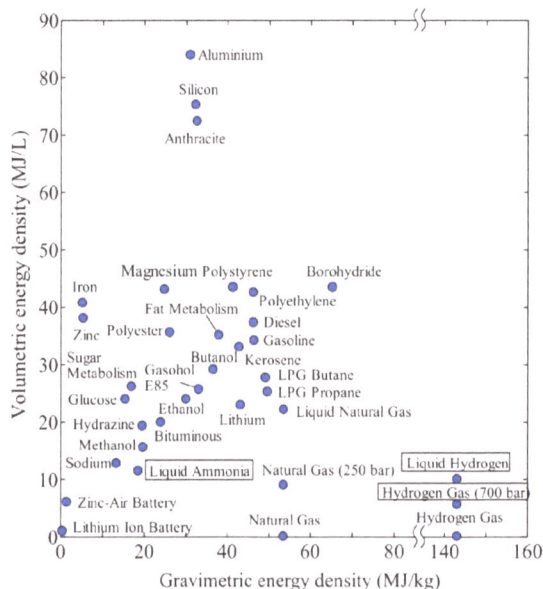

The lithium-ion battery is the highest energy density battery that we currently have for practical use. The zinc–air battery has a higher energy density than lithium-ion but is impractical except for a few specialty applications. Notice that the energy density of lithium-ion batteries is extremely low compared to any of the fossil fuels we commonly use. Lead-acid batteries are even lower than lithium-ion both in energy per kilogram and energy per liter. Batteries are much heavier and occupy much more space than any fuel sources used to power vehicles. This is one of the fundamental problems of using batteries to power vehicles.

9.2. Battery Efficiency

There are two terms related to battery efficiency: charge efficiency and energy efficiency. The charge efficiency is the amp-h charge recovered from the battery when it is discharged divided by the amp-h required to charge the battery.

$$\text{Charge Efficiency} = \frac{\text{Amp-h recovered discharging battery}}{\text{Amp-h required to charge battery}} \quad (9.1)$$

Charge efficiency for the batteries used in electric vehicles is usually very high, almost always above 90% and often near 100%. Batteries that have a liquid electrolyte, like lead-acid batteries, will use a small amount of charge in the electrolysis of water, especially when charging the battery. Charge is a fundamental quantity and must be conserved; it is a basic law of physics. But, if there is electrolysis of water, the test results will indicate that more amp-h was put into charging the battery than was released when discharging and the charge efficiency will be less than 100%. There are other secondary effects that can cause the charge efficiency to be less than 100%, and batteries will have a charge efficiency of less than 100%, though it will be near 100% for batteries we are interested in.

Energy efficiency is the watt-h energy we get discharging a battery divided by the watt-h energy required to charge the battery.

$$\text{Energy Efficiency} = \frac{\text{Watt-h recovered discharging battery}}{\text{Watt-h required to charge battery}} \quad (9.2)$$

When charging and discharging a battery, some of the energy is converted to heat. This manifests itself as a voltage difference. We charge the battery at a higher voltage and discharge the battery at a lower voltage. This is illustrated in Figure 9.2.

Plotted on the x-axis in Figure 9.2 is the percent charge in the battery such that at 0% the battery is completely discharged and at 100% the battery is completely charged. The blue line on the chart represents the open circuit voltage. If we use a high-quality voltmeter and measure the voltage across the terminals at different states of charge, we will develop the open circuit voltage curve. The open circuit voltage is drawn as a line in Figure 9.2, though it will not be exactly linear, and the exact curve depends on the type of battery. Open circuit voltage is the equilibrium voltage for the battery. Charge will not flow in or out of the battery if the voltage across the terminals is equal to the open circuit voltage. The open circuit voltage is always lower at low states of charge and higher at high states of charge. Figure 9.2 shows the open circuit voltage to vary linearly with charge, but it will not be exactly linear, though it is approximately linear except for the portions near 0 and 100% charge.

Figure 9.2 Charge and discharge of battery.

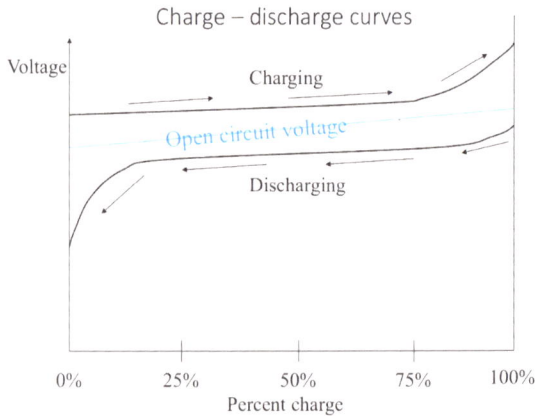

To put charge into the battery, we must increase the voltage across the terminal above the open circuit voltage. A charging path is illustrated in Figure 9.2 above the open circuit voltage curve. The voltage across the terminals is held above the open circuit voltage and charge flows into the battery until it is charged. Most batteries will see the voltage curve rise significantly as the battery approaches full charge as illustrated in Figure 9.2. We will need to reduce the charging rate as the battery approaches full charge to avoid overvoltage and damage to the battery.

In order for charge to flow out of the battery we must decrease the voltage across the terminals to a value below the open circuit voltage. A discharge path is illustrated in Figure 9.2 below the open circuit voltage curve. When

discharging a battery, the voltage will drop significantly as we approach 0% charge in the battery, as illustrated in Figure 9.2. We will need to reduce the rate of discharge as we approach 0% charge to prevent under-voltage and damage to the battery.

Energy is voltage multiplied by charge, that is, Joule = Volt × Coulomb. The area under the charging curve in the figure represents the energy required to charge the battery and the area under the discharge curve represents the energy we get back from the battery when discharging. It always takes more energy to charge the battery than what we get back when discharging. The battery is always less than 100% efficient in storing and providing energy.

Let V_{Charge} be the average voltage used in charging the battery and $V_{Discharge}$ be the average voltage when discharging the battery. If the battery had 100% charge efficiency, then the energy efficiency would be the ratio of the two voltages. As discussed earlier, because of electrolysis of water and other effects, testing of the battery will indicate a charge loss. That is, it will take more amp-h to charge the battery than what we get back in discharging the battery. The batteries we are interested in will have a high charge efficiency, but it will be less than 100%. The energy efficiency of the battery is expressed as follows:

$$\text{Energy Efficiency} = \left(\frac{V_{Discharge}}{V_{charge}} \right) \left(\text{Charge Efficiency} \right) \qquad (9.3)$$

The battery efficiency is low at low states of charge because the discharge voltage is low. The battery efficiency is low at high states of charge because the charging voltage is high. Generally, the battery is most efficient when operated

between 25 and 75% state of charge. To get the highest energy efficiency from the battery, we should keep it operating in the middle range of state of charge.

9.2.1.
Charging the Battery

To drive current into the battery, the voltage across the terminals must be higher than the open circuit voltage. Increasing the rate of charge flowing into the batteries requires higher voltage, but there is a limit to how high we can push the voltage without damaging the battery. Charging and discharging a battery causes a chemical reaction inside the battery involving the positive and negative plates and the electrolyte. The chemical reaction is reversible, going in one direction for charging and in the other direction for discharging. Applying too high of a voltage across the battery terminal will cause undesirable chemical reactions on the plates, which damages the battery. Macroscopically we specify a maximum voltage that can be applied to the battery terminals to avoid damaging the battery.

As the battery approaches full charge, most of the area on the positive and negative plates has gone through the chemical reaction, and there is a limited amount of active plate area left. The charging rate will need to be reduced as the battery approaches full charge to avoid overvoltage on the battery. All types of batteries require that the charging rate be reduced as they approach full charge.

9.2.2.
Discharging the Battery

To discharge the battery, we must reduce the voltage across the terminals below the open circuit voltage. Increasing the rate of discharge requires that the charge be drawn out at a lower voltage, but there is a limit to how fast we can discharge the battery. If we discharge too fast we will cause undesirable chemical reactions on the plates that damage the battery. Since energy is voltage multiplied by charge, it is also true that discharging at low voltages means we are not getting as much energy from the battery. It is undesirable to discharge the battery at a low voltage, and it can damage the battery.

As the battery approaches complete discharge, there is a limited amount of area on the plates that can continue the chemical reaction. The discharge rate needs to be reduced as we approach complete discharge to avoid damaging the battery.

Figure 9.3 illustrates how charging and discharging a battery quickly force the battery to operate at a lower efficiency. For electric vehicles, we want to be able to charge the batteries quickly so we can get back on the road. But this requires that we use a high charging voltage, which causes the batteries to operate at a lower efficiency. It will take more energy to charge the batteries quickly than to charge them slowly.

Figure 9.3 Effect of charging rate on battery efficiency.

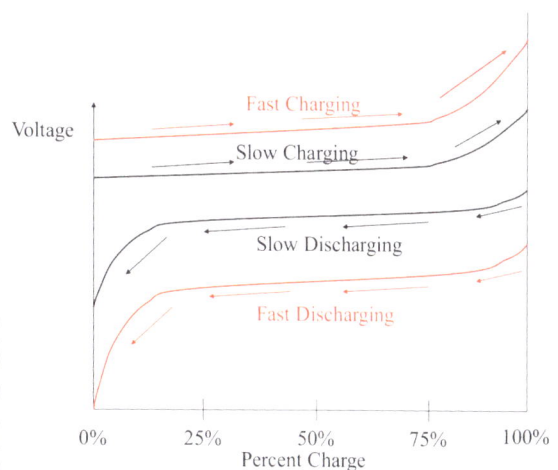

© SAE International

Figure 9.3 is a schematic that illustrates how charging rate affects the efficiency of a battery. Charging and discharging at a faster rate requires a higher charge voltage and a lower discharge voltage, which reduces the efficiency of the battery. When fast charging and discharging, a higher percentage of the energy going in and out of the battery will be converted to heat, and the battery will heat up. Too much heat will damage the battery. Charging at high voltage will damage the battery and discharging at low voltage will damage the battery. Charge and discharge rates are limited to protect the battery. Trying to charge or discharge the battery too quickly will damage the battery.

C Rating: Batteries are more efficient when charged and discharged slowly. A common way to measure the rate of charge and discharge is the "C Rating." A C/3 rating means the battery is fully discharged (or charged) over a 3-h period. 2C means it was discharged (or charged) in half an hour, etc. The C Rating allows us to compare the energy efficiency for different charge/

discharge rates, normalized to the capacity of the battery. Batteries that have the same chemistry and different capacities will have approximately the same efficiency if charged and discharged at the same C Rating.

Figure 9.4 is a typical charge and discharge curve for a lithium-ion battery used in a solar car project at Missouri S&T. Notice how the charging rate was reduced as the battery approached full charge. The maximum charge voltage used on this cell was 4.2 V. The tester was programmed to cut the charge rate current in half when the voltage approached 4.2 V. This battery also shows a 6.5% charge loss on discharge, which is what the test would show on the first cycle. When running charge–discharge cycles on a battery, the discharge voltage is limited to a value that will not damage the battery. In the first cycle, it is likely that the cutoff voltage on discharge caused the battery to stop before all the charge was drawn out of the battery. The charge efficiency will improve with the second cycle. Figure 9.4 is typical for the first cycle of the test.

Figure 9.4 Charge–discharge of lithium-ion battery.

9.2.3.
Measuring State of Charge

The open circuit voltage of a battery varies with the state of charge, so measuring open circuit voltage would be a good way to measure the state of charge of a battery. The problem with this method is that to get a true open circuit voltage measurement, the battery must stabilize at room temperature. If the battery is being charged or discharged it takes a while for the battery chemistry to stabilize enough to take an accurate open circuit voltage reading. The battery needs to sit for at least an hour to stabilize, and we often do not want to wait an hour to know the state of charge of the battery. There are amp-h meters that keep track of the charge going into and out of the battery. The amp-h meters are a much more accurate and reliable way to measure the amount of charge in the battery.

9.3.
Battery Capacity and Efficiency

Gibbs Free Energy: The Gibbs free energy is based on the chemical potential of the battery chemistry and is the maximum amount of energy that can be drawn out of the battery assuming 100% efficiency. The Gibbs free energy formula is:

$$\Delta G = nFV \qquad (9.4)$$

In the formula, n is the number of electrons involved in the chemical reaction. F is the Faraday constant, which is 96,485 C/mole, and V is the reversible (open circuit) voltage. ΔG is the amount of energy that can be drawn out of the battery, known as the Gibbs free energy.

Example 9.1

The best way to illustrate how the Gibbs free energy formula can be used to estimate battery capacity is to work through an example. The copper–zinc battery was one of the first batteries developed. There are no practical applications for the battery today, but the chemistry is simple enough that it is a good place to start talking about the specifics of battery chemistry. This example calculates the Gibbs free energy for the copper–zinc battery.

Solution: When measuring chemical potential, we can only measure the difference in chemical potential between two materials. As a reference, we define the chemical potential of hydrogen as zero. For hydrogen, the chemical potential $V^0 = 0.0$ V by definition. Based on that definition the chemical potential of copper is $V^0 = 0.337$ V and that of zinc is $V^0 = -0.763$ V. There are readily available tables of chemical potential values for many different materials [1].

The reversible (open circuit) voltage for a copper–zinc battery is the difference of the chemical potentials:

$$V = \left(0.337 \text{ V}\right) - \left(-0.763 \text{ V}\right) = 1.1 \text{ V} \qquad (9.5)$$

The chemical reaction is illustrated in Figure 9.5.

The zinc and copper cylinders in Figure 9.5 represent the zinc and copper plates in the battery. The zinc plates are in an electrolyte solution of $ZnSO_4$ and the copper plates are in an electrolyte solution of $CuSO_4$, as illustrated in the figure. There is a porous (salt) separator between the electrolytes that allows SO^{-2} ions to pass through but blocks the copper and zinc ions. The net chemical reaction is:

$$Zn + CuSO_4 \rightarrow Cu + ZnSO_4 \qquad (9.6)$$

Figure 9.5 Copper–zinc battery.

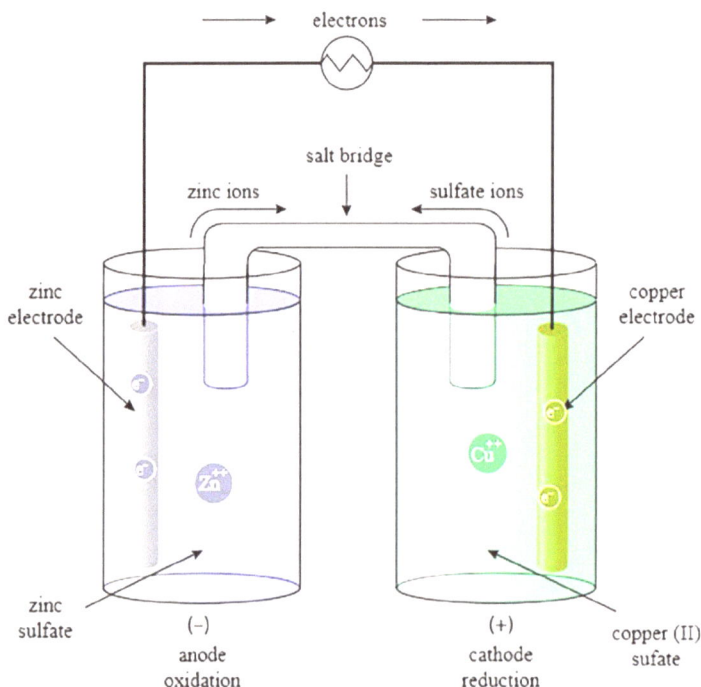

When discharging, the zinc metal goes into solution as Zn^{+2} ions, leaving two electrons in the zinc plate. To balance the charge, two electrons go up the wire, through the light bulb, and across toward the copper plate. The Cu^{+2} ion comes out of solution and is deposited on the copper plate. To balance the charge, the copper ion needs the two electrons that are coming across the wire from the zinc plate. During the reaction, the zinc plate gradually dissolves into solution and the copper plate grows as copper is added to the plate.

Charging the battery causes the chemical reaction to go the other direction with zinc being deposited on the zinc plate and copper going into solution. There are two electrons involved in the battery chemistry, so n = 2 in the Gibbs free energy equation. The Gibbs free energy for this battery can be calculated from Equation (9.4) as:

$$\Delta G = (2)(96,485)(1.1) = 212,267 \text{ J / mole} \qquad (9.7)$$

The result is that the battery stores 212,267 J of energy for each mole of material involved in the chemical reaction. One mole of zinc has a mass of 59 g, 1 mole of copper has a mass of 65 g, and 1 mole of sulfate has a mass of 96 g. A copper–zinc battery must have a mass of 59 + 65 + 96 = 220 g of active material to provide 212,267 J of energy. As an upper limit, the energy density of a copper–zinc battery is:

$$\text{Upper Limit Energy Density} = \frac{212{,}267\ J}{220\ g} = 0.965\ MJ/kg \qquad (9.8)$$

Achieving this energy density would require that other parts of the battery have a zero mass, which is not possible. The energy density of a copper–zinc battery will always be less than 0.965 MJ/kg. If a battery could be developed so that most of the mass was in the active copper, zinc, and sulfate materials, then the energy density of the battery could approach 0.965 MJ/kg. All real copper–zinc batteries have an energy density of less than 0.5 MJ/kg. For comparison, gasoline, diesel fuel, and jet fuel have an energy density of about 47 MJ/kg. So even under the most ideal conditions a petroleum fuel has 47+ times more energy per kg than a copper–zinc battery.

This type of analysis can be done for any battery chemistry. We examine the mass of the active materials and compute an upper limit for the energy density. Having active materials with a low atomic mass will allow us to make batteries that have a higher energy density. Hydrogen is element number 1 and has the lowest atomic mass. A fuel cell is a lot like a battery that uses hydrogen as one of the active materials. Fuel cells have a higher energy density than batteries, but the energy density of a fuel cell is much lower than petroleum fuel. Helium is element number 2, but helium is chemically inert, so it is impossible to make a helium battery. Lithium is element number 3 and offers the best potential for making a lightweight battery. The lithium-ion battery has a high energy density because it uses lithium as the active material.

Battery Voltage: The open circuit voltage of a battery is obtained by looking up the chemical potential V^0 of the products. For the copper–zinc battery, copper has a chemical potential of 0.337 V and zinc has −0.763 V. The open circuit voltage is calculated as 1.1 V. When we begin charging or discharging the battery, the voltage across the terminals will change because of the kinetics of the chemical reaction. The change in voltage depends on temperature and the activities of the products and reactants in the chemical reaction. The equation is normally written as:

$$V_r = V_0 - \left(\frac{R\,T}{n\,F}\right)\ln\left(\frac{\text{activities of products}}{\text{activities of reactants}}\right) \qquad (9.9)$$

In Equation (9.9), V_0 is the open circuit voltage, R is the constant for the ideal gas law, T is absolute temperature, n is the number of electrons in the chemical reaction, and F is the Faraday constant. It is almost impossible to get accurate values for the activities of the products and reactants, and the values vary with the state of charge, so

Equation (9.9) has limited practical value. The change in voltage that happens as we charge or discharge the battery is described chemically as activation or charge transfer overpotential and concentration or mass transfer overpotential, which are described below.

Activation or Charge Transfer Overpotential:
As the chemical reaction proceeds to a fully charged or fully discharged state, the active plate area decreases. The reduction in active area on the plates will cause a cell overpotential because it becomes more difficult to find areas on the plates where the chemical reaction can proceed. For example, if we are charging a battery from a low state of charge, the chemical reaction can proceed almost anywhere on the plates. As the battery approaches full charge, most of the plates have gone through the chemical reaction and there is a limited number of places on the plates for the chemical reaction to proceed. The rate of the chemical reaction will slow down, and we will measure a reduced current going into the battery. Increasing the charging voltage will cause the chemical reaction to proceed faster. If the voltage is raised too high, it will cause undesirable chemical reactions to be deposited on the plates, effectively reducing the amount of active material in the battery. The undesirable chemical reactions manifest themselves as lower battery capacity, and, in most cases, it is impossible to recover the capacity that is lost. This is why it is important to not overvoltage the battery.

Concentration or Mass Transfer Overpotential: The chemical reaction uses up ions in the electrolyte. As the chemical reaction proceeds, the ion concentration near the plates will decline as ions near the plates are absorbed and used up. This causes the ion concentration near the plates to be lower than the average ion concentration in the electrolyte. The ions must physically move through the electrolyte to participate in the chemical reaction at the plates, and this does not happen instantaneously. Macroscopically, this will manifest itself as a higher charging voltage or lower discharge voltage.

Internal Resistance Model: The cell overpotential is approximately proportional to the amount of current charging or discharging from the battery. Macroscopically, it appears that the battery has an internal resistance, though we must be careful when we model the internal resistance of the battery. Please remember that it is not physically an electrical resistance, but a way to model the chemical reaction of the battery. The internal resistance will decrease the discharge voltage and increase the charge voltage. If you simply put a resistor in series with the battery, it gives the wrong answer. The resistor needs to be inside the battery between the two terminals for charging the battery and outside the terminals when discharging the battery. If the internal resistance of the battery is R and the current is I:

$$V_{discharge} = V_0 - IR \qquad (9.10)$$

$$V_{charge} = V_0 + IR \qquad (9.11)$$

Where $V_{discharge}$ is the voltage across the battery terminals during discharge, V_{charge} is the voltage across the battery terminals when charging, and V_0 is the open circuit voltage of the battery. The energy efficiency of the battery can be expressed as:

$$\text{Energy Efficiency} = \left(\text{Charge Efficiency}\right)\left(\frac{V_0 - IR}{V_0 + IR}\right) \qquad (9.12)$$

Example 9.2

Assume a 60 amp-h deep-cycle lead-acid battery with an internal resistance of 0.07 Ω, open circuit voltage of 12.6 V, and a charge efficiency of 97%. Calculate the energy efficiency for charging the battery at 10 amps and discharging the battery at 10 amps.

Solution: Using Equation (9.12):

$$\text{Energy Efficiency} = (0.97)\left(\frac{12.6 - (10)(0.07)}{12.6 + (10)(0.07)}\right) = 86.8\% \tag{9.13}$$

If we charge the battery at 10 amps and discharge at 10 amps, we will recover 86.8% of the energy. That is, if we put 100 W-h of energy into charging the battery, we will recover 86.8 W-h when discharging. The remaining 13.2 W-h of energy will be converted to heat during the charge/discharge cycle.

Example 9.3

For the 60 amp-h battery above, it would take about 6 h to charge at 10 amps. Suppose we want to cut the charging time to 2 h by charging at 30 amps rather than 10 amps. We will still discharge the battery at 10 amps. How does this affect the energy efficiency of the battery?

Solution: Use Equation (9.12):

$$\text{Energy Efficiency} = (0.97)\left(\frac{12.6 - (10)(0.07)}{12.6 + (30)(0.07)}\right) = 78.5\% \tag{9.14}$$

This example illustrates how charging the battery faster forces it to operate at a lower efficiency. Charging faster causes a higher portion of the energy to be converted to heat, and the battery will heat up. Too much heat can damage the battery. For electric vehicles, it is desirable to charge the batteries quickly, but charging quickly will waste more of the electric energy used to charge the battery and it will be more expensive to charge the battery. Wasting energy to charge the batteries quickly takes away some of the benefits of driving electric cars.

Figure 9.6 was developed for the 60 amp-h battery in Examples 9.2 and 9.3, assuming that the charging and discharging amperage is the same. The chart shows how the energy efficiency of the battery varies with the amperage used in charging and discharging the battery.

Figure 9.6 Battery efficiency vs amperage.

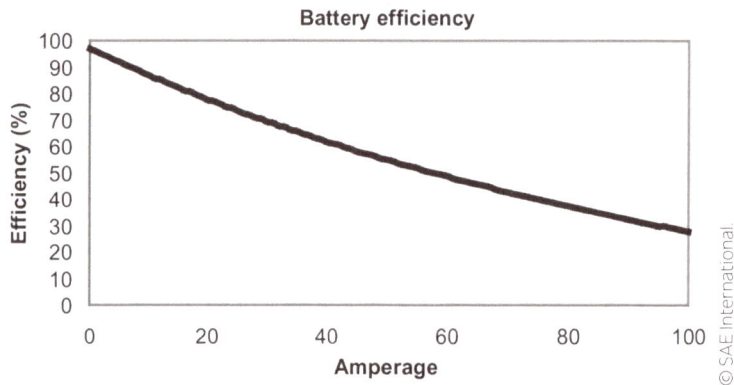

© SAE International

9.4.
Battery Chemistry

9.4.1.
Lead-Acid Batteries

Lead-acid batteries are an established technology. They are low-cost, reliable, and durable. There are many practical applications for lead-acid batteries. The starter batteries in cars and trucks are almost always lead-acid batteries. Lead acid batteries are used in wheelchairs, trolling motors, forklifts, golf carts, and other small electric vehicles. Lead-acid batteries are too heavy to be used to power electric cars and trucks, but they could be used for grid storage. Figure 9.7 shows a battery room in a power plant with racks of lead-acid batteries. The book "Batteries for Electric Vehicles" [2] is an outstanding reference for lead-acid batteries used in vehicles.

Figure 9.7 Battery room in power plant.

© Shutterstock.

The active materials in lead-acid batteries are lead on the negative plates, lead oxide on the positive plates, and sulfate ions in the electrolyte.

When discharging the battery, the following chemical reaction happens on the negative plates.

$$\text{Negative Plates}: Pb + SO_4^{-2} \rightarrow PbSO_4 + 2e^- \qquad (9.15)$$

Discharging the battery causes the negative plates to become coated with lead sulfate. The electrons on the right side of Equation (9.14) become the current the battery produces. Charging the battery causes the chemical reaction to go in the other direction. The sulfate

ions go back into solution, and the negative plates go back to being coated with lead.

The positive plates start out being coated with lead oxide. Discharging the battery causes the following chemical reaction on the positive plates.

$$\text{Positive Plates}: PbO_2 + 4H^+ + SO_4^{-2} + 2e^- \rightarrow PbSO_4 + 2H_2O \qquad (9.16)$$

The chemical reaction on the positive plates is more complex than on the negative plates. Four hydrogen ions and one sulfate ion must come in close proximity to the lead oxide molecule for the chemical reaction to take place. The oxygen atoms in the lead oxide must be stripped off to combine with the hydrogen ions and make water.

The sulfate ion combines with the lead to make lead sulfate. As the reaction proceeds, the positive plates are coated with lead sulfate. There is an excess charge of positive 2 on the left side that is balanced by the two electrons coming from the negative plates. The total chemical reaction can be written as:

$$\text{Total Reaction}: Pb + PbO_2 + 2H_2SO_4 \rightarrow 2PbSO_4 + 2H_2O \qquad (9.17)$$

As the battery is discharged, the positive and negative plates are coated with lead sulfate. The sulfate ion concentration in the electrolyte decreases as the lead sulfate is deposited on the plates. The density of the electrolyte will decrease as the battery is discharged because of the loss of sulfate ions. It is possible to measure the state of charge in a lead-acid battery by

measuring the density of the electrolyte. A schematic of the lead-acid battery cell is illustrated in Figure 9.8. In a real battery, the positive and negative plates are very close together, and a polymer separator is used to prevent the plates from touching and shorting out. The separator must be porous to allow the ions to move easily in and out of the plates.

Figure 9.8 Lead-acid battery schematic.

As the battery nears complete discharge, the plates are nearly completely coated with lead sulfate ($PbSO_4$). The reaction can proceed on the positive plate only when four free hydrogen ions and one free sulfate ion come together at an uncovered spot on the PbO_2 plate. From a probability standpoint, this becomes less likely to happen, and the chemical reaction must slow down. There is still a significant amount of energy that can be drawn out of the battery, but it must be drawn out slowly. Trying to drive the reaction too fast leads to $PbO \cdot PbSO_4$ and other undesirable chemical compounds on the positive plate. Some of the chemical reactions are almost irreversible and permanently degrade the energy storage capacity of the battery by partially coating the plates.

As the battery nears complete charge, there is very little lead sulfate ($PbSO_4$) left on the surface. Driving the chemical reaction too fast will cause adverse chemical compounds to form on the plate surface, which reduces the battery capacity. Charging and discharging rates must be reduced

as the battery approaches full charge or full discharge. This is true for all batteries.

- When charging or discharging at a fast rate, ions must move through the electrolyte at a high rate.
- The electrolyte near the plates becomes saturated with ions moving out of the plates and has a low concentration of ions moving into the plates.
- The saturation and low concentrations tend to drive the chemical reaction in the opposite direction.
- This gives an apparent increase in electrical resistance.

Deep-Cycle versus Starter Batteries:
Deep-cycle batteries are designed to provide power for electric vehicles and other electrical power applications. There are fewer plates in deep-cycle batteries and the active coating on the plates is thicker. This design works out better for providing sustained power over a long period of time, and for doing a deep discharge of the

battery before recharging. Starter batteries used in cars and trucks provide a surge of power for a few seconds to start the car and are immediately recharged by the alternator. Designing the battery to have more plates with a thinner layer of active coating provides more surface area of active material, which allows the battery to deliver higher current than deep cycle batteries. Optimizing the number of plates and the active coating thickness depends on the application of the battery.

Each cell in a lead-acid battery has a voltage of nominally 2.1 V when fully charged. Most lead-acid batteries have six cells wired in series. They have a nominal voltage of 12.6 V when fully charged and are referred to as 12 V batteries. The automotive industry settled on a 12 V system, and because of that, there are many lights, motors, and other electrical devices designed to operate at 12 V. The abundance of electrical devices available makes a 12 V system popular for low-power applications. Lead-acid batteries can be produced at any voltage that is a multiple of the 2.1 V per cell, and 6 V batteries are common. For higher operating voltages, it is common to put 12 V batteries in series. Forklifts are commonly powered by 24, 36, and 48 V lead-acid batteries.

Weakest Cell: Battery cells are put in series to generate a working voltage. There is always a weakest cell, which is the cell that has the lowest energy storage capacity. The weakest cell will always reach complete charge and discharge first when charging and discharging the battery. The weakest cell limits the capacity of the entire battery. Overcharging or discharging of the weakest cell will further reduce its capacity, which further degrades the battery. When a battery fails because its capacity is too low to be useful, it is often just one cell that has failed,

and the rest of the cells are good. Less expensive batteries are recycled when the capacity becomes too low to be useful. Expensive batteries, like the ones used in forklifts can be refurbished by wiring around and/or replacing bad cells in the battery.

When charging a lead-acid battery, some of the electrons going into the negative plates will combine with hydrogen ions in the electrolyte to form hydrogen gas. Some of the electrons being released into the negative plates will come from oxygen ions in solution, which will cause the release of oxygen gas near the positive plates. Figure 9.9 illustrates the electrolysis of water in a lead-acid battery.

Electrolysis of Water: For batteries that have a water-based electrolyte like sulfuric acid or potassium hydroxide, there is always some electrolysis of water when charging and discharging the battery. This electrolysis of water gives an apparent charge loss when charging and discharging the battery. It will take more amp-h to charge the battery than will be recovered when discharging the battery. There is always some charge leakage in batteries, which also causes an apparent charge loss. Batteries that sit idle for long periods of time will have significant charge leakage, but leakage is a small percentage of the charge loss for batteries that are used every day. For batteries that are used daily, most of the charge loss is from electrolysis of water. An excessive amount of electrolysis will deplete the electrolyte. Older lead-acid batteries required that water be added to the electrolyte occasionally to make up for the loss of water. The newer lead-acid batteries (sealed batteries) have been designed so that the hydrogen and oxygen created from electrolysis are recombined into water and put back in the battery. It is not normally necessary to add water to modern

Figure 9.9 Electrolysis of water in lead-acid battery.

Electrolysis of water

lead-acid batteries. But even though the water is put back in the battery, there is still an apparent charge loss due to the electrolysis process. The chemistry of electrolysis will be studied later in the chapter when talking about hydrogen production for fuel-cell applications.

9.4.2.
Silver–Zinc Batteries

Silver–zinc batteries were one of the first batteries developed. They are still being used for military applications because the electrolyte can be stored separately and then added to the battery when the battery is needed. This gives the silver–zinc battery a long shelf life, which makes it suitable for certain military applications. These batteries are expensive because of the silver on the positive plates.

The batteries can hold a full charge almost indefinitely until the electrolyte is added. Once the electrolyte is added the batteries have a short

life compared to lead-acid or lithium-ion batteries. Some silver–zinc batteries can only be charged and discharged 10 or 20 times before they are worn out. There are applications where a battery will need to sit idle for many years before it is used. There is usually a mechanism to break a vial and add the electrolyte when it is needed. The battery will be able to provide a full cycle of energy. In some applications, one cycle is all that is needed.

The negative plates in a silver–zinc battery are coated with zinc and the positive plates are coated with silver oxide. The electrolyte is a potassium hydroxide solution. Figure 9.10 illustrates the chemistry of the silver–zinc battery.

The chemical reaction on the positive plate is complex. It can be written in three steps as shown in Equations (9.18), (9.19), and (9.20).

Figure 9.10 Silver–zinc battery schematic.

The chemical reaction on the negative plate is:

$$2K^+ + AgO + 2e^- \rightarrow K_2O + Ag \tag{9.18}$$

$$K_2O + Ag \rightarrow 2K^+ + O^{-2} + Ag \tag{9.19}$$

$$O^{-2} + 2H^+ \rightarrow H_2O \tag{9.20}$$

The chemical reaction on the negative plate is:

$$Zn + 2(OH)^- \rightarrow Zn(OH)_2 + 2e^- \tag{9.21}$$

The net chemical reaction is:

$$Zn + AgO + H_2O \rightarrow Zn(OH)_2 + Ag \tag{9.22}$$

9.4.3.
Nickel–Cadmium Batteries

Nickel–cadmium batteries have been used in many applications, and there are many sizes of batteries. They have an energy density that is higher than lead-acid, but lower than lithium-ion. They are reliable and can be cycled many times. The nickel–cadmium battery uses cadmium as the negative plate and the positive plates are coated with nickel oxyhydroxide. The electrolyte is potassium hydroxide. A schematic illustrating the battery chemistry is shown in Figure 9.11. The chemistry is similar to the silver–zinc battery.

Figure 9.11 Nickel–cadmium battery schematic.

The chemical reaction on the positive plate is complex. It can be written in three steps as:

$$2K^+ + OH^- + NiO(OH) + e^- \rightarrow K_2O + Ni(OH)_2 \tag{9.23}$$

$$K_2O + Ni(OH)_2 \rightarrow 2K^+ + O^{-2} + Ni(OH)_2 \tag{9.24}$$

$$O^{-2} + 2H^+ \rightarrow H_2O \tag{9.25}$$

The chemical reaction on the negative plate is:

$$Cd + 2(OH)^- \rightarrow Cd(OH)_2 + 2e^- \tag{9.26}$$

The net chemical reaction is:

$$Cd + 2NiO(OH) + 2H_2O \rightarrow Cd(OH)_2 + 2Ni(OH)_2 \tag{9.27}$$

9.4.4.
Nickel–Hydrogen Batteries

Nickel–hydrogen batteries have been used in satellites. The energy density is significantly higher than lead-acid batteries, but lower than lithium-ion batteries. The batteries are capable of many charge/discharge cycles, which is important for space applications. The primary applications for the batteries are space applications. They can be charged and discharged in more cycles than any type of battery. A pressure vessel is required to store the hydrogen gas.

The hydrogen gas is the negative plate in the battery, so a porous electrode must be developed to allow the hydrogen gas to flow in and out of the negative plate. The negative terminal is similar to a fuel cell. The hydrogen gas is used up when the battery is discharged and replenished when the battery is charged. The pressure inside the battery can be used to indicate the state of charge. The positive plates in the battery are nickel oxyhydroxide, and the electrolyte is potassium hydroxide. The chemistry is similar to the nickel–cadmium battery. Figure 9.12 is a schematic of the nickel–hydrogen battery.

Figure 9.12 Nickel–hydrogen battery schematic.

The chemical reaction on the positive plate is complex. It can be written in three steps as:

$$2K^+ + OH^- + NiO(OH) + e^- \rightarrow K_2O + Ni(OH)_2 \tag{9.28}$$

$$K_2O + Ni(OH)_2 \rightarrow 2K^+ + O^{-2} + Ni(OH)_2 \tag{9.29}$$

$$O^{-2} + 2H^+ \rightarrow H_2O \tag{9.30}$$

The chemical reaction on the negative plate is:

$$H_2 + 2(OH)^- \rightarrow 2H_2O + 2e^- \tag{9.31}$$

The net chemical reaction is:

$$H_2 + 2NiO(OH) \rightarrow 2Ni(OH)_2$$

(9.32)

9.4.5.
Nickel–Metal Hydride Batteries
Nickel–metal hydride batteries are used in many applications. They are reliable and durable and have a higher energy density than lead-acid batteries. The chemistry is the same as the nickel–hydrogen batteries, but the hydrogen is stored in a metal hydride rather than in a pressure vessel. The catalyzed electrode in the previous figure is replaced with a metal hydride electrode.

9.4.6.
Lithium Batteries
Lithium is a lightweight element, and people have tried for many years to make lithium batteries because of the potential for lightweight batteries. Lithium is very reactive and explosive, and batteries made using lithium metal have always been regarded as too dangerous to be used in practice. A lot of interesting work has been done, but no practical batteries have been developed.

9.4.7.
Lithium-Ion Batteries
Lithium-ion batteries are very different from other battery chemistries. There is no metallic lithium in the batteries. The charge moves in the batteries by transporting lithium-ions across an organic or polymer membrane electrolyte. Lithium is a small enough atom to fit into the crystal lattice of larger atoms. The negative electrode is made of carbon graphite, which is a layered material, and the lithium atoms fit between the layers in the graphite. The structure of graphite is illustrated in Figure 9.13.

Because of the way the lithium atoms fit into the graphite structure, there needs to be a minimum of six carbon atoms for each lithium atom. Carbon has an atomic mass of 12, and lithium has an atomic mass of 6.9. With a minimum of six carbon atoms required for each lithium atom, the mass of the carbon in the negative terminal will be at least 6*12/6.9 = 10.4 times the mass of the lithium. The negative terminal will be much heavier than the active lithium in the battery.

The positive terminal is made from a metal oxide and is the heaviest part of the battery. The earliest lithium-ion batteries used cobalt oxide for the positive terminal, and cobalt oxide is still used in batteries today. Nickel oxide and manganese oxide have been used, and a blend of metal oxides has been used. Cobalt is heavy and expensive, and much research has gone into finding other materials that will work for the positive terminal in the batteries. If cobalt oxide is used for the positive terminal, the way the lithium atoms fit into the crystal lattice requires a minimum of two cobalt oxide molecules for each lithium atom. The atomic mass of cobalt oxide is 91, so the mass of cobalt oxide will be at least 2*91/6.9 = 26.4 times the mass of lithium. Cobalt oxide is more than twice as heavy as lithium and carbon combined. The best way to reduce the weight of the lithium-ion battery is to find a lighter-weight material for the positive electrode.

The chemical reaction for lithium-ion batteries is shown assuming cobalt oxide is used for the positive terminal. The positive plates follow the chemical reaction:

Figure 9.13 Graphite structure.

142 pm

335 pm

$$2CoO_2 + Li^+ + e^- \rightarrow Li(CoO_2)_2 \tag{9.33}$$

The negative plate reaction is:

$$LiC_6 \rightarrow C_6 + Li^+ + e^- \tag{9.34}$$

The overall reaction is:

$$LiC_6 + 2CoO_2 \rightarrow C_6 + Li(CoO_2)_2 \tag{9.35}$$

A schematic of the lithium-ion battery is illustrated in **Figure 9.14**.

The lithium-ions move across the organic or polymer electrolyte, which is also called the separator. The electrolyte (or separator) should be very thin to reduce internal resistance in the battery. The early lithium-ion batteries had problems with dendritic growth through the electrolyte. As the lithium ions moved back and forth between the plates during the charge/discharge cycles, an occasional bridge of lithium metal would connect the positive and negative

Figure 9.14 Schematic of lithium-ion battery.

plates shorting the battery out. The result was catastrophic failure and the battery would ignite. Kyocera had a major recall of products in the early 2000s because of this type of battery failure. The more recent lithium-ion batteries have been designed in such a way as to prevent dendritic bridging and shorting out and are much safer.

Lithium-ion batteries seem to be replacing other battery technologies in almost all applications. Lead-acid batteries are still less expensive than lithium-ion batteries for the same amount of storage capacity, but the price gap is narrowing. Lithium-ion batteries are lighter in weight than lead-acid batteries and have a higher charge/discharge efficiency. They generate less heat than lead-acid batteries. Lithium has an atomic mass of 6.9 g/mole, so 1 kg of lithium is 145 moles. The batteries operate at a nominal voltage of 3.8 V and there is one electron involved in the process, so the Gibbs free energy associated with 1 kg of lithium is calculated as:

$$\text{Gibbs Free Energy} = (1)(96,485)(3.8)(145) = 53.1 \, \text{MJ} / \text{kg} \tag{9.36}$$

Petroleum fuels have an energy density of 47 MJ/kg, so the active lithium in the battery has an energy density comparable to petroleum fuels. But for the battery to work, there must be a minimum of 12 carbon atoms in the negative terminal for each active lithium atom, and there must be a minimum of two cobalt oxide molecules for each active lithium atom in the positive terminal. To have 145 moles of active lithium in the battery requires a minimum of 6*145 moles of carbon for the negative terminal and 2*145 moles of cobalt oxide for the positive terminal. The minimum weights for the negative and positive terminals are:

$$\text{Negative Terminal} = \frac{(6)(12)(145)}{1000} = 10.44 \, \text{kg Carbon} \tag{9.37}$$

$$\text{Positive Terminal} = \frac{(2)(91)(145)}{1000} = 26.39 \,\text{kg Cobalt Oxide} \qquad (9.38)$$

To make a battery with 53.1 MJ of energy storage requires 1 kg of lithium, 10.44 kg of carbon, and 26.39 kg of cobalt oxide. The total mass of the battery would be 37.83 kg. The energy density would be 53.1 MJ/37.83 kg = 1.4 MJ/kg. This does not include the mass in the polymer electrolyte (separator), the electrical wiring and terminals, and the battery case. It is an upper limit for lithium-ion batteries using cobalt oxide as the positive terminal. Real lithium-ion batteries will have an energy density less than 1.4 MJ/kg.

9.4.8.
Solid-State Lithium Batteries

Solid-state lithium batteries have been in development for many years. The basic idea is to replace the carbon negative terminal in a lithium-ion battery with lithium metal, eliminating most of the weight of the negative terminal. The problem has always been that dendritic growth from the lithium metal through the electrolyte (separator) causes the battery to short out and catch fire. The current approach is to use a ceramic electrolyte (separator) that will prevent the dendritic growth. It is unknown whether we will be able to solve the dendritic growth problem or not. If successful, most of the carbon in the negative terminal can be eliminated, improving the energy density of the battery. Using the numbers in the paragraph above, if the 10.44 kg of carbon is eliminated, the mass of the battery would be 27.39 kg and the energy density would be 53.1/27.39 = 1.9 MJ/kg. This would be a significant improvement in energy density compared to lithium-ion batteries.

9.4.9.
Lithium Iron Phosphate Batteries

The cobalt used in traditional lithium-ion batteries is expensive. Iron phosphate is less expensive and can be used to replace the cobalt oxide for the positive terminal. The iron phosphate is reported to be chemically a little more stable than the cobalt oxide, making the battery safer than traditional lithium-ion batteries [3]. Two iron phosphate molecules are required for each active lithium atom in the chemical reaction:

$$\text{Li} + 2\text{FePO}_4 \rightarrow \text{Li}\left(\text{FePO}_4\right)_2 \qquad (9.39)$$

Lithium iron phosphate batteries operate at a nominal voltage of 3.3 V, so the Gibbs free energy for 1 kg of active lithium is 46.2 MJ rather than the 53.1 MJ for traditional lithium-ion batteries. One kg of lithium is 145 moles. The mass of the negative (carbon) terminal is 10.44 kg since it is unchanged from the traditional lithium-ion battery. The positive terminal made from iron phosphate will have a mass of 2*145*151/1000 = 43.79 kg. The maximum energy density of the lithium iron phosphate battery is 46.2 MJ/55.23 kg = 0.837 MJ/kg. The maximum energy density is 60% of the traditional lithium-ion batteries. Lithium iron phosphate batteries are heavier than traditional lithium-ion batteries, but they are less expensive and may be a better solution overall. Iron, phosphorous, carbon, and oxygen are common elements in the earth's crust. Cobalt, nickel, and manganese are much less common elements. It is questionable that we can mine enough cobalt, nickel, and manganese to make large quantities of batteries required for electric vehicles and other applications.

9.4.10.
Flow Batteries

Flow batteries are being considered for utility grid electric storage. The batteries offer the potential benefits of low cost and very large storage capacity. There is a relatively small battery portion where the chemical reaction takes place and the electricity is generated. The active materials for the battery are stored in solution in tanks and pumped to the battery as needed. Adding more or larger tanks of active

material increases the capacity of the battery. There is essentially no limit on battery capacity. The power the battery can provide is limited by the size of the battery portion. Figure 9.15 is a schematic of a vanadium flow battery. In the schematic, the catholyte is the material that is pumped to the redox flow cell as the cathode for the battery, and the anolyte is the material for the anode of the battery.

Because the active materials are stored in tanks, the battery can remain at low or high states of charge for long periods of time with essentially no electrical leakage and very little degradation of the active materials. They can be charged and discharged many times before needing to

Figure 9.15 Vanadium redox flow battery.

be refurbished. Active materials can be recycled and replaced when refurbishing the battery. The primary disadvantage is that the batteries have a low energy density, requiring a much larger mass and volume than lithium-ion or lead-acid batteries for the same energy storage. Flow batteries are not suitable for vehicle applications, but may be suitable for utility grid storage when size and weight are less important [4].

Vanadium is the most common flow battery today, but other chemistries are being considered. Flow batteries are early in the development stage, and it is difficult to say at this point which chemistry, if any, will be successful. The chemical reaction on the positive and negative electrode and the overall reaction are as follows:

$$\text{Positive Electrode}: VO_2^+ + 2H^+ + e^- \rightarrow VO_2^{+2} + H_2O \tag{9.40}$$

$$\text{Negative Electrode}: V^{+2} \rightarrow V^{+3} + e^- \tag{9.41}$$

$$\text{Overall Reaction}: VO_2^+ + V^{+2} + 2H^+ \rightarrow VO_2^{+2} + V^{+3} \tag{9.42}$$

The theoretical working voltage is 1.259 V per cell. In practice, there will be internal resistance and the actual working voltage will be lower. Following is an illustration of the calculations associated with sizing a 1 MW-h vanadium flow battery.

The chemical reaction for the vanadium flow battery transfers one electron. Assuming a voltage of 1.259 V, the Gibbs free energy is:

$$\Delta G = (1)(96,485)(1.259) = 121,475\,J\,/\,mole \tag{9.43}$$

Examining the left side of the overall reaction, the molar weight of the active material to transfer one electron is:

$$\text{Molar Weight Active Material} = (51)(2) + (16)(2) + (2)(1) = 136\,g\,/\,mole \tag{9.44}$$

This means that 136 g of active material is required to transfer 1 mole of electrons and produce 121,475 J of electrical energy. One kg of active material will transfer 1000/136 = 7.353 moles of electrons, yielding an energy density of 7.353*121,475 = 893,199 J/kg active material.

The active material is dissolved in a sulfuric acid solution, and only a small percentage of the solution is active material. For this example, assume that the active material ($VO_2^+ + V^{+2} + 2H^+$) is 10% of the solution by mass. With this assumption, the energy density is 89,319.9 J/kg solution. To provide enough solution in the storage tanks for 1 MW-h of energy:

$$\left(1{,}000{,}000 \text{ W-h}\right)\left(\frac{3600 \text{ J}}{\text{W-h}}\right)\left(\frac{\text{kg solution}}{89{,}319.9 \text{ J}}\right) = 40{,}300 \text{ kg solution} \qquad (9.45)$$

A typical sulfuric acid solution would have a density of 1800 kg/m^3. Storing 40,300 kg of solution would require a volume of 40,300/1800 = 22.4 m^3. To put this into perspective, a 55 gal barrel holds 0.2082 m^3 of volume. It would require 107.5 barrels of solution to provide 1 MW-h of electric energy. Grid storage will require hundreds of MW-h of capacity. The mass and volume of solution for flow batteries will be very large. Any battery used for grid storage will need to be large in volume and mass. Flow batteries will be larger and more massive than lithium-ion or lead-acid batteries.

9.4.11.
Polymer Electrolyte Membrane (PEM) Fuel Cells

Hydrogen is being considered as a potential fuel for vehicles because it produces no carbon dioxide gas when burned or used in fuel cells. Hydrogen can be used to replace fossil fuels in an internal combustion engine or jet engine. The thermal efficiency of an engine burning hydrogen should be similar to engines burning gasoline, diesel, or jet fuel. Fuel cells can be used to convert the chemical energy in the hydrogen to electricity at a much higher efficiency than internal combustion engines. Fuel cells were originally developed in the 1800s, and there are many types of fuel cells. NASA used fuel cells to provide electricity for the Apollo missions. The Polymer Electrolyte Membrane (PEM), or sometimes called Proton Exchange Membrane, fuel cell seems to have the most promise today. Toyota uses a PEM fuel cell to power the Murai.

Figure 9.16 is a schematic showing how the PEM fuel cell works. Hydrogen gas enters on the right side through a gas diffusion layer. A catalyst helps separate the electron from the nucleus of the hydrogen atom in the anode. The electron flows up and is what produces the electric current. The hydrogen nucleus (proton) moves through the PEM to the cathode. In the cathode, the hydrogen nucleus (proton) comes with oxygen and the electrons flowing in from the electrical current to form water. A catalyst is used to help with the chemical reaction in the cathode. The oxygen is provided by using air rather than pure oxygen. The vehicle will not need to carry tanks of oxygen. Platinum is used as the catalyst. The most common polymer membrane is Nafion, made by Dupont.

The chemical reaction takes hydrogen and oxygen gas and makes water. The theoretical voltage is 1.18 V, but a more practical operating voltage is 0.6 to 0.7 V per cell. The efficiency of fuel cells is (0.6 to 0.7)/1.18, which is 50 to 60% efficiency. Cells are stacked in series to produce a suitable working voltage. Like batteries, fuel cells will have an apparent internal resistance so that efficiency decreases with higher current. Gaseous hydrogen has an energy density of 115 MJ/kg. The hydrogen must be compressed to a high pressure to get it to a small enough volume that it will fit on the vehicle. The energy required to compress the hydrogen is significant and reduces the overall energy efficiency of using hydrogen fuel.

Figure 9.16 Figure 9.16 PEM fuel cell.

9.4.12.
Production of Hydrogen Fuel

The least expensive way to produce hydrogen is by steam reforming of methane. Hydrogen produced this way is called blue hydrogen. Gray hydrogen is produced using coal, and green hydrogen is produced using electrolysis of water. The chemical reaction for steam reforming of methane is:

$$CH_4 + 2H_2O\,(gas) + Heat \rightarrow CO_2 + 4H_2 \qquad (9.46)$$

This is an endothermic reaction, so heat must be added to keep the chemical reaction going. The heat is normally provided by burning methane, which creates additional water and carbon dioxide not shown in Equation (9.46).

Methane has 810 kJ/mole heat energy. It requires 40.7 kJ/mole to convert liquid water to steam. The minimum heat energy required is 164.8 kJ/mole of methane converted to overcome the endothermic nature of the chemical reaction. The energy required (left side of the equation) is:

$$\frac{810\,\text{kJ}}{\text{mole}} + 2\left(\frac{40.7\,\text{kJ}}{\text{mole}}\right) + \frac{164.8\,\text{kJ}}{\text{mole}} = \frac{1056.2\,\text{kJ}}{\text{mole}} \tag{9.47}$$

Hydrogen gas releases 231 kJ/mole when combined with oxygen to form water. The energy in 4 moles of hydrogen is 924 kJ. The maximum possible energy efficiency associated with steam reforming of methane is 924/1056.2, which is 87.5%. But adding only 164.8 kJ/mole heat energy will cause the chemical reaction to proceed at a very slow rate, which is impractical. Typical steam reforming processes add an additional 300–400 kJ/mole to speed the process, yielding a process that is approximately 65% efficient. The result is that the energy in the hydrogen generated is only about 65% of the energy in the methane used to make the hydrogen.

Steam reforming of methane generates carbon dioxide and hydrogen gas as shown in Equation (9.46). The heat required to convert the liquid water to steam and the additional heat required to keep the chemical reaction going comes from burning methane, which creates more carbon dioxide. The additional heat energy is 2(40.7 kJ) for the water, 164.8 kJ for the minimum theoretical heat energy for the chemical reaction, and an additional 300–400 kJ to keep the chemical reaction going at a reasonable rate. The total is 546.2 to 646.2 kJ of heat energy that must be added to generate 4 moles of hydrogen. Methane generates 810 kJ/mole, so an additional 0.67 to 0.80 moles of methane must be burned to generate the heat to produce 4 moles of hydrogen. If an average value of 0.736 moles of methane is assumed for the heat, the chemical reaction can be revised to include the additional methane as:

$$(1.736)CH_4 + (1.472)O_2 + 2H_2O\,(\text{gas}) \rightarrow (1.736)CO_2 + 4H_2 + (1.472)H_2O \tag{9.48}$$

The carbon footprint of the hydrogen gas generated using steam reforming of methane can be calculated knowing that the 4 moles of hydrogen gas have 924 kJ of energy and produce 1.736 moles of carbon dioxide. In Chapter 6, it was calculated that burning methane gives about 19 BTU of energy for each gram of carbon dioxide generated. For a comparison:

$$\left(\frac{924\,\text{kJ}}{(1.736)(42\,\text{g CO}_2)}\right)\left(\frac{\text{BTU}}{1.055\,\text{kJ}}\right) = \frac{12.0\,\text{BTU}}{\text{g CO}_2} \tag{9.49}$$

The carbon footprint of the hydrogen produced by steam reforming of methane is higher than the methane used to produce the hydrogen. Methane provides 19 BTU of heat energy for each gram of CO_2 produced, whereas hydrogen provides only 12 BTU/g CO_2. Using methane as the fuel would have a lower carbon footprint than using the hydrogen gas produced by steam reforming of methane. Blue hydrogen does not make sense as a long-term solution. The energy efficiency is lower than using methane as the fuel source, and the carbon emissions are higher. A similar process is used to produce gray hydrogen from coal. The gray hydrogen produced from coal has a lower energy efficiency and higher carbon footprint than using coal as the fuel source. Blue hydrogen and gray hydrogen can be used as a bridge to allow us to develop fuel-cell-powered vehicles, but they do not help reduce our carbon footprint. Vehicles that run on natural gas will have a lower carbon footprint than vehicles using blue hydrogen.

In order to reduce our carbon footprint, hydrogen gas needs to be produced using electrolysis of water, and the electricity needs to come from renewable sources that do not produce carbon dioxide. The Department of Energy [5] estimates that 53.4 to 70.1 kW-h energy is required to produce 1000 g of hydrogen using electrolysis. In Chapter 5, it was estimated that in the United States we generate an average of 419 g of carbon dioxide for each kW-h of electricity produced. Hydrogen gas has a molecular weight of 2 and 231 kJ/mole of energy. These can be combined as follows:

$$\left(\frac{1000\,\text{g H}_2}{53.4\,\text{to}\,70.1\,\text{kW-h}}\right)\left(\frac{\text{kW-h}}{419\,\text{g C})_2}\right)\left(\frac{\text{mole H}_2}{2\,\text{g H}_2}\right)\left(\frac{231\,\text{kJ}}{\text{mole H}_2}\right)\left(\frac{1.055\,\text{BTU}}{\text{kJ}}\right) = \frac{5.45\,\text{to}\,4.15\,\text{BTU}}{\text{g CO}_2} \tag{9.50}$$

In Chapter 6, we calculated that burning natural gas yields about 19 BTU/g CO_2, petroleum yields about 14 BTU/g CO_2, and coal yields about 9 BTU/g CO_2. Blue hydrogen yields about 12 BTU/g CO_2 [Equation (9.49)]. The hydrogen produced by electrolysis (green hydrogen) has a higher carbon footprint than any of the fossil fuels or blue hydrogen. In order for green hydrogen to have a low carbon footprint, it must be produced using electricity that comes from renewable sources. We can use electricity from the grid to produce hydrogen and develop our production capabilities of producing hydrogen from electrolysis in the hope that in the future electricity will be produced using renewable energy. But in the near term, blue hydrogen has a lower carbon footprint than green hydrogen because of the carbon dioxide generated by producing the electricity. Burning petroleum or natural gas has a lower carbon footprint than blue hydrogen.

9.4.13.
Storing and Transporting Hydrogen

The standard 40-L steel pressure vessel has a mass of 54 kg and is capable of storing 0.37 kg of hydrogen at 125 bar (1800 psi). The tank is 145 times as heavy as the hydrogen it stores. Vehicles will need much lighter weight storage tanks for the hydrogen fuel. The tanks on the Toyota Murai are reported to have a mass of 85 kg and store 5 kg of hydrogen.

Hydrogen gas could be transported in pipelines similar to how natural gas is transported. It can be pressurized and transported on trucks, or it can be transported as ammonia and reformed on-site.

9.5.
Homework

1. A 5-kW-h, lithium-ion battery system has a weight of 80 lb. A 5-kW-h, lead-acid battery system has a weight of 300 lb.

   (a) Find the energy density of the two battery systems in kilowatt-hours per kilogram (kW-h/kg) and in mega-joules per kilogram (MJ/kg).

   (b) Use Figure 9.1 to compare the energy density of gasoline to the energy density of the two battery systems. (1) How many kg of lithium-ion batteries would it take to provide the same amount of energy as 1 kg of gasoline? (2) How many kg of lead-acid batteries would it take to provide the same amount of energy as 1 kg of gasoline? (Partial Answer: about 350 kg lead-acid batteries.)

(c) Assume that 25% of the energy in the gasoline will be converted to useful mechanical energy by the engine in the car. Also assume that 90% of the electric energy in the battery system will be converted to useful mechanical energy by the electric motor in the car. (1) How many kg of lithium-ion batteries would it take to provide the same amount of useful mechanical energy as 1 kg of gasoline? (2) How many kg of lead-acid batteries would it take to provide the same amount of useful mechanical energy as 1 kg of gasoline? (Partial Answer: about 100 kg of lead-acid batteries.)

2. Figure 9.17 is a rough illustration of what happens when discharging a lead-acid battery at different rates. Assume that the data are for a 12 V 60 amp-h capacity lead-acid battery.

The lowest curve on the graph is labeled 3.0 C, which means the discharge amperage was 3 × 60 = 180 amps. The second lowest curve labeled 2.0 C means the battery is discharged at a rate of 120 amps. There are several other curves on the graph. Using the graph, estimate the amount of energy that can be drawn from the battery at the different discharge rates (3.0 C, 2.0 C, 1.0 C, 0.4 C, 0.2 C, 0.05 C, 0.01 C).

[Process: First use five to seven points along the curve to find the average discharge voltage. Point 1 should be at 1 min, that is, do not use the first minute in finding the average voltage. Use the x-axis on the graph to get the number of hours of discharge and multiply the discharge amperage (180 amps for the lowest curve) by the hours to get the amp-hours discharge. The energy drawn from the battery is the average voltage multiplied by the amp-hours. Use Excel to plot the energy that can be drawn out of the battery vs the discharge amperage. (Partial Answer: At 1.0 C there is about 480 W-h energy.)

Figure 9.17 Lead-acid battery discharge curves.

© Ricktek

3. Data are collected for charging a small lithium-ion cell. For most of the charging cycle, the battery is charged at a steady 1.3 amps. After 10 min of charging, the voltage is observed to be 3.85 V. After 90 min of charging, the voltage is observed to be 4.16 V. (Raising the charging voltage of a lithium-ion cell above 4.2 V will damage the battery, so after 90 min the charging amperage must decrease to avoid damaging the battery.)

(a) Assume that the battery voltage varies linearly between 10 and 90 min. Calculate the energy put into the battery (W-h) and the amp-hours of charge put into the battery during this 80 min of the charging cycle. Assume 100% charge efficiency.

(b) Assume that the amp-hours charge calculated above will be drawn out of the battery at an average of 3.7 V. Calculate the energy that will be drawn out of the battery. What is the energy efficiency of the battery? (Partial Answer: 92.4% efficient.)

4. This problem is related to calculating the maximum theoretical energy density for a lead-acid battery. Suppose we want the lead-acid battery to be able to store 1 kW-h of energy. What is the minimum possible weight for a lead-acid battery that can store 1 kW-h of energy?

(a) Use the Gibbs free energy analysis to calculate the amount of energy per mole of active materials in the battery. Assume the cell voltage is 2.1 V. (Answer: 405,237 J/mole.)

(b) Assume lead has an atomic weight of 207.2 g/mole, oxygen has 16 g/mole, sulfur has 32 g/mole, and hydrogen has 1 g/mole. Calculate the mass for 1 mole of active material in the battery. (Answer 642.4 g.)

Following is the chemistry associated with a lead-acid battery:

$$Pb + PbO_2 + 2H_2SO_4 \rightarrow 2PbSO_4 + 2H_2O \qquad (9.51)$$

(c) Calculate the number of moles of active material required for 1 kW-h of energy. (8.88 moles.)

(d) Calculate the minimum possible weight of the battery. (5.7 kg or 12.5 lb)

(e) A real lead-acid battery capable of storing 1 kW-h of energy will weigh approximately 80 lb. Based on your calculations above, what percentage of the battery is active material and what percentage is the other components required to make the battery functional?

5. Assume that a large lead-acid forklift battery is at a 50% state of charge. The battery is charged briefly at a rate of 50 amps, and the voltage is observed to be 14.3 V. The battery is then discharged briefly at 50 amps, and the voltage is observed to be 11.0 V. Use these data to estimate the internal resistance of the battery. [Hint: $V_0 + (50 \text{ amps}) \times R = 14.3$ V.]

Calculate the energy efficiency for this battery if it is charged at 30 amps and discharged at 20 amps. (Answer: 87.9%)

6. Assume that a lead-acid cell is put on a trickle charger to maintain full charge on the battery. There is no real increase in charge in the battery. The energy going into the battery contributes to the electrolysis of the water in the electrolyte. Assume the voltage across the cell is maintained at 2.2 V and 1 mole of water is converted to hydrogen and oxygen gas.

(a) How much energy in Joules is used to convert the water to hydrogen and oxygen gas?

(b) How much energy in W-h? (Answer: 118 W-h)

(c) How many grams of water in the electrolyte will be lost for each kW-h of energy?

7. The goal of this homework problem is to have you think about the battery chemistry of lead-acid batteries. Lead-acid batteries will be important in small electric vehicles like forklifts and golf carts for a long time because they are inexpensive and reliable. Review the notes on lead-acid batteries and do a brief Google search on lead-acid battery chemistry. Make a hand-drawn sketch of a lead-acid battery similar to the figure in the notes.

(a) Write a short paragraph explaining the chemistry on the positive plates.

(b) Write a short paragraph explaining the chemistry on the negative plates.

8. The most important battery technology in the near future is the lithium-ion battery. Lithium-ion batteries show the most promise for us in electric cars. Review the notes on lithium-ion batteries and do a brief Google search on lithium-ion battery chemistry.

(a) Make a hand-drawn sketch of a lithium-ion battery similar to the figure in the notes.

(b) Write a short paragraph explaining the chemistry on the positive plates.

(c) Write a short paragraph explaining the chemistry on the negative plates.

(d) Write a short paragraph explaining how the electrolyte works.

9. Section 9.4.7 shows how to calculate the maximum possible energy density for lithium-ion batteries, and the result was 1.4 MJ/kg. For this problem estimate the maximum energy density of solid-state lithium-ion batteries subject to the assumptions below.

(a) Calculate the energy in 1 kg of active lithium, assuming the average discharge voltage is 3.8 V per cell. (Answer: 53.2 MJ)

(b) Assume that the negative terminal is made of pure lithium. Most of the metal in the negative terminal is active lithium, but in order to allow the battery to charge and discharge correctly, we cannot use all the lithium in the negative terminal as active material. Assume a 20% surcharge of lithium so that if there is 1 kg of active lithium, there will be 0.2 kg of lithium that stays at the negative terminal and is not used in the chemical reaction.

(c) Assume that cobalt oxide CoO_2 is used for the positive terminal, and assume that for each lithium atom in the positive terminal there will need to be at least two CoO_2 molecules. That is, make the same assumption that was made for the lithium-ion batteries. Calculate the mass of the positive terminal required for 1 kg of active lithium. (Answer: 26.4 kg)

(d) Neglecting everything else, calculate the maximum possible energy density based on the mass in the positive terminal, the active lithium, and the negative terminal. (Answer: 1.93 MJ/kg) How does this compare to the energy density of gasoline?

10. Suppose we have a vanadium flow battery that follows the chemical reaction:

$$VO_2^+ + V^{+2} + 2H^+ \rightarrow VO^{+2} + H_2O + V^{+3}$$

(9.52)

The goal of this problem is for you to make an estimate of how much solution is required to store 300 MW-h of energy. Vanadium flow batteries are envisioned to be used for grid storage, so a battery would need to hold a large amount of energy to be effective.

(a) The chemical reaction transfers one electron. Assume that the voltage for the cell is the theoretical 1.259 V. Use the Gibbs free energy equation to calculate the energy released on discharge for 1 mole of material. (Answer: 121,475 J/mole)

(b) Assuming vanadium has a molar mass of 51 g, calculate the maximum energy density for the battery using only the materials in the above formula. (Answer: 893,200 J/kg)

(c) To make the battery work, the active materials are dissolved in a sulfuric acid solution. As a first approximation, a typical solution would add about 5 sulfuric acid molecules (H_2SO_4) and 50 water molecules (H_2O) to both sides of the equation. The sulfuric acid solution does not participate in the chemical reaction, but is necessary to keep the ions in solution. Calculate the energy density of the battery solution including the sulfuric acid molecules and water molecules. (Answer: 79,600 J/kg)

(d) Convert 300 MW-h to Joules and use the answer in (c) above to calculate the number of kg of solution required to allow the battery to store 300 MW-h of energy.

(e) Assume that the solution has a density of 1830 kg/m³. Calculate the number of cubic meters of solution required to store 300 MW-h of energy. (Answer: 7414 m³.) If we were to build a cubic-shaped tank, what would the dimensions be? (Answer 19.5 m or about 64 ft.)

(f) Assuming 264.2 gal/m³, and that a barrel holds 55 gal, how many barrels of solution will it take to allow the battery to store 300 MW-h of energy? (Answer: 35,600 barrels.)

11. Explore the efficiency of steam reforming of methane. Steam reforming of methane follows the chemical reaction below.

$$CH_4 + 2H_2O\,(gas) + heat \rightarrow CO_2 + 4H_2$$

(a) Assume methane has 810 kJ/mole heat energy.

(b) Assume that it requires 40.7 kJ/mole to convert liquid water to steam.

(c) Assume it requires a minimum of 164.8 kJ heat energy for each mole of methane to keep the chemical reaction going. This is an endothermic reaction; we must add heat to keep it going.

(d) Assume that hydrogen gas releases 231 kJ/mole when burned with oxygen.

Use these numbers to estimate the maximum possible energy efficiency in the steam reforming process. (Answer: 87.5%. The energy in the hydrogen generated will never be more than 87.5% of the energy in methane used to create hydrogen. The chemical reaction will proceed very slowly with only 164.8 kJ/mole heat energy added.

This is not a practical process for steam reforming of methane because it is too slow.)

For a more practical methane reforming process, assume 400 kJ/mole of heat added instead of only 164.8 kJ/mole. This will make the chemical reaction happen much faster. Calculate the efficiency for this methane reforming process. (Answer: 71.60%. The energy in the hydrogen will be 71.6% of the energy in the methane used to create the hydrogen.)

References

1. ch302.cm.utexas.edu/echem/echem-all.php.

2. Rand, D.A.J., Woods, R., and Dell, R.M., *Batteries for Electric Vehicles* (Warrendale: Society of Automotive Engineers, Inc., 1998).

3. Lithium iron phosphate battery—Wikipedia.

4. Vanadium redox battery—Wikipedia.

5. Technology Brief: Analysis of Current-Day Commercial Electrolyzers: National Renewable Energy Laboratory (Fact Sheet) (nrel.gov).

Unit 03: Detailed Modeling of Vehicle Efficiency

In Unit 1, we studied the energy consumption and fuel economy of vehicles at their cruising or operating speed. The energy consumed and fuel economy of the vehicle depend on the type of vehicle, its size and weight, and the thermal efficiency of the engine and drive system. When operating at the cruising or operating speed, the following thermal efficiency ranges are typical for the internal combustion and jet engines in different types of vehicles:

(a) Cars and light trucks	25–30%
(b) Larger trucks	30–35%
(c) Trains and ships	30–35%
(d) Airplanes	25–30%
(e) Jets	35%

Electric motors used in electric vehicles have a much higher efficiency than the internal combustion engines used in most vehicles. Electric motors are typically 85 to 90% efficient in converting electric energy to mechanical energy. The batteries required to store the electric energy are heavy, and historically, it has been the weight of the batteries that held back the development and adoption of electric vehicles. Lead-acid batteries were suitable for electric bicycles and scooters, wheelchairs and other mobility vehicles, forklifts and vehicles used in manufacturing plants, and other small electric vehicles that do not require a lot of driving range. The development of lithium-ion batteries approximately tripled the amount of energy per kg that can be stored in batteries, compared to lead-acid batteries. This made it possible for more vehicles to be powered by batteries. Today, there are many electric cars and light trucks in service, and we are experimenting with electrifying large trucks and small aircraft.

In some countries, trains are powered directly from the electric power grid and do not need to carry batteries. With the right infrastructure, trains can be electric-powered using the technology available today. Battery-powered trains would require very large batteries, which seems impractical when they can be run directly off the grid. The navy has used nuclear-powered electric ships for many years, so it is possible in principle to electrify the shipping industry. Nuclear-powered commercial ships would be expensive, and it would be difficult to convince the public they are safe. Battery-powered ships would require extremely large batteries, which would reduce the amount of cargo they can carry. Now, battery-powered ships seem impractical. Electrifying aircraft will be difficult. Jet fuel is a significant portion of the weight of a jet, and it is hard to imagine any battery with an energy density comparable to jet fuel. Battery-powered airplanes can work for short flights, but the batteries are too heavy for long flights. Now, hydrogen fuel seems like the best possibility for eliminating carbon emissions from aircraft. Hydrogen fuel could be used in a jet engine or could be used in a fuel cell to provide electricity for the motors driving the propellers on an airplane.

In Unit 2, we learned that the power plants that produce electricity from fossil fuels are approximately 33% efficient overall in converting the primary energy sources into electricity. If we use fossil fuels to produce electricity, then electric vehicles have a significant carbon emission because of the carbon dioxide produced in generating the electricity. In order to reduce our carbon emissions, we will need to convert vehicles to electric power and produce electricity from sources other than fossil fuels.

Most vehicles spend their life operating at cruising or operating speed. The methods taught in Unit 1 work well for estimating the energy consumption and fuel economy for large trucks, trains, ships, airplanes, and jets. Cars and light trucks spend a large percentage of their time in city traffic, and city driving has a large impact on their energy consumption and fuel economy. The thermal efficiency of the engine in city driving is lower than for highway driving. Electric-powered and hybrid-electric-powered cars and light trucks can be much more efficient in city driving than those powered by internal combustion engines. In Unit 3, we will do detailed modeling of cars and light trucks in city driving and suburban driving and gain a better understanding of how it impacts the energy consumption and fuel economy of the vehicles.

Internal combustion (IC) engines operate over a range of revolutions per minute (RPMs) and torques. The engines can be tested, and a mapping of the engine performance as shown in _____ can be produced. The RPMs are on the x-axis and the torques are on the y-axis. The fuel consumption is measured in pound per horse-power-hour (lb/hp-h) in US customary units and in grams per kilowatt-hour (g/kW-h) in metric units. The highest thermal efficiency for this engine is the dark red region in the upper middle part of the graph. For the line encircling this region, the engine uses 0.411 lb of gasoline for each hp-h of energy produced by the engine. In the metric system, the engine uses 250 g of gasoline for each kW-h of energy produced.

The numbers on the graph can be used to obtain the thermal efficiency of the engine at different RPM and torque values. Gasoline has an energy density of 0.128 lb/hp-h or in the metric system 77.85 g/kW-h. Dividing by the numbers on the graph gives the thermal efficiency of the engine. For example, dividing 0.128 lb/hp-h by 0.411 lb/hp-h yields a thermal efficiency of 31.1%. When operating the engine on the line bordering the red region, the engine is operating at a thermal efficiency of 31.1%. This means that 31.1% of the

Figure 10.1 Engine map for Saturn 1.9 L engine.

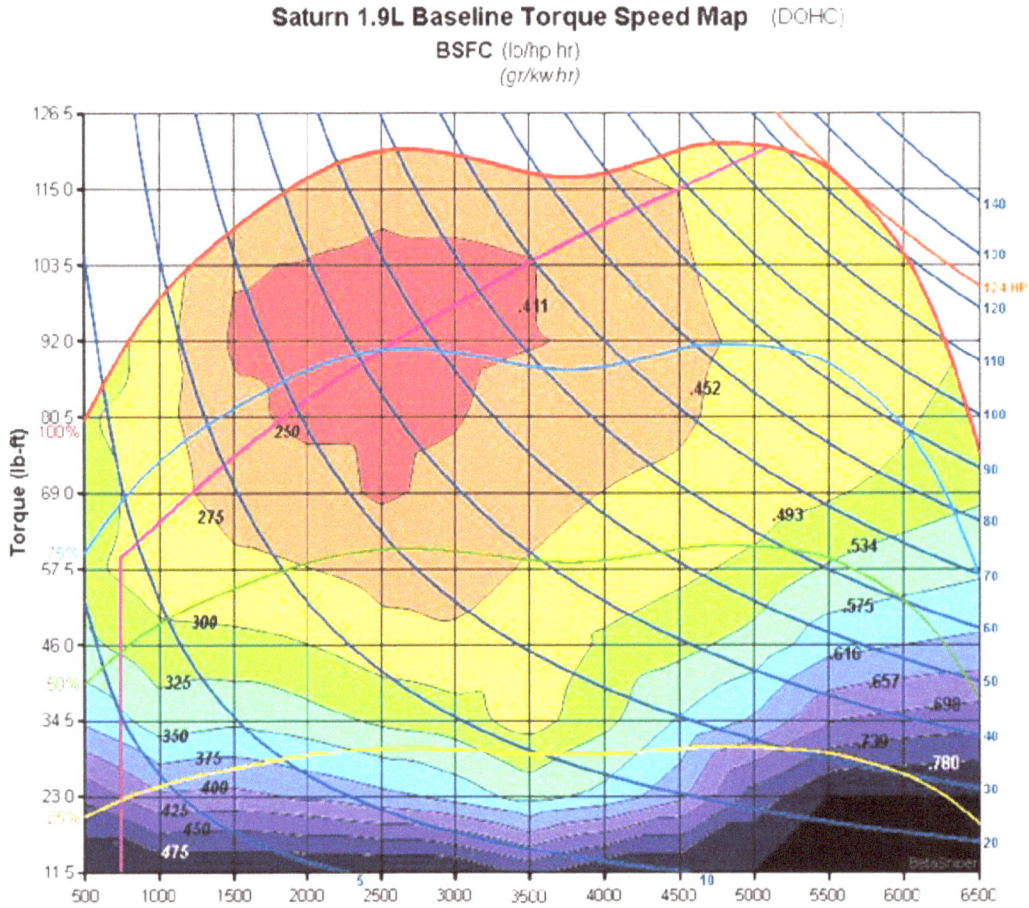

Saturn 1.9L Baseline Torque Speed Map (DOHC)
BSFC (lb/hp hr)
(gr/kw hr)

heat energy in the gasoline is being converted into useful mechanical energy and the other 68.9% is being converted into heat. In the metric system, divide 77.85 g/kW-h by 250 g/kW-h to get 31.1% thermal efficiency of the engine. Let FC be the fuel consumption in lb/hp-h or g/kW-h.

$$\text{Thermal Efficiency} = \frac{0.128}{FC} \qquad \left(\text{with FC in lb / hp-h}\right) \qquad (10.1)$$

$$\text{Thermal Efficiency} = \frac{77.85}{FC} \qquad \left(\text{with FC in g / kW-h}\right) \qquad (10.2)$$

Example 10.1

Assume the engine is producing 23.0 ft-lb of torque at 1500 RPM and calculate the thermal efficiency of the engine.

Solution: From Figure 10.1, locate the point 1500 RPM and 23.0 ft-lb on the graph. The fuel consumption is between the 400 and 425 curves; a little closer to the 400 curve. Estimate the fuel consumption (FC) to be 410 g/kW-h when operating at 23.0 ft-lb torque and 1500 RPM and use Equation (10.2). Thermal efficiency is 77.85/410 = 19.0%. In city driving, the engine will spend a lot of time operating at low torque and RPM and the thermal efficiency of the engine will be lower than when the vehicle is on the highway.

The blue lines in Figure 10.1 are lines of constant horsepower, with the numbers on the right side of the graph. For example, the engine can produce 20 hp by generating 34.5 ft-lb of torque at 3045 RPM, or it can generate 69.0 ft-lb of torque at 1522 RPM. Engine power is defined as the torque multiplied by the angular velocity. Assume the torque is to be in ft-lb, the angular velocity in RPM, and the power in hp. The following equation is a convenient way to relate the three quantities:

$$\text{Power}\left(\text{hp}\right) = \frac{\left[\text{Torque}\left(\text{ft-lb}\right)\right]\left(\text{RPM}\right)}{5252} \tag{10.3}$$

There is a pink line on the graph that shows how to get maximum efficiency out of the engine for the required power output. The pink line should be approximately perpendicular to the horsepower curves through the high-efficiency part of the mapping down to idle speed and then drops vertically. The transmission will be designed to follow this line; however, it will not be able to exactly follow the line, especially for low power requirements. The transmission operates most efficiently in high gear, so overall efficiency will be a function of engine and transmission efficiency.

Three-speed transmissions were common for many years. As fuel economy and efficiency became more important more gears were added to the transmission. Having more gears allows the transmission access to the required power and has the engine operate at a higher thermal efficiency. Having the engine operate at a higher thermal efficiency will cause the vehicle to get better fuel economy. The newest cars have a continuous variable transmission (CVT). The electronics and software controlling the CVT will choose a gear ratio that allows the engine to operate at maximum thermal efficiency for the amount of horsepower it needs to deliver. For low power requirements, the CVT will push the engine to low RPMs to get the best thermal efficiency possible.

Example 10.2

Assume that the Saturn vehicle has a weight of 3000 lb, a rolling resistance coefficient of 0.010, a frontal area of 1.5 m², and a drag coefficient of 0.31. Assume level ground and the vehicle traveling at 40 mph. What is the ideal RPM for the engine to maximize the thermal efficiency of the engine? (Assume the transmission would be designed to shift and allow the engine to operate at maximum thermal efficiency for the required power.) Calculate the thermal efficiency of the engine using Figure 10.1. Assume that the transmission and final drive for the vehicle are 90% efficient in getting the engine power to the wheels and calculate the fuel economy of the vehicle in mpg.

Solution: When cruising at a steady speed on level pavement, the car must overcome rolling resistance and aerodynamic drag. From Unit 1, the power in hp is calculated as:

$$\text{Power (hp)} = \frac{(0.002667)\text{W V Crr}\left(1+\frac{V}{200}\right)+(7.184\times10^{-5})V^3\text{A Cd}}{\text{Transmission / Drive Efficiency}} \tag{10.4}$$

Plugging the given values into the equation yields:

$$\text{Power} = \frac{(0.002667)(3000)(40)(0.010)\left(1+\frac{40}{200}\right)+(7.184\times10^{-5})40^3(1.5)(0.31)}{0.90} = 6.642\text{ hp} \tag{10.5}$$

The engine must provide 6.642 hp to push the car along at 40 mph on level pavement. According to Figure 10.1, at 6.642 hp the engine will operate most efficiently at 750 RPM, because that is where the 6.642 hp line would cross the pink line on the graph. The torque requirement is calculated as:

$$\text{Torque (ft-lb)} = \frac{(6.642\text{ hp})(5252)}{750\text{ RPM}} = 46.5\text{ ft-lb} \tag{10.6}$$

The point for 750 RPM and 46.51 ft-lb is in the green region in Figure 10.1 between the 300 and 325 lines, but closer to the 325 line. Estimate the value to be 320 g/kW-h.

$$\text{Thermal Efficiency} = \frac{77.85}{320} = 24.3\% \tag{10.7}$$

This is the thermal efficiency of the engine when delivering 6.642 hp at 750 RPM and 46.51 ft-lb torque. Gasohol will have an energy density of about 121,000 BTU/gal, and 23.4% of the energy in the fuel will be converted to useful mechanical energy to power the car. The rate of fuel consumption, or fuel power, is calculated as:

$$\text{Fuel Power} = \frac{6.642\text{ hp}}{0.243} = 27.33\text{ hp} \tag{10.8}$$

For unit conversions, 1 hp is equal to 2544 BTU/h. Putting this together:

$$\left(27.33\,\text{hp}\right)\left(\frac{2544\,\text{BTU}\,/\,\text{h}}{\text{hp}}\right)\left(\frac{\text{gal}}{121,000\,\text{BTU}}\right) = 0.575\,\text{gph} \tag{10.9}$$

Driving on level ground at 40 mph, the car would use 0.575 gal of gas each hour. The fuel economy for the car is the speed divided by the gph fuel consumption.

$$\text{Fuel Economy} = \frac{40\,\text{mph}}{0.575\,\text{gph}} = 69.6\,\text{mpg} \tag{10.10}$$

This seems like a very high fuel economy for the car. The Environmental Protection Agency (EPA) rating is less than 69.6 mpg for city or highway driving. But if the car were operated at 750 RPM and 40 mph steady speed, it would get close to the 69.6 mpg calculated. GM did not program the transmission to have the engine operate at 750 RPM when providing 6.642 hp. Having the engine operate at 750 RPM is the maximum thermal efficiency for the engine and provides the best fuel economy, but the car would be unpleasant to drive if the engine were only turning at 750 RPM.

Example 10.3

Operating the car at 750 RPM will make the car feel sluggish if you want to accelerate or if you start up a slight incline in the road. Looking at Figure 10.1, the maximum possible horsepower for the engine at 750 RPM is about 10 hp. When you mash the gas pedal, you want the engine to produce more power so you can speed up, but the engine cannot produce much more power at 750 RPM. You will feel like you are driving a very underpowered vehicle, and most people would be very dissatisfied with the car. The car would be very fuel-efficient, but not many people would buy the car because it is unpleasant to drive.

Solution: To alleviate the sluggishness, assume that the transmission causes the engine to operate at 1500 RPM and calculate the fuel economy for steady cruising at 40 mph on level pavement. At 1500 RPM, the engine could produce a little over 30 hp and that will make the car accelerate better and be more pleasant to drive.

The power the engine must produce to push the car along at 40 mph is the same 6.642 hp regardless of the engine RPM. The torque for 1500 RPM is calculated as:

$$\text{Torque}\left(\text{ft-lb}\right) = \frac{\left(6.642\,\text{hp}\right)\left(5252\right)}{1500\,\text{RPM}} = 23.26\,\text{ft-lb} \tag{10.11}$$

Locating the point 1500 RPM and 23.26 ft-lb torque in Figure 10.1 puts us in the lavender region between the 400 and 425 lines, but closer to the 400 line. Assume that the fuel consumption is 410 g/kW-h. The thermal efficiency is calculated as:

$$\text{Thermal Efficiency} = \frac{77.85}{410} = 19.0\% \tag{10.12}$$

Notice that the thermal efficiency of the engine is reduced from 24.3% at 750 RPM to 19.0% at 1500 RPM. This increases the amount of fuel that must be consumed to generate the 6.642 hp required to push the car along at 40 mph.

$$\left(\frac{6.642 \text{ hp}}{0.190}\right)\left(\frac{2544 \text{ BTU}/\text{h}}{\text{hp}}\right)\left(\frac{\text{gal}}{121{,}000 \text{ BTU}}\right) = 0.735 \text{ gph} \tag{10.13}$$

$$\text{Fuel Economy} = \frac{40 \text{ mph}}{0.735 \text{ gph}} = 54.4 \text{ mpg} \tag{10.14}$$

This is still a good fuel economy, and the car will be much more pleasant to drive if it is geared like this. If you were to take the actual production car and put it on a level track driving at 40 mpg, it would deliver a fuel economy near the 54.4 mpg calculated.

Example 10.4

Rework the problem for highway mileage assuming the car is traveling at 70 mph. What is the ideal RPM for the engine and what is the fuel economy?

Solution: The power required to push the car along at 70 mph on the level pavement is calculated as:

$$\text{Power} = \frac{(0.002667)(3000)(70)(0.010)\left(1+\frac{70}{200}\right)+(7.184\times10^{-5})70^3(1.5)(0.31)}{0.90} = 21.13 \text{ hp} \tag{10.15}$$

The 21.13 hp curve crosses the pink line on the graph at about 1550 RPM. The torque required to produce 22.24 hp at 1550 RPM is 71.6 ft-lb. This falls in the orange region on the graph between the 250 and 275 lines, but closer to the 250 line. Assume the fuel consumption is 260 g/kW-h. This yields a thermal efficiency of 77.85/260 = 29.9%, fuel consumption of 1.484 gph, and fuel economy of 47.2 mpg. If the transmission were geared so that the engine operated at 1550 RPM at 70 mph, the car would get approximately 47.2 mpg fuel economy. But this will make the car feel a little sluggish on the highway if you want to pass someone. To make the car more pleasant to drive, the transmission will be geared to a higher RPM on the highway.

Example 10.5

To balance drivability with fuel economy, the manufacturer (GM/Saturn) chose to have the engine turn 2500 RPM at 70 mph. Calculate the fuel economy for the car traveling at 70 mph and the engine turning 2500 RPM.

Solution: The power required is 21.13 hp for driving at 70 mph. If the engine is operating at 2500 RPM, Equation (10.3) is used to calculate the torque to be 44.39 ft-lb. This is in the yellow region of the graph approximately halfway between 275 and 300 lines. Assume a fuel consumption of 287 g/kW-h. The thermal efficiency is 77.85/287 = 27.12%. Fuel consumption is 1.638 gph, and fuel economy is 42.7 mpg.

The EPA fuel economy estimate for this vehicle was 40 mpg for highway driving. The purpose of the four examples was to illustrate how to use the engine map (Figure 10.1) to calculate the fuel economy of the vehicle. In city driving, the car will accelerate and decelerate, and the engine will operate over a range of RPMs and torques. To calculate fuel economy for city driving, we calculate the instantaneous fuel economy and then numerically integrate over a typical city driving pattern. The approach illustrated in the four examples above will be used later in the unit to calculate the fuel economy for city driving. Before we get to the detailed modeling, there are a few more important topics to be discussed.

The size of the engine is an important factor in the fuel economy of the vehicle. In general, putting a larger engine in a vehicle will yield a lower fuel economy. It is important to understand why this is true. Steady speed driving, even at highway speeds, does not use enough horsepower for us to get into the highest efficiency range for the vehicle. If the engine were used to power a generator, we could gear it to operate at the highest efficiency, but for a car, this is not possible.

Example 10.6

What if we use a smaller engine to power the car, so we can operate at a higher thermal efficiency when driving on the highway?

Solution: The short answer is that the car will get better fuel economy on the highway because the engine will operate at a higher thermal efficiency. Car performance would suffer, and people would not enjoy driving the car with the smaller engine, but the fuel economy would improve.

Examining the engine map (Figure 10.1) for the Saturn car, the maximum thermal efficiency is in the middle of the red range at about 2500 RPM and 92.0 ft-lb of torque. The maximum horsepower the engine can produce is 124 hp, but at maximum thermal efficiency, the engine is producing 43.8 hp.

$$\frac{(2500\,\text{RPM})(92.0\,\text{ft-lb})}{5252} = 43.8\,\text{hp} \qquad (10.16)$$

From previous examples, we calculated that it requires 21.13 hp to push the car along on level pavement at 70 mph. If we were to cut the engine size in half, the maximum rated horsepower would be 62 hp, and the engine would deliver maximum thermal efficiency at half of 43.8 hp, which is 21.4 hp. The conclusion is that using an engine half the size of the one Saturn put in the car would allow the car to operate at maximum thermal efficiency when driving on the highway. The smaller engine would operate at 2500 RPM and produce 44.39 ft-lb of torque. The fuel consumption when operating at that torque and RPM would be approximately 230 g/kW-h. The thermal efficiency of the engine would be 77.85/230 = 33.85%. Fuel consumption and economy are calculated as follows:

$$\left(\frac{21.13 \text{ hp}}{0.190}\right)\left(\frac{2544 \text{ BTU}/\text{h}}{\text{hp}}\right)\left(\frac{\text{gal}}{121,000 \text{ BTU}}\right) = 1.313 \text{ gph} \tag{10.17}$$

$$\text{Fuel Economy} = \frac{70 \text{ mph}}{1.313 \text{ gph}} = 53.3 \text{ mpg} \tag{10.18}$$

Using an engine half the size of the one that Saturn actually used would raise the fuel economy from 42.7 to 53.3 mpg, an approximate 25% improvement in fuel economy. This is a large improvement in fuel economy, but driving a car with an engine half the size will not be much fun. Part of the reason Saturn chose to put a 124 hp engine in the car instead of a 62 hp engine is to provide adequate acceleration of the vehicle.

Acceleration of the car is important. The engine must have enough horsepower to accelerate the car properly around town and on the highway. Testing indicates that around town people expect to be able to accelerate at about 7 ft/s². Sports cars will be able to accelerate faster than this, but for most people, an acceleration of 7 ft/s² will feel solid and allow the car to keep up in traffic. Car acceleration is commonly measured as the time it takes to accelerate from 0 to 60 mph. Since 60 mph is equal to 88 ft/s, the time for a 7 ft/s² acceleration is calculated as:

$$0-60 \text{ mph time} = \frac{88 \text{ ft}/\text{s}}{7 \text{ ft}/\text{s}^2} = 12.6 \text{ s} \tag{10.19}$$

Sports cars will accelerate 0 to 60 mph in less than half this amount of time, which means they can provide at least 14 ft/s² acceleration. If you mash the gas pedal to the floor, the Saturn will be able to accelerate faster than 7 ft/s², but for normal driving, people do not mash the pedal to the floor. A 7 ft/s² acceleration is acceptable for normal city driving. We sometimes measure acceleration in g's, and the acceleration of gravity is 32.2 ft/s². Another way to look at the acceleration is:

$$\text{Acceleration} = \frac{7 \text{ ft}/\text{s}^2}{32.2 \text{ ft}/\text{s}^2} = 0.22 \text{ g acceleration} \tag{10.20}$$

The engine must be able to provide the horse-power required for acceleration. The weight of the Saturn car we have been discussing is 3000 lb. The mass of the car is 3000 lb/32.2 ft/s² = 93.2 slug. The force required to accelerate at 7 ft/s² is mass multiplied by acceleration:

$$\text{Force} = (93.2 \text{ slug})(7 \text{ ft}/\text{s}^2) = 652.2 \text{ lb} \tag{10.21}$$

The engine power required for acceleration is force multiplied by speed. One hp is 550 ft-lb/s, so the power requirements at 15, 30, and 60 mph can be calculated as:

(a) 15 mph = 22 ft/s, Power = (652.2)(22)/(550) = 26.1 hp

(b) 30 mph = 44 ft/s, Power = (652.2)(44)/(550) = 52.2 hp

(c) 60 mph = 88 ft/s, Power = (652.2)(88)/(550) = 104.3 hp

The calculations illustrate that the engine horsepower must increase as speed increases to provide the same acceleration. Accelerating the car at highway speed requires a lot more horse-power than accelerating in city traffic. Most cars do not have enough horsepower to accelerate at 7 ft/s^2 when traveling at highway speed. Drivers do not expect the car to be able to accelerate at 7 ft/s^2 on the highway. An acceleration of 4 ft/s^2 is acceptable on the highway.

The calculations above are the horsepower required for acceleration. The power to overcome rolling and aerodynamic drag must be added and the efficiency of the drive system must also be factored in. The engine will need to produce more horsepower than the numbers above. For a solid acceleration of 7 ft/s^2, most of the engine horsepower goes to provide the acceleration, but a significant percentage goes for rolling and aerodynamic drag and inefficiencies in the drive system. The hp calculations in (a), (b), and (c) above are reasonable, but low estimates of the engine horsepower are required.

Around town, the 62 hp engine will have enough horsepower to accelerate the car up to approxi-mately 25 mph, and then the car will feel sluggish and unable to keep up in traffic. The driver will mash the pedal to the floor all the time trying to keep up in traffic. After a test-drive, not many people will be willing to buy the car. The 124 hp engine that Saturn selected is adequate to provide acceleration and allow the car to keep up in traffic. It is not a racecar, but it is adequate.

If the car were fitted with a larger engine, the acceleration would improve, but acceleration is also limited by the friction between the tires and the pavement. If the engine is powerful enough to spin the tires, then acceleration is limited by friction rather than by engine horsepower. As the car accelerates, the weight shifts to the rear tires to some extent. The friction force is equal to the weight of the tires multiplied by the coefficient of friction. Since there is more weight on the rear tires when accelerating, rear-wheel-drive cars can accelerate faster than front-wheel-drive cars. All the really fast cars are rear-wheel drive, including NASCAR, Indy Car, Dragsters, Corvette, Camaro, Mustang, Charger, Porsche, Ferrari, etc. When designing for high acceler-ation, rear-wheel drive is the best choice.

In theory, four-wheel-drive cars should be able to provide more acceleration than rear-wheel-drive cars, but there are control issues associated with dividing the power between the front and rear wheels and not causing the front wheels to slip. For traditional IC engine cars, the control issues have been difficult to solve, and from a practical standpoint, all the fastest IC engine cars have been rear-wheel drive. Some of the newer electric cars are four-wheel drive, and with electric cars, companies have been able to solve the control issues. Some of the new electric cars have 0 to 60 mph times of less than 5 s, which is an average acceleration of 17.6 ft/s^2. The focus of this book is on energy efficiency, but at this point, it is important to do the dynamic analysis associated with the acceleration of a car.

Example 10.7

Assume a 3000 lb car with the center of gravity (CG) 2 ft above the pavement and centered between the front and rear axle as illustrated in Figure 10.2. Assume that the car is to accelerate at 13 ft/s². Find the reactions on the front and rear (N_F and N_R) and the traction force T required to provide the acceleration.

$$\Sigma F_y = 0 = 1211\,\text{lb} - T \tag{10.22}$$

$$\Sigma F_y = 0 = N_F + N_R - 3000\,\text{lb} \tag{10.23}$$

$$\Sigma M_{N_F} = 0 = N_R\left(10\right) - 3000\left(5\right) - 1211\left(2\right) \tag{10.24}$$

Solution: Solving the equations, the traction force T = 1211 lb, N_R = 1742 lb, and N_F = 1258 lb. Front-wheel-drive cars have the traction force on the front tires, and rear-wheel-drive cars have the traction force on the rear wheels. The minimum coefficient of friction required to prevent slipping is equal to the traction force divided by the normal force.

$$\text{Front-Drive Minimum Coefficient of Friction} = \frac{1211}{1258} = 0.963 \tag{10.25}$$

$$\text{Rear-Drive Minimum Coefficient of Friction} = \frac{1211}{1742} = 0.695 \tag{10.26}$$

A conservative estimate for the coefficient of friction between tires and the dry pavement is 0.80. From the calculations above, the rear-drive car would be able to achieve an acceleration of 13 ft/s² without slipping the tires because the required coefficient of friction is 0.695 and the actual coefficient of friction is 0.80. A front-wheel-drive car would not be able to achieve an acceleration of 13 ft/s² because the required coefficient of friction is 0.963 and the actual coefficient of friction is 0.80. The front-wheel-drive car can only achieve an acceleration of about 11 ft/s² before the front tires begin to slip. The rear-wheel-drive car can achieve an acceleration of about 15 ft/s² before the rear wheels begin to slip.

Figure 10.2 Dynamics of acceleration for Example 10.7

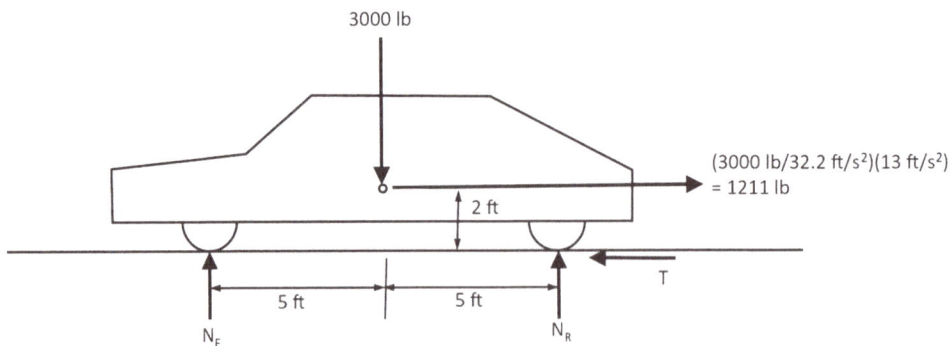

3000 lb

(3000 lb/32.2 ft/s²)(13 ft/s²)
= 1211 lb

2 ft

T

5 ft   5 ft

N_F N_R

© SAE International

The problem above can be set up in more general terms as illustrated in Figure 10.3:

$$\text{Front-Drive Maximum Acceleration} = \frac{\mu b g}{(a + b + \mu c)} \qquad (10.27)$$

$$\text{Rear-Drive Maximum Acceleration} = \frac{\mu b g}{(a + b - \mu c)} \qquad (10.28)$$

where μ is the coefficient between the tires and pavement, which is typically assumed to be 0.80 for dry pavement, and g is the acceleration of gravity. If the values a = b = 5 ft, c = 2 ft are substituted into the equations above, the maximum acceleration for front-drive cars is 11.1 ft/s^2 and for rear-drive cars it is 15.3 ft/s^2. Equations (10.27) and (10.28) form a nice solution to the problem, but it is very difficult to get accurate values for the a, b, and c distances on production cars, and the 0.80 coefficient of friction between the tires and the pavement depends on the tires and the pavement. Sports cars will use tires that have a higher coefficient of friction than the tires used on typical sedans.

Figure 10.3 Dynamics of acceleration, general case.

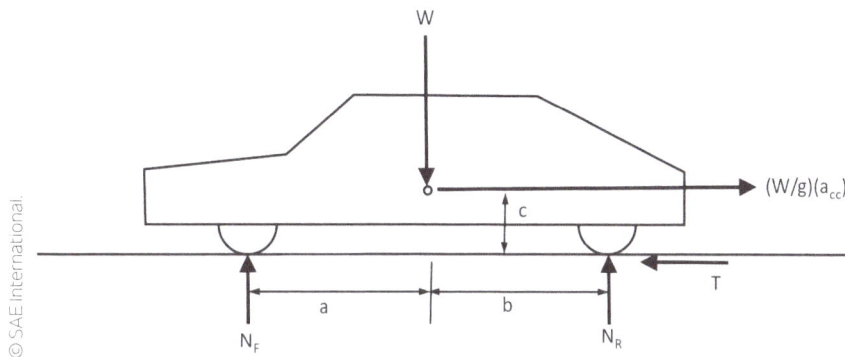

Since many people are familiar with the 0 to 60 mph acceleration times, it seems prudent at this point to calculate the 0 to 60 mph times for front-wheel-drive and rear-wheel-drive cars using the values above (60 mph = 88 ft/s).

$$\text{Front Drive}: \frac{88 \text{ ft / s}}{11.1 \text{ ft / s}^2} = 7.93 \text{ s} \qquad (10.29)$$

$$\text{Rear Drive}: \frac{88 \text{ ft / s}}{15.3 \text{ ft / s}^2} = 5.75 \text{ s} \qquad (10.30)$$

There are very few production cars that can accelerate from 0 to 60 mph in 5.75 s. There are, however, a few expensive sports cars that can accelerate 0 to 60 mph in less than 5.75 s because they have very powerful engines and tires that provide excellent traction. There are a few expensive front-drive cars that can come close to the 0 to 60 mph time of 7.93 s. Most rear-drive cars can accelerate at 15 ft/s² at low speeds and most front-drive cars can accelerate at 11 ft/s² at low speeds. As speed increases, the horsepower required to sustain the acceleration increases and most production cars do not have enough horsepower to sustain the acceleration up to 60 mph. The relatively inexpensive four-cylinder cars and small sport utility vehicles (SUVs) that many people drive can accelerate from 0 to 60 mph in about 12 s. That is an acceptable acceleration for most people.

To model the acceleration of a car, we need to consider that the acceleration is limited by the traction between the tires and the pavement and by the horsepower of the engine. For lower speeds, the engine will have enough horsepower to slip the tires and traction will be what limits the acceleration of the car. For this unit, we will assume front-wheel-drive cars are limited to 11 ft/s² acceleration and rear-wheel-drive cars are limited to 15 ft/s² acceleration. Actual acceleration limits vary from car to car as illustrated in Figure 10.3 and Equations (10.27) and (10.28). Using limits of 11 and 15 ft/s² will be close to the correct values and will allow us to talk about the differences in front- and rear-drive cars.

As the car speeds up, there comes a point where the engine cannot provide enough horsepower to slip the tires, and at that point, it is the horsepower of the engine that limits the acceleration of the car. From a practical standpoint, most people will not push the engine to the maximum possible horsepower. For example, the Saturn engine is rated at 124 hp, but to get 124 hp from the engine, the engine would need to be turning 5700 RPM, and most people are not going to push the engine that hard. At 4500 RPM, the engine will produce 90 hp, which is about 75% of the rated horsepower. From a practical standpoint, most people will never have the engine producing more than 75% of the rated horsepower.

We will use Excel to develop a model for the acceleration of the car. In the model, there will be a logic statement such that:

(a) Acceleration cannot be more than 11 ft/s² for front-wheel-drive or 15 ft/s² for rear-wheel-drive cars. This is the traction limit for acceleration.

(b) Acceleration cannot be more than $(0.75\,P_{max})/(mass*speed)$. P_{max} is the maximum rated horsepower for the engine, and the driver will not use more than 75% of the rated horsepower. This is the power limitation for acceleration.

Example 10.8

Model acceleration of the Saturn car starting from rest for 20 s of acceleration. Assume the weight is W = 3000 lb, rolling coefficient is Crr = 0.010, drag area is A = 1.5 m², drag coefficient is Cd = 0.31, maximum rated engine power is P_{max} = 62 hp, maximum traction acceleration is 11 ft/s² (front-wheel drive), and transmission/drive efficiency is 0.88.

Solution: In a previous example, we discovered that using a small 62 hp engine in the Saturn car would allow it to achieve about 53.3 mpg fuel economy on the highway, which is significantly better than the 42.7 mpg highway fuel economy the car achieves with the 124 hp engine that came in the car. The purpose of this example is to illustrate quantitatively the sluggish performance with the small engine and help you understand why we would not want to put such a small engine in the car. Table 10.1 illustrates the acceleration spreadsheet model for the Saturn car for the first 2.6 s.

Table 10.1 Acceleration of Saturn car.

Weight	3000	lb		Rated power	62	hp
Crr	0.01			Max acceleration	11	ft/s^2
A	1.5	m^2		75% power	46.5	hp
Cd	0.31			Transmission efficiency	0.88	

Time (s)	Speed (ft/s)	Speed (mph)	Rolling and aerodynamic Power (hp)	Available acceleration Power (hp)	Max acceleration (ft/s^2)	Cleaned (ft/s^2)
0	0.000	0.000	0.000	40.920	Infinity	11.000
0.1	1.100	0.750	0.060	40.860	219.281	11.000
0.2	2.200	1.500	0.121	40.799	109.477	11.000
0.3	3.300	2.250	0.182	40.738	72.875	11.000
0.4	4.400	3.000	0.245	40.675	54.573	11.000
0.5	5.500	3.750	0.307	40.613	43.591	11.000
0.6	6.600	4.500	0.371	40.549	36.269	11.000
0.7	7.700	5.250	0.436	40.484	31.038	11.000
0.8	8.800	6.000	0.502	40.418	27.114	11.000
0.9	9.900	6.750	0.569	40.351	24.061	11.000
1.0	11.000	7.500	0.637	40.283	21.619	11.000
1.1	12.100	8.250	0.706	40.214	19.620	11.000
1.2	13.200	9.000	0.777	40.143	17.953	11.000
1.3	14.300	9.750	0.849	40.071	16.542	11.000
1.4	15.400	10.500	0.923	39.997	15.332	11.000
1.5	16.500	11.250	0.998	39.922	14.283	11.000
1.6	17.600	12.000	1.075	39.845	13.365	11.000
1.7	18.700	12.750	1.154	39.766	12.553	11.000
1.8	19.800	13.500	1.235	39.685	11.832	11.000
1.9	20.900	14.250	1.318	39.602	11.186	11.000
2	22.000	15.000	1.403	39.517	10.604	10.604
2.1	23.060	15.723	1.487	39.433	10.095	10.095
2.2	24.070	16.411	1.568	39.352	9.651	9.651
2.3	25.035	17.069	1.648	39.272	9.260	9.260
2.4	25.961	17.701	1.727	39.193	8.912	8.912
2.5	26.852	18.308	1.804	39.116	8.599	8.599
2.6	27.712	18.895	1.880	39.040	8.316	8.316

In the first row, we start with time and speed equal to zero. Since the speed is zero, the rolling and aerodynamic power are both zero too. The maximum power that can be delivered to the wheels is 75% of the maximum rated engine power multiplied by the transmission/drive efficiency. The equation for maximum power acceleration has the speed term in the denominator, so for zero speed, the equation yields infinity. The logic statement makes the maximum acceleration equal to the traction acceleration, which in this case is 11 ft/s².

(a) In this model, the time step was chosen as 0.1 s. The column for the time was incremented at 0.1 to 20 s to model the first 20 s of acceleration.

(b) In the second column for speed, the speed is equal to the initial speed at the beginning of the time step plus the acceleration multiplied by the time step. For the second row, the speed is 0 ft/s plus (11 ft/s²) (0.1 s) = 1.1 ft/s. For the third row, the speed is 1.1 ft/s plus (11) (0.1) = 2.2 ft/s.

(c) In the third column, the speed is converted to mph because we think of speed in terms of mph and it will make more sense when we graph it. We also need the speed in mph to be able to use the equations for rolling and aerodynamic drag. To convert speed in ft/s to mph we multiply by (3600/5280).

(d) In the fourth column, the rolling and aerodynamic drag are calculated using equations from Unit 1. Rolling + Aerodynamic = (0.002667) WVCrr (1 + V/200) + (7.184 × 10⁻⁵) V³ACd. Rolling + Aerodynamic is small and insignificant at low speed as illustrated by the numbers in the table, but as the car speeds up, it becomes more significant.

(e) In the fifth column, the available power for acceleration is the 40.92 hp the engine can deliver to the wheels minus the power to overcome rolling and aerodynamic drag.

(f) In the sixth column, the maximum power acceleration is calculated as: Maximum Power Acceleration = (available power) (550)/ (mass*speed). The 550 term converts horsepower to ft-lb/s and the speed should be in ft/s. At low speeds, the engine has plenty of power to spin the wheels, but as the car speeds up, the maximum power acceleration will become less than 11 ft/s² and will be what limits the acceleration of the car.

(g) The seventh column is a logic statement that chooses the smaller value of the maximum power acceleration in column 6 and the maximum traction acceleration of 11 ft/s². The acceleration will be 11 ft/s² until the car reaches a speed where the engine no longer has enough horsepower to continue accelerating at 11 ft/s², and then the acceleration will decrease. For this example, the transition from traction-limited acceleration to hp-limited acceleration happens at 2.0 s.

Once the first three lines of the Excel program are developed, it is possible to select the third line and copy it to the bottom of the spreadsheet at time = 20 s. There are details of indexing in Excel, and it is assumed that the reader has a good understanding of how to use Excel. The results for the 62 hp engine are illustrated in Figure 10.4.

Figure 10.4 Acceleration with 62 hp engine.

Speed vs Time

The graph in Figure 10.4 shows that it will take 20 s for the car to accelerate to 60 mph, that is, the 0 to 60 mph time is 20 s. This will seem very sluggish and underpowered compared to the cars we drive. Looking at the numbers in the spreadsheet at a time of 2 s the engine no longer has enough horsepower to accelerate at 11 ft/s^2. As the car speeds up, the power to overcome rolling and aerodynamic drag increases, which reduces the power available for acceleration and the acceleration of the car. The maximum top speed for this vehicle would be when the rolling and aerodynamic power totals to the 40.92 hp the engine can provide to the wheels. In this case,

the maximum top speed is 96 mph, and it takes a long time to get to that speed.

The car comes with a 124 hp engine, and if the parameters in the spreadsheet model are changed, the car can accelerate to 60 mph in 10.7 s, which is typical of the cars and trucks that we drive. If the spreadsheet was developed correctly, it is only necessary to change the rated power to 124 hp and the results should be recalculated. The 124 hp engine has enough hp to accelerate the car at 11 ft/s^2 for 3.8 s, and then acceleration is limited by engine hp. The results are illustrated in Figure 10.5.

Figure 10.5 Acceleration with 124 hp engine.

Speed vs Time

Hill Climbing: Another aspect of normal driving is that the car will need to have adequate power to drive uphill. For hill climbing, the power to overcome gravity is equal to the gravitational force pushing the car downhill multiplied by the vertical speed component of the car (Figure 10.6).

If the car has a weight W and is traveling at speed V up a hill of angled slope θ, the power to overcome gravity is:

$$\text{Gravitational Power} = W\, V \sin(\theta) \tag{10.31}$$

For highways, we normally measure the slope in grade such that the grade is the sine of the angle. For example, a 6% grade means the sine of the angle θ is 0.06. Most interstate highways limit the grade to 4%, though in mountainous areas the grades may be as high as 7%. Local roads may have grades as high as 11%, though it is rare. An 11% grade is very steep.

Figure 10.6 Car traveling uphill.

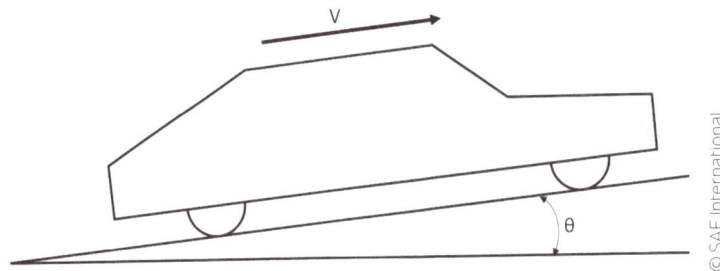

Example 10.9

Calculate the horsepower required to overcome gravity for a 5000 lb car traveling 70 mph up a 4% grade (70 mph = 102.67 ft/s).

Solution: Use Equation (10.31).

$$\text{Gravitational Power} = \frac{(5000\ \text{lb})(102.67\ \text{ft-lb / s})(0.04)}{\dfrac{550\ \text{ft-lb / s}}{\text{hp}}} = 37.33\ \text{hp} \tag{10.32}$$

Example 10.10

Assume that the Saturn car is accelerating uphill. Modify the spreadsheet (Table 10.1) to account for the gravity loading in addition to the aerodynamic and rolling power.

Solution: The spreadsheet was set up so that the grade of the hill could be adjusted. An additional column was added for the hill-climbing power, which was subtracted to reduce the available power for acceleration. A 4% grade was used for the example below with the small 62 hp engine (Table 10.2).

Table 10.2 Acceleration of Saturn car uphill.

Weight	3000	lb			Rated power	62	hp
Crr	0.01				Max acceleration	11	ft/s²
A	1.5	m²	Grade		75% power	46.5	hp
Cd	0.31		4	%	Transmission efficiency	0.88	

Time (s)	Speed (ft/s)	Speed (mph)	Rolling and aerodynamic Power (hp)	Hill climbing Power (hp)	Available acceleration Power (hp)	Max acceleration (ft/s²)	Cleaned (ft/s²)
0	0.000	0.000	0.000	0.000	40.920	Infinite	11.000
0.1	1.100	0.750	0.060	0.240	40.620	217.993	11.000
0.2	2.200	1.500	0.121	0.480	40.319	108.189	11.000
0.3	3.300	2.250	0.182	0.720	40.018	71.587	11.000
0.4	4.400	3.000	0.245	0.960	39.715	53.285	11.000
0.5	5.500	3.750	0.307	1.200	39.413	42.303	11.000
0.6	6.600	4.500	0.371	1.440	39.109	34.981	11.000
0.7	7.700	5.250	0.436	1.680	38.804	29.750	11.000
0.8	8.800	6.000	0.502	1.920	38.498	25.826	11.000
0.9	9.900	6.750	0.569	2.160	38.191	22.773	11.000
1.0	11.000	7.500	0.637	2.400	37.883	20.331	11.000
1.1	12.100	8.250	0.706	2.640	37.574	18.332	11.000
1.2	13.200	9.000	0.777	2.880	37.263	16.665	11.000
1.3	14.300	9.750	0.849	3.120	36.951	15.254	11.000
1.4	15.400	10.500	0.923	3.360	36.637	14.044	11.000
1.5	16.500	11.250	0.998	3.600	36.322	12.995	11.000
1.6	17.600	12.000	1.075	3.840	36.005	12.077	11.000
1.7	18.700	12.750	1.154	4.080	35.686	11.265	11.000
1.8	19.800	13.500	1.235	4.320	35.365	10.544	10.544
1.9	20.854	14.219	1.315	4.550	35.055	9.923	9.923
2	21.847	14.895	1.391	4.767	34.762	9.393	9.393
2.1	22.786	15.536	1.465	4.972	34.484	8.934	8.934
2.2	23.679	16.145	1.537	5.166	34.217	8.530	8.530
2.3	24.532	16.727	1.607	5.353	33.961	8.172	8.172
2.4	25.350	17.284	1.675	5.531	33.714	7.851	7.851
2.5	26.135	17.819	1.742	5.702	33.476	7.562	7.562
2.6	26.891	18.335	1.807	5.867	33.246	7.298	7.298

Figure 10.7 is a graph showing the speed of the Saturn car for accelerating up a 4% grade for the first 20 s. Notice that it takes 20 s to accelerate to 50 mph. It would take a very long time for the car to accelerate to highway speed if the ramp for entering the highway had a 4% uphill grade.

Figure 10.8 is the same graph as Figure 10.7 except the engine size was changed to 124 hp. With the larger engine, the car could accelerate up a 4% grade to highway speed in less than 20 s. This is much more reasonable performance for the car.

Figure 10.7 Uphill acceleration with 62 hp engine.

Figure 10.8 Uphill acceleration with 124 hp engine.

The examples used for the Saturn car comparing the 62 hp engine and the 124 hp engine that comes in the car are meant to help the reader understand why the small engine is inadequate. The smaller engine will deliver better fuel economy, but it will not be able to provide the acceleration and hill-climbing capability we expect from our cars and trucks. Very few people would buy such an underpowered car. The manufacturer must balance fuel economy with performance. We all know that in the abstract. The purpose of this unit is to teach you how to do the calculations and understand them quantitatively.

The next step is to look at the engine map for a typical diesel engine. The mapping in Figure 10.9 is for a Volkswagen 2.0 TDI diesel engine. The colored curves are of constant horsepower, and the numbers are in grams of fuel per kW-h

energy from the engine. Diesel fuel has an energy density of 69.5 g/kW-h. The maximum efficiency is at about 2250 RPM and 84 hp, where the engine is using 196 g/kW-h fuel consumption. The maximum thermal efficiency of the engine is calculated as $69.5/196 = 35.5\%$ thermal efficiency.

The y-axis in the engine map (Figure 10.9) is given in brake mean effective pressure (BMEP). Plotting BMEP on the y-axis normalizes the mapping with respect to the size of the engine. When torque is plotted on the y-axis, large engines will have much higher torque values than small engines, and it is harder to compare the efficiency of large and small engines. BMEP will be the same for large and small engines and makes it easier in some ways to compare large and small engines. Many engineers prefer BMEP plotted on the y-axis rather than torque.

Figure 10.9 Volkswagen 2.0 TDI diesel engine map.

BMEP can be converted to torque values if you know the displacement volume (DV) of the engine and whether it is a two-cycle or four-cycle engine. All cars and trucks use four-cycle engines, so Equation (10.33) is developed assuming a four-cycle engine. The 4π factor in the equation would be 2π if it were a two-cycle engine.

$$\text{Engine Torque} = \frac{(DV)(BMEP)}{4\pi} \quad (\text{assuming four-cycle engine}) \qquad (10.33)$$

A bar is defined as $100\,\text{kPa}$. In the engine map (Figure 10.9), the torque of the engine in ft-lb is equal to the bar reading on the y-axis multiplied by 11.74. A 2.0-L engine has a DV of $0.002\,\text{m}^3$.

$$\frac{\left(0.002\,\text{m}^3\right)\left(100,000\,\text{Pa / bar}\right)\left(0.73756\,\text{ft-lb / N-m}\right)}{4\pi} = \frac{11.74\,\text{ft-lb}}{\text{bar}} \qquad (10.34)$$

At points where the y-axis has a reading of 10 bar, the engine is producing 117.4 ft-lb of torque. The conversion factor will be different for different sizes of engines. For example, a 4.0 L engine will have a conversion factor twice that shown in Equation (10.34). This illustrates how the BMEP approach can be used to compare engines of different sizes on the same graph. BMEP is used for gasoline engines too. It is not just for diesel engines.

Students sometimes ask why we do not use two-cycle engines in cars and trucks. The advantage of two-cycle engines is that they produce more horsepower for their weight than four-cycle engines. Weed eaters, chain saws, and blowers used by lawn crews use two-cycle engines because it makes them lighter in weight and weight is very important in those applications. Two-cycle engines inherently have a lower thermal efficiency than four-cycle engines. A car with a two-cycle engine would have poor fuel economy. From an emissions standpoint, two-cycle engines are filthy compared to four-cycle engines. If all cars used two-cycle engines, the air would be unbreathable. The four-cycle engine is much cleaner, more efficient, less expensive to operate, and better in every way except that it is heavier than the two-cycle engine.

Every engine will have a different mapping. The engineers who work out the details of the transmission design will fit the shift points of the transmission to optimize the efficiency of the engine, with a compromise to make the car drivable. For drivability we want the transmission to shift and allow the engine to make horsepower when we press down on the gas pedal. It is hard to balance thermal efficiency and drivability. Optimizing too much for thermal efficiency can make the car seem sluggish to drive.

Homework

1. If 77.85 g of gasoline fuel provides 1 kW-h of heat energy, what is the maximum thermal efficiency illustrated in the engine map (Figure 10.6)? (Answer: 31.8%)

2. Reading from the engine map (Figure 10.10):

   (a) Assume that the engine provides 48 hp at 1200 RPM. What is the thermal efficiency? (Answer: 29.9–30.3%)

Figure 10.10 Engine map homework #1.

(b) Assume that the engine provides 48 hp at 2200 RPM. What is the thermal efficiency? (Answer: 25.5–26.4%)

Please remember to calculate the torque so you can locate the (RPM, torque) point on the graph. Use a ruler to help you locate the points on the graph. Each printer may scale the graph a little differently, and I will not know exactly how the graph will be scaled in the book. For my printer 388.5 ft-lb corresponded to 84.5 mm on the y-axis. Then knowing the ft-lb torque (T), the y-distance on the graph is calculated as y = (T*84.5/388.5). For the x-axis, 1000 RPM corresponded to 0 and 4500 RPM corresponded to 145 mm. Then knowing the RPM, the x-distance as x = (RPM – 1000)*145/3500. Following this process allows for a little better accuracy using the graph.

3. Assume that the engine powers an older large SUV that has a weight of 5400 lb, a rolling resistance coefficient of 0.012, a frontal area of 3.2 m², and a drag coefficient of 0.48. Assume that the transmission and final drive for the SUV are 88% efficient in getting the engine power to the wheels.

Assume that the vehicle is traveling on level ground at 45 mph. What is the ideal RPM for the engine to maximize fuel economy? Assume that the minimum RPM for the engine is 1000 RPM. Use the engine map to get the thermal efficiency of the engine and calculate the fuel economy of the SUV in mpg. (Answer: 25.9–26.9 mpg)

4. Operating the car at 1000 RPM will make the SUV feel sluggish. Suppose that the transmission causes the SUV in problem 3 to operate at 1500 RPM. Calculate the fuel economy for steady cruising at 45 mph and 1500 RPM. (Answer: 21.8–22.5 mpg)

5. Assume that the SUV in problems 3 and 4 is geared to operate at 2200 RPM at a highway speed of 70 mph. Calculate the fuel economy of the SUV in mpg on the highway. (Answer: 15.3–15.4 mpg)

6. For this problem, please neglect the rolling and aerodynamic power and calculate only the power required for acceleration. Assume that the weight of the vehicle is 4500 lb.

  (a) Assume that the vehicle is to accelerate from 0 to 30 mph at an acceleration rate of 12 ft/s². What is the maximum horsepower (at 30 mph) required for the vehicle?

  (b) Assume that the vehicle is to accelerate from 60 to 80 mph at an acceleration rate of 6 ft/s² to pass another vehicle on the highway. What is the maximum horsepower required (at 80 mph) for this acceleration? (Answer: 178.9 hp)

7. In this problem, you are to develop an acceleration spreadsheet like the one developed in Table 10.1 for the acceleration of the older large SUV in problems 3, 4, and 5. Assume that the acceleration is limited to 13 ft/s² at lower speeds and is limited by the horsepower of the engine at higher speeds. Assume that the engine will never provide more than 75% of the rated horsepower.

  (a) Develop a graph of the speed of the vehicle vs time for a 30-s acceleration. Assume that the rated horsepower for the engine is 175 hp.

  (b) What is the 0–60 mph time for the SUV? (Answer: 13.3 s)

  (c) What is the time required for the SUV to accelerate from 60 to 80 mph?

  (d) Develop a graph of the speed of the vehicle vs time for a 30-s acceleration. Assume that the rated horsepower for the engine is 275 hp.

  (e) What is the 0–60 mph time for the SUV with the 275 hp engine?

  (f) What is the time required for the SUV to accelerate from 60 to 80 mph with the 275 hp engine? (Answer: approximately 7.3 s)

Gasoline engines are typically 25–30% efficient in converting the chemical energy in gasoline into useful mechanical energy when driving on the highway. The thermal efficiency is zero when idling. At lower speeds, the thermal efficiency is less than that on the highway. The engine is most efficient when operating at a specific revolution per minute (RPM) and is usually geared so that it is most efficient in high gear at 70 mph, with consideration for acceleration performance.

The ideal RPM for the engine depends on the amount of power required to push the vehicle along. When the vehicle requires a lot of power (accelerating), it is more efficient to operate at a higher RPM than when it requires less power (cruising at low speed). Variable valve timing on modern engines has broadened the efficiency curve significantly, and this allows engines to operate more efficiently over a range of RPMs. Transmissions have several gears or are continuously variable. The electronics controlling the transmission will try to find the optimum gear ratio to allow the engine to operate at the most efficient RPM for the required power output. This sometimes creates a lag when you want to accelerate as the transmission must find a new gear to deliver the power. Drivers find the lag to be frustrating, but it provides better fuel economy.

The graph (Figure 11.1) shows data from fueleconomy.gov and is something to work with, but not a perfect model. It is for gasoline engines and typical midsize sedan cars built around the year 2000. New cars get better fuel economy at all speeds. Most gasoline engine and hybrid gas–electric cars get their best fuel economy when driving 55 to 60 mph. The thermal efficiency is higher at 70 mph, but it requires more power to drive at 70 mph. Fuel economy depends on the thermal efficiency of the engine and the power required and is typically best at 55 or 60 mph. Electric cars will have the most range when driven slower. Though not practical, the maximum driving range for an electric car is achieved by driving at a steady speed of about 30 mph. Range decreases as the electric car is driven faster.

Figure 11.1 Overall fuel economy for cars.

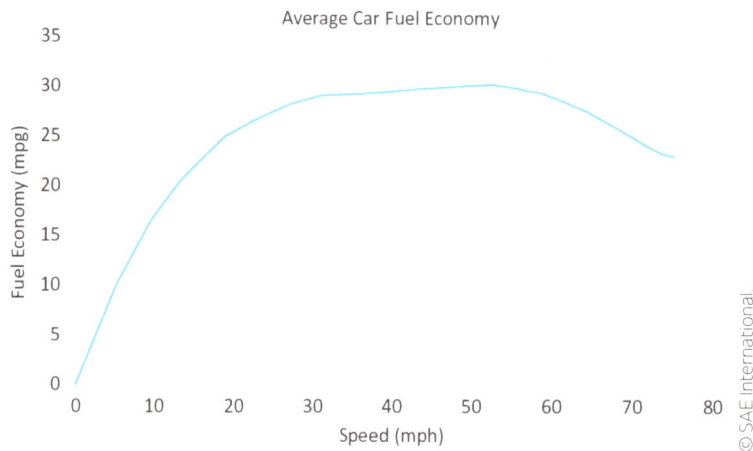

There have been a number of improvements made to engines over the years that have contributed to more power, better fuel economy, and lower emissions for internal combustion (IC) engines. During the 1960s, mechanical mechanisms were added to the distributor to advance the ignition timing when the engine was turning at high RPM, improving the power and efficiency of the engine. Vacuum advance was added to reduce fuel consumption and emissions when idling. Cars emitted a lot of unburned fuel, carbon monoxide, and nitrous oxide in the 1960s, which led to air quality and smog problems, especially in California. During the 1970s, we developed methods of reducing the emissions from IC engines, which led to the catalytic converter in the late 1970s. The catalytic converter is effective in reducing the pollutants from gasoline engines. Cars today produce a tiny fraction of pollution compared to cars of the 1960s. Electronic ignition and fuel injection were added in the 1980s, which helped increase the power and fuel economy of the engine and reduced the emissions. Variable valve timing was added in the 1990s, which allowed the engines to operate at higher efficiency over a broader range of RPM. Since 2000, electronic controls have allowed us to continue to improve and refine engine performance. Transmission and drive systems have gradually become more efficient.

Companies are experimenting with variable compression ratios, which will allow engines to operate more efficiently when generating low amounts of power. Companies are also experimenting with direct injection of gasoline into the combustion chamber, which potentially improves the combustion process.

When driving at a steady speed on level pavement, the power required will be the power to overcome rolling resistance and aerodynamic drag. In developing Figure 11.2, points were taken from Figure 11.1 for the fuel economy of the vehicle at different speeds. It was also assumed that the vehicle was a 4000 lb car with a coefficient of rolling resistance of Crr = 0.10, an

aerodynamic drag area of 1.9 m², a drag coefficient Cd of 0.33, and a drive efficiency of 90%. Figure 11.2 shows how the efficiency of the engine/drive varies with speed. This is an imperfect model, but it illustrates how the thermal efficiency of the gasoline engine varies with the overall average speed of the vehicle. For highway speeds, it works out best to assume a constant thermal efficiency of about 28%. For city driving, the thermal efficiency will be lower and depends on the power provided by the engine and the RPM as illustrated in the engine map. Figure 11.2 shows the overall thermal efficiency of the engine as a function of vehicle speed for an average car built around the year 2000.

Figure 11.2 Gasoline engine thermal efficiency at different speeds.

Looking at the engine map for the Saturn 1.9 L engine (Figure 10.1), it is clear the engine can produce power over a range of RPMs and that the efficiency of the engine depends on the RPM as selected by the transmission ratio. In general,

choosing a lower RPM (higher gear) allows the engine to operate at higher efficiency in producing the required power. Data were collected from the engine map (Figure 10.1) and presented in Figure 11.3.

Figure 11.3 Thermal efficiency of Saturn 1.9 L engine as a function of power.

Efficiency vs Watts Power

On the x-axis of the graph is the power the engine is producing in watts. The y-axis shows the corresponding thermal efficiency. The blue dots on the graph are the data points collected from the engine map (Figure 10.1) and illustrate that the engine operates over a range of thermal efficiencies for any required power output. The higher blue dots correspond to the power being produced at low RPMs (higher gears), and the lower blue dots on the graph correspond to the power being produced at higher RPMs (lower gears).

The red dots on the graph and the dotted line illustrate the model that will be used for modeling city and suburban driving. For city driving, the transmission will choose a mid-range gear to produce the required power. If the transmission always chooses the highest possible gear for the engine to produce the required power the car will feel sluggish when the driver wants to accelerate. The driver will not be happy with the car. Choosing a mid-range gear as illustrated in the model will make the car more lively and fun to drive around town. The trend line through the data follows the curve efficiency = 0.0384 ln (power + 25.9) – 0.1249 for this engine.

To use the model for other gasoline-powered cars and trucks, it is necessary to normalize to the size of the engine. The model was developed for a 124 hp engine. To normalize, the power term in the equation is multiplied by the quantity $(124/P_{max})$, where P_{max} is the rated hp of the engine. For city driving, we will use Equation (11.1) to estimate the thermal efficiency of a gasoline engine:

$$\text{Thermal Efficiency} = (0.0384)\ln\left(\frac{124\,P}{P_{max}} + 25.9\right) - 0.1249 \qquad \text{(11.1)}$$

There is a mix of units for P and P_{max}. The 124 in the equation has units of hp, and to normalize it, the P_{max} term needs to be the rating of the engine in hp. In the development that follows, it is convenient to have the power produced by the engine in W, so the P term is in W.

(a) P_{max} = Maximum rated horsepower of the engine in hp.

(b) P = Power the engine is producing in W.

IC engines continue to use fuel when the car stops at a stoplight because they must be idle. Larger engines will use more fuel at idle than smaller engines. Including the idle power is a small but necessary correction for the models to be developed for city driving. For this analysis, we will assume the power at idle is $(0.002)P_{max}$, and the idle power will need to be converted to W.

$$\text{Idle Power} = (0.002)P_{max} \qquad \text{(11.2)}$$

The Saturn 1.9 L engine that the model is based on is rated at a maximum of 124 hp. Multiplying the power term in the log function by the ratio of $(124/P_{max})$ scales the model to different sizes of engines based on their power output capabilities. The assumption is that all gasoline engines will have an engine map similar to the Saturn 1.9 L. Section 11.2 gives a more general approach if the reader has a specific car and engine in mind. Newer gasoline engines will be more efficient than the Saturn 1.9 L. The model has been tested on many production vehicles, and it yields a reasonable estimate of the fuel economy when compared to the Environmental Protection Agency (EPA) estimate for city driving. The newest cars will get slightly better fuel economy than predicted by the model because their engines have better thermal efficiency than the Saturn 1.9 L engine the model is based on.

The model underestimates the fuel economy of vehicles for highway driving. When driving on the highway, the transmission will shift to a higher gear and allow the car to be a little more sluggish to get better fuel economy. The thermal efficiency predicted by the model will be lower than what is achieved in highway driving. For highway driving, it is best to assume a constant thermal efficiency of about 28% for gasoline-powered cars.

Suburban driving is a mix of city and highway driving. Overall, the model used in this class is reasonably accurate in modeling suburban driving. In most cases, the car will get a slightly better fuel economy that is predicted by the model, but it is reasonably close.

11.2.
General Approach to Model Engine Thermal Efficiency

To be able to model fuel economy for a car in city driving, it is necessary to have a model for the thermal efficiency of the engine as a function of the power it produces. We will use Equation (11.1) for all the examples in the book, but if the reader has a specific car and engine in mind, and has a good engine map, it is possible to develop a more precise equation to replace Equation (11.1).

Start with an engine map similar to Figure 10.1. Follow the hp curves and take a data point each place it crosses a fuel consumption line. For example, looking at Figure 10.1, following the 30 hp curve on the graph starting on the right side of the graph and proceeding leftward, it crosses

475 g/kW-h, then 450, 425, 400, 375, 350, 325, 300, 275, 250. These values can be converted to thermal efficiency by taking 77.85 and dividing by the numbers above. The engine can produce 30 hp at all these different thermal efficiencies. Similar data can be taken by following the other hp curves. The data points can be plotted to yield a plot similar to the blue dots in Figure 11.3. A trend line is fit through the data points so that it falls in the middle of the data points. The trend line should have the form:

$$\text{Thermal Efficiency} = A\ln(P+B)+C \qquad (11.3)$$

Start with A = 0.0384, B = 25.9, and C = −0.1249 since those numbers work for the Saturn engine. In the spreadsheet, take the absolute value of the difference between the trend line and each data point as the "error" and total the error. Adjust the parameters A, B, and C to minimize the total error, with the requirement that the trend line should go through (0,0). Since there are three parameters to adjust, it will be impossible to know you have achieved the ideal optimal solution for their values, but it is possible to achieve a good solution. The trend line can then be used to estimate the thermal efficiency of the engine in city driving.

For highway driving, the transmission will shift to allow the engine to operate near the higher thermal efficiency points in the graph. The trend line will be a low estimate of the thermal efficiency for highway driving because it fits through the middle of the data. For highway driving, it is probably close enough to use a constant thermal efficiency for the engine as was done in Unit 1 of the book. If the reader wanted to develop a formula for highway driving, it would make sense to fit the trend line

closer to the high values of thermal efficiency in Figure 11.3.

11.3.
Efficiency of Electric Motors

The engine map in Figure 11.4 is for a typical electric motor that would be used to power an electric vehicle.

The efficiency of an electric motor is good over a very large operating range. The efficiency is zero at 0 RPMs. Above 10% of the rated RPMs and 10% of the rated torque, the efficiency becomes very good. The transmission and drive systems will absorb some energy making the overall motor/drive system 85–90% for an electric car. Solar cars and electrathon cars attach the wheel directly to the motor, avoiding the losses in the transmission and drive systems.

For modeling electric cars in city driving, we will assume an overall efficiency for the motor/drive system. The efficiency does not vary a lot, and this is an acceptable approximation.

Figure 11.4 Electric engine map.

One of the main advantages of hybrid-electric cars is that we can turn the gasoline engine off when we stop and save the cost of idling. The electric motor is used to get the car started rolling and then the gasoline engine will start and provide the power for most of the driving. The small battery system is charged by the gasoline engine. All the energy for driving comes from the gasoline engine. The electric motor is used for low speed where it is much more efficient than the gasoline engine. The gasoline engine operates in a speed/power range where it is more efficient in powering the car and charging the battery. The result is better fuel economy for city driving.

11.4.
Simulation of Fuel Economy

With the models for the efficiency of gasoline and electric motors, we can now begin modeling fuel economy for cars. The SAE J1082

specification was developed for road testing of vehicles in urban, suburban, and highway driving. We will develop mathematical models of vehicles driving the different cycles and use the results to compare fuel economy for different vehicles. The driving cycles are listed in the tables below.

SAE J1082 was developed for the actual driving of cars on a test track, and it required a lot of practice for drivers to be able to perform the tasks listed in the tables. Results from the tests were dependent on the ability of the driver, and the specification is no longer used. But for our purposes, we will use the computer to simulate the driving in the specification, and the computer will not have the issues of driver performance. There was a lot of thought put into developing the SAE J1082 specification and the driving patterns that would simulate urban, suburban, and highway driving. The specification is useful in simulating the fuel economy performance of cars in different types of driving (Tables 11.1 through 11.4).

Table 11.1 SAE J1082 urban cycle.

Distance (mi)	Operation
0.0	Start the fuel meter and timing device, idle 15 s, accelerate to 15 mph at 7 ft/s². Proceed at 15 mph to the 0.2-mi marker.
0.2	Stop at 4 ft/s², accelerate to 15 mph at 7 ft/s². Proceed at 15 mph to the 0.3-mi marker.
0.3	Decelerate to 5 mph at 4 ft/s², accelerate to 15 mph at 7 ft/s². Proceed at 15 mph to the 0.5-mi marker.
0.5	Stop at 4 ft/s², record fuel temperature and idle 15 s, accelerate to 20 mph at 7 ft/s². Proceed at 20 mph to the 0.7-mi marker.
0.7	Stop at 4 ft/s², accelerate to 20 mph at 7 ft/s². Proceed at 20 mph to the 0.8-m marker.
0.8	Decelerate to 10 mph at 4 ft/s², accelerate to 20 mph at 5 ft/s². Proceed at 20 mph to the 1.0-mi marker.
1.0	Stop at 4 ft/s², record fuel temperature and idle 15 s, accelerate to 15 mph at 7 ft/s², then to 25 mph at 5 ft/s². Proceed at 25 mph to the 1.2-mi marker.
1.2	Stop at 4 ft/s², accelerate to 15 mph at 7 ft/s², then to 25 mph at 5 ft/s². Proceed at 25 mph to the 1.3-mi marker.
1.3	Decelerate to 15 mph at 4 ft/s², accelerate to 25 mph at 5 ft/s². Proceed at 25 mph to the 1.5-mi marker.
1.5	Stop at 4 ft/s², record fuel temperature and idle 15 s, accelerate to 15 mph at 7 ft/s², then to 30 mph at 5 ft/s². Proceed at 30 mph to the 1.7-mi marker.
1.7	Stop at 4 ft/s², accelerate to 15 mph at 7 ft/s², and then to 30 mph at 5 ft/s². Proceed at 30 mph to the 1.8-mi marker.
1.8	Decelerate to 20 mph at 4 ft/s², accelerate to 30 mph at 5 ft/s². Proceed at 30 mph until it is time to begin braking for the 2.0-mi marker.
2.0	Begin braking at 4 ft/s² so the car will stop at the 2.0-mi marker. Stop the fuel meter and timing device at stop, and record fuel consumed, elapsed time, and fuel temperature.

© SAE International.

Table 11.2 SAE J1082 suburban cycle.

Distance (mi)	Operation
0.0	Approach the starting line at 40 mph. At the line start fuel measuring and timing devices, accelerate to 60 mph at 3 ft/s². Proceed at 60 mph to the 0.7-mi marker.
0.7	Decelerate to 30 mph at 4 ft/s². Accelerate to 50 mph at 3 ft/s². Proceed at 50 mph to the 2.0-mi marker.
2.0	Stop at 4 ft/s², record the fuel temperature and idle 7 s. Accelerate to 15 mph at 7 ft/s². Continue accelerating to 25 mph at 5 ft/s². Continue accelerating to 40 mph at 3 ft/s². Proceed at 40 mph to the 2.6-mi marker.
2.6	Accelerate to 50 mph at 3 ft/s². Proceed at 50 mph to the 3.3-mi marker.
3.3	Stop at 4 ft/s², record the fuel temperature and idle 7 s. Accelerate to 15 mph at 7 ft/s². Continue accelerating to 25 mph at 5 ft/s². Continue accelerating to 40 mph at 3 ft/s². Proceed at 40 mph to the 5.2-mi marker.
5.2	Stop the fuel measuring and timing devices while driving at 40 mph at the 5.2-mi marker. Record fuel consumed, elapsed time, and fuel temperature.

© SAE International.

Table 11.3 SAE J1082 interstate cycle (55 mph).

Distance (mi)	Operation
0.0	Approach the starting line at 55 mph. Record fuel temperature at the line and start fuel measuring and timing devices. Proceed at 55 mph to the 0.2-mi marker.
0.2	Accelerate to 60 mph at 1 ft/s². Immediately decelerate to 50 mph at 1 ft/s². Immediately accelerate to 55 mph at 1 ft/s². Proceed at 55 mph to the 1.2-mi marker.
1.2	Accelerate to 60 mph at 1 ft/s². Immediately decelerate to 50 mph at 1 ft/s². Immediately accelerate to 55 mph at 1 ft/s². Proceed at 55 mph to the 2.2-mi marker.
2.2	Accelerate to 60 mph at 1 ft/s². Immediately decelerate to 50 mph at 1 ft/s². Immediately accelerate to 55 mph at 1 ft/s². Proceed at 55 mph to the 3.2-mi marker.
3.2	Accelerate to 60 mph at 1 ft/s². Immediately decelerate to 50 mph at 1 ft/s². Immediately accelerate to 55 mph at 1 ft/s². Proceed at 55 mph to the 4.7-mi marker.
4.7	Stop the fuel measuring and timing devices while driving at 55 mph at the 4.7-mi marker. Record fuel consumed, elapsed time, and fuel temperature.

© SAE International.

Table 11.4 SAE J1082 interstate cycle (70 mph).

Distance (mi)	Operation
0.0	Approach the starting line at 70 mph. Record fuel temperature at the line and start fuel measuring and timing devices. Proceed at 70 mph to the 0.2-mi marker.
0.2	Accelerate to 75 mph at 1 ft/s². Immediately decelerate to 65 mph at 1 ft/s². Immediately accelerate to 70 mph at 1 ft/s². Proceed at 70 mph to the 1.2-mi marker.
1.2	Accelerate to 75 mph at 1 ft/s². Immediately decelerate to 65 mph at 1 ft/s². Immediately accelerate to 70 mph at 1 ft/s². Proceed at 70 mph to the 2.2-mi marker.
2.2	Accelerate to 75 mph at 1 ft/s². Immediately decelerate to 65 mph at 1 ft/s². Immediately accelerate to 70 mph at 1 ft/s². Proceed at 70 mph to the 3.2-mi marker.
3.2	Accelerate to 75 mph at 1 ft/s². Immediately decelerate to 65 mph at 1 ft/s². Immediately accelerate to 70 mph at 1 ft/s². Proceed at 70 mph to the 4.7-mi marker.
4.7	Stop the fuel measuring and timing devices while driving at 70 mph at the 4.7-mi marker. Record fuel consumed, elapsed time, and fuel temperature.

© SAE International.

For these models, we will assume the road is flat so that gravitational energy is not a factor. The car will accelerate and decelerate, which means that the kinetic energy of the car will change. The change in kinetic energy between steps i and i + 1 is:

$$\text{Change in Kinetic Energy} = \frac{1}{2}m\left(V_{i+1}^2 - V_i^2\right) \tag{11.4}$$

If the change in kinetic energy happens over a time step T, the average kinetic power over that time step is:

$$\text{Kinetic Power} = \frac{\frac{1}{2}m\left(V_{i+1}^2 - V_i^2\right)}{T} \tag{11.5}$$

If the weight of the vehicle (W) is in pounds and the speed (V) is in mph, there will be unit conversions to get the energy and power into sensible units. It is assumed the time step T used in the spreadsheet is in seconds. The following equations are useful in developing the models. In all equations, it is assumed the vehicle weight is in lb, the speed in mph, and the aerodynamic drag area in m². Several of the equations are repeated from Unit 1 for convenience.

$$\text{Kinetic Energy} = (0.0454)WV^2 \ \text{J} \tag{11.6}$$

$$\text{Kinetic Energy} = (0.0334)WV^2 \ \text{ft-lb} \tag{11.7}$$

$$\text{Kinetic Power} = \frac{(0.0454)W\left(V_{i+1}^2 - V_i^2\right)}{T} \ \text{W} \tag{11.8}$$

$$\text{Kinetic Power} = \frac{(0.0334)W\left(V_{i+1}^2 - V_i^2\right)}{T} \ \text{ft-lb / s} \tag{11.9}$$

$$\text{Rolling Power} = (1.989)WV\left(1 + \frac{V}{200}\right) \ \text{W} \tag{11.10}$$

$$\text{Rolling Power} = (1.467)WV\left(1 + \frac{V}{200}\right) \ \text{ft-lb / s} \tag{11.11}$$

$$\text{Aero Power} = (0.05357)V^3 \ \text{A Cd W} \tag{11.12}$$

$$\text{Aero Power} = (0.03931)V^3 \ \text{A Cd ft-lb / s} \tag{11.13}$$

If the road is flat, then there is no change in gravitational energy, and the total power required from the batteries for electric vehicles, or the fuel for IC engine vehicles is:

$$\text{Total Power} = \frac{\text{Kinetic Power} + \text{Rolling Power} + \text{Aero Power}}{\text{Motor / Drive Efficiency}} \tag{11.14}$$

Electric Vehicle Model: The easiest model to develop for city driving is the electric vehicle model, so it will be developed first. For the electric vehicle, we will assume that the motor/drive efficiency is constant.

Example 11.1

Model an electric car driving the urban cycle. Assume that there is no regenerative braking for this case. Parameters for the car are: W = 3800 lb, Crr = 0.011, A = 1.9 m², Cd = 0.32, motor/drive efficiency = 85%, battery capacity = 30,000 W-h. This example was developed by using parameters for the 2017 Nissan Leaf electric car.

1. Determine the average W-h per mi energy consumption for urban driving.
2. What is the range of the vehicle for urban driving?

Solution: We start with modeling an electric car because it is easier than modeling a gasoline or diesel-powered car. The W-h of electric energy used per mi traveled can be related to the cost per mi of driving the car by multiplying by the cost of the electricity used to charge the batteries. The cost of electricity per mi for an electric car is significantly less than the cost of gasoline per mi for a gasoline-powered car. The more critical factor for an electric car is the driving range.

Looking at the urban cycle, the car idles for the first 15 s, simulating being stopped at a stoplight. An electric car will consume no energy when at the stoplight, so the energy consumed in the first 15 s is 0. Table 11.5 is used to illustrate how the model was developed.

The time step chosen for this model was 1 s. The values in all the other columns are zero for the first 15 s, and Table 11.5 starts at time 16 s. At the beginning of the 16th s, the car accelerates at 7 ft/s². The speeds and power are all zero at the beginning of the time step. The way the model is developed, each row represents what is happening at the beginning of the time step.

In column 2, the car accelerates at 7 ft/s² until it reaches 15 mph, which is 22 ft/s. The speed in ft/s is shown in the third column and converted to mph in the fourth column. Notice that at time step 19 s, the acceleration is reduced to 1 ft/s² because that is what is needed to make the speed 15 mph at the beginning of time step 20. The first 4 columns of the spreadsheet are developed using Table 11.1 so that the car has the correct speed and acceleration according to the specification SAE J1082.

Table 11.5 Urban driving for an electric car.

Time (s)	Acc (ft/s²)	Speed (ft/s)	Speed (mph)	Dist (ft)	Dist (mi)	Roll (W)	Aero (W)	Kin (W)	Total (W)	Clean (W)	W-h
16	7	0	0	0	0	0	0	0	0	0	0
17	7	7	4.772	3.5	0.000663	406.3	3.54	3930	5105	5105	0.709
18	7	14	9.545	14	0.002652	831.5	28.3	11,780	14,882	14,882	3.485
19	1	21	14.32	31.5	0.005966	1276	95.6	19,649	24,730	24,730	8.987
20	0	22	15	53	0.010038	1341	109.9	3449	5764	5764	13.22

Column 5 is the distance traveled by the car. Distance is the average velocity during the time step multiplied by the time step (T = 1 s in this case). Using D for the distance in column 5 and V for the speed in column 3, the distances (ft) are calculated as:

$$D_{i+1} = D_i + \frac{1}{2}(V_{i+1} + V_i)(T)$$

(11.15)

The distances are converted to mi in column 6. The first six columns are the same for any vehicle traveling on the urban cycle. None of the terms depend on the weight of the vehicle, the rolling or aerodynamic losses, or other car parameters. The columns on the right side of the table are used to estimate the amount of electric energy used during each time step and then total it for the cycle.

Columns 7, 8, 9, and 10 calculate the rolling, aero, and kinetic power and then total the power. To be sure you understand how to do the calculations, you should go through the calculations and make sure the numbers match those in the table. The calculations are done using Equations (11.8), (11.10), and (11.12) to calculate rolling, aero, and kinetic power. The total power is calculated in column 10 using Equation (11.16).

$$\text{Total Power} = \frac{\text{Rolling} + \text{Aero} + \text{Kinetic}}{\text{Motor / Drive Efficiency}}$$

(11.16)

A logic statement is required for column 11 because there are times in the spreadsheet where the car is braking. Braking will yield a negative number for the kinetic power and for the total power. For the first example, we are assuming there is no regenerative braking, so all the braking energy will be lost as heat in the brakes.

The total power is positive for rows 16–20, but there will be places in the spreadsheet where the total power is negative. Column 11 is the "cleaned" power, which will zero the power requirements when the total power is negative. In Excel, the logic statement is:

$$\text{If}(\text{Total Power} > 0, \text{Total Power}, 0)$$

(11.17)

In the next example, we will include regenerative braking and assess how much difference regenerative braking makes in extending the range of the vehicle. This logic statement [Equation (11.17)] will be modified to include regenerative braking.

The last column in the spreadsheet totals the number of watt-h energy used from the batteries.

The energy used is the average power during the time step multiplied by the time step. The 3600 factor in the equation converts the units from Joules to watt-h. Let P be the cleaned power in column 11, and E be the energy used in column 12. The total energy used in column 12 is calculated from Equation (11.18):

$$E_{i+1} = E_i + \frac{\frac{1}{2}(P_{i+1} + P_i)(T)}{3600}$$

(11.18)

The reader should go through enough calculations to feel confident with the equations. The model can then be developed as shown in Table 11.6. The actual spreadsheet goes to a time of 480 s.

Table 11.6 shows the first 26 s of the urban simulation. Readers can use the numbers in Table 11.6 to ensure the spreadsheet they are developing is working correctly.

Table 11.6 Urban driving model for an electric car.

Car weight	3800	lb		Battery	30,000	W-h				W-h/mi	254.2
Crr	0.011			eff	0.85					Range	118.0
Area	1.9	m²								Regen eff	0
Cd	0.32										

Time (s)	Acceleration (ft/s²)	Speed (ft/s)	Speed (mph)	Distance (ft)	Distance (mi)		Roll (W)	Aero (W)	Kinetic (W)	Total (W)	Cleaned (W)	Total (W-h)
0	0	0	0	0	0		0	0	0	0	0	0
1	0	0	0	0	0		0	0	0	0	0	0
2	0	0	0	0	0		0	0	0	0	0	0
3	0	0	0	0	0		0	0	0	0	0	0
4	0	0	0	0	0		0	0	0	0	0	0
5	0	0	0	0	0		0	0	0	0	0	0
6	0	0	0	0	0		0	0	0	0	0	0
7	0	0	0	0	0		0	0	0	0	0	0
8	0	0	0	0	0		0	0	0	0	0	0
9	0	0	0	0	0		0	0	0	0	0	0
10	0	0	0	0	0		0	0	0	0	0	0
11	0	0	0	0	0		0	0	0	0	0	0
12	0	0	0	0	0		0	0	0	0	0	0
13	0	0	0	0	0		0	0	0	0	0	0
14	0	0	0	0	0		0	0	0	0	0	0
15	0	0	0	0	0		0	0	0	0	0	0
16	7	0	0	0	0		0	0	0	0	0	0
17	7	7	4.77	3.5	0.00066		406.3	3.54	3930	5105	5105	0.709
18	7	14	9.55	14	0.00265		831.5	28.33	11,789	14,882	14,882	3.485
19	1	21	14.32	31.5	0.00597		1275.6	95.61	19,649	24,730	24,730	8.987
20	0	22	15	53	0.01004		1340.6	109.93	3449	5764	5764	13.222
21	0	22	15	75	0.01420		1340.6	109.93	0	1707	1707	14.259
22	0	22	15	97	0.01837		1340.6	109.93	0	1707	1707	14.733
23	0	22	15	119	0.02254		1340.6	109.93	0	1707	1707	15.207
24	0	22	15	141	0.02670		1340.6	109.93	0	1707	1707	15.682
25	0	22	15	163	0.03087		1340.6	109.93	0	1707	1707	16.156
26	0	22	15	185	0.03504		1340.6	109.93	0	1707	1707	16.630

A portion of the spreadsheet where the total power goes negative is shown in Table 11.7 to illustrate how the logic statement in column 11 works when the car is braking.

In column 2 notice that the car is braking at a deceleration of −4 ft/s² during the first part of the table. The kinetic power in column 10 goes negative when the car is braking. The "cleaned" power column sets the negative power to zero because without regenerative braking the car cannot recover the braking energy. The total energy used from the batteries in the far right column does not increase during braking. In the next example, we will add a factor for regenerative braking and allow the car to recover some of the braking energy. When we include regenerative braking, the total energy used from the batteries in the far right column will decrease slightly as energy is put back into the batteries.

The number at the bottom of column 6 (at 480 s) is the total distance traveled by car in mi. The number at the bottom of column 12 is the total number of W-h of energy used from the batteries. Dividing the total W-h used by the total distance traveled gives the 254.2 W-h/mi reading shown in the top right corner of the spreadsheet. The range of the vehicle is calculated as the battery capacity divided by the 254.2 W-h/mi and is 118 mi in this case.

The simulation shows that the electric car would have a range of 118 mi in city driving. If the electricity to charge the batteries cost 14¢/kW-h, the cost to charge the batteries would be $4.20 and the cost per mile would be about $0.036/mi. This would be a typical cost for people living in the Midwest charging their car at home. A fast-charging station might charge 25¢/kW-h, bringing the cost to $0.064/mi. People living in

Table 11.7 Portion of an urban driving model for electric cars illustrating what happens during braking.

Time (s)	Acceleration (ft/s²)	Speed (ft/s)	Speed (mph)	Distance (ft)	Distance (mi)	Roll (W)	Aero (W)	Kinetic (W)	Total (W)	Cleaned (W)	Total (W-h)
65	0	22	15.00	1043	0.19754	1340.6	109.93	0	1707	1707	35.117
66	−4	22	15.00	1065	0.20170	1340.6	109.93	0	1707	1707	35.591
67	−4	18	12.27	1085	0.20549	1083.0	60.21	−12,832	−13,752	0	35.828
68	−4	14	9.55	1101	0.20852	831.5	28.33	−10,266	−11,066	0	35.828
69	−4	10	6.82	1113	0.21080	586.2	10.32	−7699	−8356	0	35.828
70	−4	6	4.09	1121	0.21231	347.1	2.23	−5133	−5628	0	35.828
71	−2	2	1.36	1125	0.21307	114.1	0.08	−2566	−2885	0	35.828
72	7	0	0.00	1126	0.21326	0.0	0.00	−321	−377	0	35.828
73	7	7	4.77	1129.5	0.21392	406.3	3.54	3930	5105	5105	36.537
74	7	14	9.55	1140	0.21591	831.5	28.33	11,789	14,882	14,882	39.313
75	1	21	14.32	1157.5	0.21922	1275.6	95.61	19,649	24,730	24,730	44.815
76	0	22	15.00	1179	0.22330	1340.6	109.93	3449	5764	5764	49.050

California may pay 25¢/kW-h when charging at home, and higher prices at a fast-charging station. If we compare to a small gasoline-powered car that gets 35 mpg in the city and burns fuel that costs $3.00 per gallon the fuel cost is $0.086/mi. The energy cost of driving an electric car is less than for a similar-sized gasoline-powered car if you can charge the car at home, but the savings are less when charging at a fast-charging station. A small hybrid-electric car might get 60 mpg in the city, yielding a cost of $0.05/mi if the gasoline cost is $3.00 per gallon.

The range of the electric car is the main issue. Everyone would drive electric cars if the cost was comparable to gasoline-powered cars and if they had adequate range. Getting an adequate range

for the car has always been a problem. One of the things that can be done to improve the range in city driving is to put regenerative braking on the car. Then when the car brakes some of the braking energy can be put back into the batteries and the range of the car will be extended. Safety considerations prevent us from being able to put all the braking energy back into the batteries. It is more important to be able to stop the car in a controlled manner than to recover the braking energy. Realistically it is possible to recover about 30% of the braking energy using a regenerative braking system. To accommodate this in the spreadsheet, we allow a fraction of the braking energy to be put back in the batteries. The logic statement in column 11 is modified to be:

$$\text{If}(\text{Total Power} > 0, \text{Total Power}, \text{regen eff} * \text{Total Power}) \quad (11.19)$$

Example 11.2

Modify the spreadsheet in the previous example to illustrate how regenerative braking affects the efficiency and range of the electric car, with the regenerative braking efficiency set to 30%.

Solution: The "cleaned" power column is modified using Equation (11.19). The result is that the average energy consumption per mi is reduced to 215.3 W-h/mi and the range is extended to 139.4 mi. Being able to recover 30% of the braking energy increases the range of the car from 118 to 139.4 mi: a very significant increase. The first 26 s of the model are shown in Table 11.8 to allow readers who are developing spreadsheets to compare numbers.

Table 11.8 Urban driving for an electric car with regenerative braking.

Car weight	3800	lb		Battery	30,000	W-h				W-h/mi	215.3	
Crr	0.011			eff	0.85					Range	139.4	
Area	1.9	m²								Regen eff	0.3	
Cd	0.32											

Time (s)	Acceleration (ft/s²)	Speed (ft/s)	Speed (mph)	Distance (ft)	Distance (mi)		Roll (W)	Aero (W)	Kinetic (W)	Total (W)	Cleaned (W)	Total (W-h)
0	0	0	0	0	0		0	0	0	0	0	0
1	0	0	0	0	0		0	0	0	0	0	0
2	0	0	0	0	0		0	0	0	0	0	0
3	0	0	0	0	0		0	0	0	0	0	0
4	0	0	0	0	0		0	0	0	0	0	0
5	0	0	0	0	0		0	0	0	0	0	0
6	0	0	0	0	0		0	0	0	0	0	0
7	0	0	0	0	0		0	0	0	0	0	0
8	0	0	0	0	0		0	0	0	0	0	0
9	0	0	0	0	0		0	0	0	0	0	0
10	0	0	0	0	0		0	0	0	0	0	0
11	0	0	0	0	0		0	0	0	0	0	0
12	0	0	0	0	0		0	0	0	0	0	0
13	0	0	0	0	0		0	0	0	0	0	0
14	0	0	0	0	0		0	0	0	0	0	0
15	0	0	0	0	0		0	0	0	0	0	0
16	7	0	0	0	0		0	0	0	0	0	0
17	7	7	4.77	3.5	0.00066		406.3	3.54	3930	5105	5105	0.709
18	7	14	9.55	14	0.00265		831.5	28.33	11,789	14,882	14,882	3.485
19	1	21	14.32	31.5	0.00597		1275.6	95.61	19,649	24,730	24,730	8.987
20	0	22	15	53	0.01004		1340.6	109.93	3449	5764	5764	13.222
21	0	22	15	75	0.01420		1340.6	109.93	0	1707	1707	14.259
22	0	22	15	97	0.01837		1340.6	109.93	0	1707	1707	14.733
23	0	22	15	119	0.02254		1340.6	109.93	0	1707	1707	15.207
24	0	22	15	141	0.02670		1340.6	109.93	0	1707	1707	15.682
25	0	22	15	163	0.03087		1340.6	109.93	0	1707	1707	16.156
26	0	22	15	185	0.03504		1340.6	109.93	0	1707	1707	16.630

A portion of the spreadsheet illustrating what happens during regenerative braking is shown in Table 11.9.

Notice that during regenerative braking, the total W-h drawn from the batteries in the right column decreases slightly because energy is being put into the batteries. During this regenerative braking cycle, the energy drawn from the batteries goes from 35.591 W-h down to 32.339 W-h, indicating that 3.252 W-h of energy was recovered.

Table 11.9 Portion of an urban driving model for electric cars illustrating what happens during regenerative braking.

Time (s)	Acceleration (ft/s²)	Speed (ft/s)	Speed (mph)	Distance (ft)	Distance (mi)	Roll (W)	Aero (W)	Kinetic (W)	Total (W)	Cleaned (W)	Total (W-h)
65	0	22	15.00	1043	0.19754	1340.6	109.93	0	1707	1707	35.117
66	−4	22	15.00	1065	0.20170	1340.6	109.93	0	1707	1707	35.591
67	−4	18	12.27	1085	0.20549	1083.0	60.21	−12,832	−13,752	−4125	35.255
68	−4	14	9.55	1101	0.20852	831.5	28.33	−10,266	−11,066	−3320	34.221
69	−4	10	6.82	1113	0.21080	586.2	10.32	−7699	−8356	−2507	33.412
70	−4	6	4.09	1121	0.21231	347.1	2.23	−5133	−5628	−1688	32.829
71	−2	2	1.36	1125	0.21307	114.1	0.08	−2566	−2885	−865	32.475
72	7	0	0.00	1126	0.21326	0.0	0.00	−321	−377	−113	32.339
73	7	7	4.77	1129.5	0.21392	406.3	3.54	3930	5105	5105	33.032
74	7	14	9.55	1140	0.21591	831.5	28.33	11,789	14,882	14,882	35.808
75	1	21	14.32	1157.5	0.21922	1275.6	95.61	19,649	24,730	24,730	41.310
76	0	22	15.00	1179	0.22330	1340.6	109.93	3449	5764	5764	45.545

The model shows that adding regenerative braking to the car will extend the range from 118 to 139.4 mi, an improvement of 18%. Once the model is developed, we can use it to help us understand how the different car parameters impact the range of the vehicle. We know that reducing the weight of the car will give the car a better range in the city. The model allows us to quantify how much improvement in the range can be achieved.

Example 11.3

Suppose we consider replacing some of the steel components in the car with aluminum, and estimate it will reduce the weight of the car by 100 lb. How much does this impact the range of the vehicle?

Solution: Change the weight in the top left corner to 3700 and the spreadsheet will automatically recalculate everything. The range of the vehicle improves to 142.85 mi. Reducing the weight by 100 lb will improve the range by 3.5 mi. There will be a cost associated with replacing the steel components with aluminum, and we will need to decide if the additional 3.5 mi in the range is worth the additional cost.

Example 11.4

Suppose we find better tires for the car that have a rolling resistance coefficient of 0.009 rather than the 0.011 of the tires in the model. Changing the Crr to 0.009 increases the range of the vehicle to 152.04 mi, an improvement of about 12.5 mi. From the analysis, it is clear that using low-rolling-resistance tires is important for electric vehicles in city traffic.

Example 11.5

Suppose we work on the aerodynamics of the car and reduce the aerodynamic drag coefficient to 0.30. Making that change in the model causes the range of the vehicle to improve to 140.00 mi, an improvement of 0.64 mi. This analysis shows that aerodynamics is not very important for city driving. When we model the car in the highway driving cycle, the aerodynamic drag will be more important, but for city driving it is not very important.

Example 11.6

For another trade-off, consider what happens by adding more batteries. Suppose adding 70 lb of batteries increases the battery capacity by 4000 W-h. This would be approximately correct for adding lithium-ion batteries to the car. Adding 70 lb to the weight of the car and 4000 W-h to the battery capacity improves the range to 155.29 mi, which is an improvement of about 16 mi in range. It also adds cost to the car, and the batteries will take up some of the storage space in the car. The designer will need to decide if it is worth adding the additional batteries.

The models help guide the design of the car. We want to spend development money and effort on things that will help most in the performance of the car, and for an electric vehicle, range is one of the most important objectives of the design. There are many possibilities for improving the range of the electric car, and nearly all of them involve added cost. Adding regenerative braking improved the range of the car by 18%. It is almost certain that the added cost of regenerative braking is well spent. Reducing the weight of the vehicle using aluminum components is of less benefit. It may or may not be worth it. The model allows the designer to quantify how much improvement in the range is associated with different design alternatives and is an important design tool.

Gasoline Vehicle Model: The next step is to model a gasoline car in city driving. The process is very similar to the electric vehicle. Equation (11.1) will be used to estimate the thermal efficiency of the engine and Equation (11.2) will be used to estimate the power the engine consumes while idling.

Example 11.7

Assume a pickup has a weight of 5500 lb and the tires have a rolling resistance coefficient Crr = 0.010. The power rating for the engine is 375 hp. The frontal area is 2.77 m^2 and the drag coefficient is Cd = 0.50. Assume the power associated with idling is given by Equation (11.2) (0.002*P$_{max}$). The drive efficiency is 90%. This example is developed using specifications for the 2018 Ford F-150 truck.

Solution: For this model, we can start with the model for the electric car. The rolling, aerodynamic, kinetic, and total power columns are calculated the same way as for the electric car. The truck is a larger and heavier vehicle, so the numbers are larger, but the formulas used are the same. When we do the "cleaned" column after the total power, we need to recognize that the minimum power the engine provides is idle. The logic statement will need to say that if the total power is less than the idle power, then the truck is using idle power. Idle power for the truck is estimated as:

$$\text{Idle Power} = (0.002)(375)(745.7) = 559 \text{ W} \qquad (11.20)$$

The truck is stationary at the stoplight for the first 15 s, so the power requirement is idle power through the beginning of the 16th step. As with the electric car, the first 15 steps are omitted in the portion of the spreadsheet shown in Table 11.10. The reader should work through the calculations to ensure understanding.

Table 11.10 Urban driving for a gasoline vehicle.

Time (s)	Acc (ft/s^2)	Speed (ft/s)	Speed (mph)	Dist (ft)	Dist (mi)	Roll (W)	Aero (W)	Kin (W)	Total (W)	Clean (W)
16	7	0	0	0	0	0	0	0	0	559
17	7	7	4.772	3.5	0.000663	534.6	8.06	5688	6923	6923
18	7	14	9.545	14	0.002652	1094	64.5	17,064	20,247	20,247
19	1	21	14.32	31.5	0.005966	1678	218	28,439	33,706	33,706
20	0	22	15	53	0.010038	1764	250	4991	7784	7784

There are too many columns in this spreadsheet to fit everything on one-page width. Added to the right of Table 11.10 will be four more columns used to calculate the amount of fuel used as the truck goes through the urban driving model. The four columns are in Table 11.11.

Once we know the power the engine must produce, we can use the efficiency formula to calculate the thermal efficiency of the engine. Remember that P in Equation (11.1) is the required engine power in watts and P$_{max}$ is the rated engine power in horsepower, which in this case is 375 hp. The first column in Table 11.11 is calculated using Equation (11.21), where P is from the far-right column in Table 11.10.

$$\text{Efficiency} = (0.0384)\ln\left[\frac{124\,P}{375} + 25.9\right] - 0.1249 \qquad (11.21)$$

The fuel power in the second column of Table 11.11 is equal to the engine power divided by the efficiency as shown in Equation (11.22).

Table 11.11 Efficiency and fuel consumption of gasoline vehicles in urban driving.

Efficiency	Watt	BTU/s	Gallon
0.08058	6941	6.579	0.0000544
0.17259	40,111	38.02	0.000369
0.21352	94,826	89.88	0.001111
0.23303	144,643	137.10	0.002244
0.17705	43,967	41.67	0.002589

$$\text{Fuel Power} = \frac{P}{\text{Efficiency}} W \tag{11.22}$$

Since there are 1055 J in 1 BTU, the fuel power can be converted to BTU/s by dividing by 1055 as shown in column 3 of Table 11.11. Assuming 121,000 BTU/gal, the fuel used during the time step is the BTU/s fuel power divided by 121,000.

This result is multiplied by the time step T, which in this case is 1 s. We want to keep a running total in the gallons of fuel used in the fourth column of Table 11.11.

$$\text{Gallons}_{i+1} = \text{Gallons}_i + \frac{\left(\frac{\text{BTU}}{s}\right)_{i+1}(1\,s)}{121,000} \tag{11.23}$$

The reader should be able to calculate the numbers in the tables using the formulas given. The distance traveled in miles is at the bottom of the distance column, and the gallons of gasoline used are at the bottom of the gallons column. Dividing the miles traveled by the gallons used gives the average fuel economy of the truck in city traffic. The EPA estimate for the F-150 truck is 18 mpg in the city. The results of the model show the fuel economy to be 18.073 mpg. Table 11.12 shows the spreadsheet for the first 30 s of the urban gasoline vehicle model.

Ford has gone through considerable effort to lighten the truck by using aluminum body

components. If the weight of the vehicle is reduced by 500 lb, the fuel economy improves from 18.073 to 19.228 mpg, which is a 6.4% improvement in fuel economy. Other studies can be performed to see how changing other car parameters impacts the overall fuel economy of the truck in city driving.

Many cars now have an instantaneous fuel economy meter on the dashboard showing the instantaneous fuel economy while driving. The instantaneous fuel economy is the distance traveled divided by the fuel used over a brief time period. We can plot this in the model. The plot is shown in Figure 11.5.

Table 11.12 Urban model for a gasoline vehicle.

Truck weight	5500	lb		P_{max}	375		hp			Fuel energy	121,000	BTU/gal	
Crr	0.01			Eff	0.9								
Area	2.77	m²		Idle	559		W			Econ	18.073	mpg	
Cd	0.5												

Time (s)	Acce-leration (ft/s²)	Speed (ft/s)	Speed (mph)	Dist (ft)	Dist (mi)	Rolling (W)	Aero (W)	Kinetic (W)	Total (W)	Clean (W)	Thermal efficiency	Fuel (W)	Fuel (BTU/s)	Fuel (gal)
0	0	0	0	0	0	0	0	0	0	559	0.081	6941	6.58	0.00000
1	0	0	0	0	0	0	0	0	0	559	0.081	6941	6.58	0.00005
2	0	0	0	0	0	0	0	0	0	559	0.081	6941	6.58	0.00011
3	0	0	0	0	0	0	0	0	0	559	0.081	6941	6.58	0.00016
4	0	0	0	0	0	0	0	0	0	559	0.081	6941	6.58	0.00022
5	0	0	0	0	0	0	0	0	0	559	0.081	6941	6.58	0.00027
6	0	0	0	0	0	0	0	0	0	559	0.081	6941	6.58	0.00033
7	0	0	0	0	0	0	0	0	0	559	0.081	6941	6.58	0.00038
8	0	0	0	0	0	0	0	0	0	559	0.081	6941	6.58	0.00043
9	0	0	0	0	0	0	0	0	0	559	0.081	6941	6.58	0.00049
10	0	0	0	0	0	0	0	0	0	559	0.081	6941	6.58	0.00054
11	0	0	0	0	0	0	0	0	0	559	0.081	6941	6.58	0.00060
12	0	0	0	0	0	0	0	0	0	559	0.081	6941	6.58	0.00065
13	0	0	0	0	0	0	0	0	0	559	0.081	6941	6.58	0.00071
14	0	0	0	0	0	0	0	0	0	559	0.081	6941	6.58	0.00076
15	0	0	0	0	0	0	0	0	0	559	0.081	6941	6.58	0.00082
16	7	0	0	0	0	0	0	0	0	559	0.081	6941	6.58	0.00087
17	7	7	4.77	3.5	0.0007	535	8	5688	6923	6923	0.173	40,111	38.02	0.00118
18	7	14	9.55	14	0.0027	1094	65	17,064	20,247	20,247	0.214	94,826	89.88	0.00193
19	1	21	14.32	31.5	0.0060	1678	218	28,439	33,706	33,706	0.233	144,643	137.10	0.00306
20	0	22	15	53	0.0100	1764	250	4991	7784	7784	0.177	43,967	41.67	0.00340
21	0	22	15	75	0.0142	1764	250	0	2238	2238	0.130	17,201	16.30	0.00354
22	0	22	15	97	0.0184	1764	250	0	2238	2238	0.130	17,201	16.30	0.00367
23	0	22	15	119	0.0225	1764	250	0	2238	2238	0.130	17,201	16.30	0.00381
24	0	22	15	141	0.0267	1764	250	0	2238	2238	0.130	17,201	16.30	0.00394
25	0	22	15	163	0.0309	1764	250	0	2238	2238	0.130	17,201	16.30	0.00408
26	0	22	15	185	0.0350	1764	250	0	2238	2238	0.130	17,201	16.30	0.00421
27	0	22	15	207	0.0392	1764	250	0	2238	2238	0.130	17,201	16.30	0.00435
28	0	22	15	229	0.0434	1764	250	0	2238	2238	0.130	17,201	16.30	0.00448

Figure 11.5 Instantaneous fuel economy.

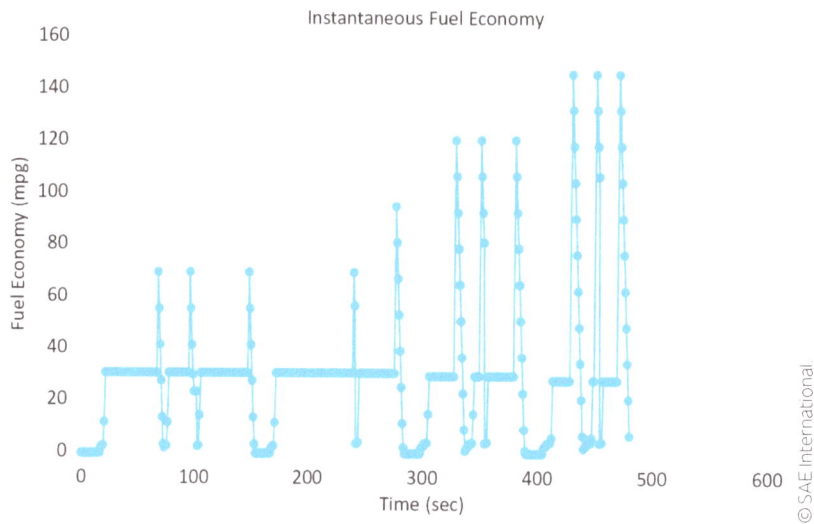

Examining **Figure 11.5** shows that the fuel economy is zero when idling. Not much fuel is consumed when idling, so idling does not have a huge impact on the overall fuel economy. The fuel economy is low during hard acceleration and significant amounts of fuel are consumed. Fuel economy levels out at 31 mpg when the truck is at a constant speed in this model because we are assuming it all takes place on level pavement. If small grades were added in to make the model more like a city driving, the constant regions shown in the chart would not be constant. The high mpg readings on the chart happen when the truck is braking. The instantaneous fuel economy varies from 0 to 150 mpg, but the average fuel economy is 18 mpg.

One of the main disadvantages of an IC engine car is that it burns fuel while idling at stoplights or any time the car stops and is not shut off. It is a waste of fuel and it lowers the fuel economy of the vehicle. There are regular gasoline cars that turn the engine off when the car stops and then automatically start the engine when the driver

presses the gas pedal. This results in significant fuel savings in city driving, but it also creates a lag in performance when taking off at a stoplight. Many drivers find the lag frustrating. It is not a popular option. This can be modeled for the F-150 pickup in the previous example by setting the idle power to zero rather than the $(0.002)P_{max}$. Making this change in the spreadsheet improves the fuel economy from 18 to 19.35 mpg, which is a significant improvement. But very few trucks are sold with the feature of turning off when stopped because drivers do not like it.

Another problem with the IC engine cars is that the thermal efficiency of the engine is low at low speeds or low power requirements, such as when taking off at a stoplight or stop sign or cruising at low speeds. With a hybrid-electric car, we will use the electric motor when the IC engine is operating at low thermal efficiency. The IC engine will power the car and charge the battery when it is operating at a higher thermal efficiency. The hybrid vehicle requires an IC engine and an electric motor. It is more complex than gasoline vehicles, but there is a potential for significant improvement in fuel economy. The improvement in fuel economy comes from being able to turn the gasoline engine off when the vehicle is stopped, and by allowing the gasoline engine to operate at a higher thermal efficiency when it is on. The electric motor is used to power the vehicle when taking off from stoplights, or other instances when the gasoline engine would be operating at low thermal efficiency. Another benefit to the hybrid vehicle is that we can recover some of the braking energy with regenerative braking. The electric motor in the car is used as a generator, and the wheels and drivetrain power the generator to produce the electric energy and charge the batteries.

All the energy for powering the car comes from the IC engine because it charges the batteries that provide the power for the electric motor. If the battery is large enough to provide power for a few minutes, the electric motor can also supplement the power of the IC engine when the car is climbing a steep hill or accelerating. This gives the car more power than can be provided by the IC engine alone. Since the electric motor can supplement the power of the IC engine, we can use a smaller IC engine than would be required without the electric motor. Using a smaller engine will allow us to further increase the overall thermal efficiency of the engine. In Chapter 10, there is a section illustrating how using a smaller engine will allow the vehicle to get better fuel economy.

The control logic of when to use the gasoline engine and when to use the electric motor is complex. The car companies are still working out how to optimize the hybrid-electric car. The battery should be large enough to power the car for a few minutes. Larger batteries provide more flexibility in deciding when the gasoline engine needs to be started. Optimizing the size of the battery for the car is also a difficult problem that is not fully understood at this point in time. The hybrid-electric cars are complex and expensive, but they offer the potential for a large improvement in urban fuel economy compared to a gasoline-powered car or truck. The improvement in fuel economy comes without any sacrifice in the performance of the vehicle. Hybrid vehicles are filled with gas like a gasoline-powered vehicle, which makes them more convenient than electric vehicles when making a long trip. Electric vehicles have limited range and it takes a while to charge the batteries.

The gasoline engine will need to run continuously on the highway to provide the necessary power. The electric motor can assist when the gasoline engine needs to produce more power, and this allows a smaller gasoline engine to be used than would be required for a standard gasoline-powered car. A hybrid-electric car will deliver an improvement in the highway fuel economy, but the improvement will not be as large as the improvement in urban fuel economy.

Example 11.8

Consider what would happen if we make a hybrid version of a pickup truck (modeled after the F-150). The weight is 5500 lb, Crr is 0.01, the drag area is 2.77 m², and the aero drag coefficient is 0.5. Assume the transmission and drive are 90% efficient in transmitting power from the engine to the wheels. The truck will use the electric motor in conjunction with the gas motor, so a smaller gas engine can be used. Assume the gas engine is rated at 150 hp. Using a smaller gas engine improves thermal efficiency.

Discussion: We will start with the spreadsheet developed for the truck (**Table 11.12**). The idle power can be set to zero because the gas engine will be turned off when the truck stops. The rest of the modifications to the spreadsheet depend on how the control system is programmed to decide between using the gas engine and the electric motor. Ideally, the electric motor will be used when the truck is traveling at low speed or with a low power requirement. The gas motor will be started to charge the batteries and power the car at the same time, which will allow the gas engine to operate at a higher thermal efficiency on average. The amount of time the truck can run on electric power depends on the size of the battery system.

I hope the reader can appreciate the complexity of optimizing the performance and fuel economy of hybrid vehicles. We spent decades optimizing gasoline and diesel vehicles. Having two sources of power (fuel and electric) and a finite-sized battery makes optimizing the hybrid vehicle an order of magnitude more complex. We want to deliver adequate power and maximize fuel economy. It will require decades of research and experimentation to fully optimize the hybrid vehicle. Simulations will be part of the development effort. The models will be modified according to the strategy used

in deciding when to use the gas engine or the electric motor. The solution below is one possibility in how the hybrid vehicle could be used in urban traffic.

Solution: For this model assume that the control logic will be set so that the gas engine will operate at a minimum thermal efficiency of 15%. Assume that 30% of the braking energy will be recovered and put into the batteries. Assume that the battery is large enough to allow the vehicle to accomplish these two things. Developing the software and control system to accomplish this would be difficult but assume that it is possible. The previous model for the truck can be modified to generate the model for the hybrid-electric vehicle. The changes are listed below.

1. The total power column is rolling + aerodynamic + kinetic. It should not be divided by the efficiency of the gas engine because it will not be all gas-powered. Portions will be electrically powered.

2. The "cleaned" power column should have a logic statement saying that if the total power is positive use the total power divided by the drive efficiency. If the total power is negative it should be multiplied by the regenerative braking efficiency.

$$\text{If}\left(\text{Total} > 0, \frac{\text{Total}}{\text{drive eff}}, \text{Total} * \text{regen eff}\right) \tag{11.24}$$

3. The thermal efficiency needs to be separated into two columns because of the way Excel works. We need to use two logic statements to work it out. For the first column, if the cleaning

power is negative or zero, set the thermal efficiency to zero. Otherwise, set it to the logarithmic formula used for gasoline engines.

If you leave out this step, the negative powers will cause problems in the logarithmic equation.

$$\text{If}\left(P > 0, (0.0384)\ln\left[\frac{124\,P}{375} + 25.9\right] - 0.1249, 0 \right)$$

(11.25)

4. A column is added for minimum efficiency. In the example, the minimum efficiency is set to 15%. It is assumed in this example that the gas engine will only be turned on when it can operate at an efficiency of at least 15%. It is assumed the electric motor will need to power the car in all cases when the gas engine would need to operate at less than 15% thermal efficiency. The logic statement here is that if the thermal efficiency in the previous column is less than the minimum, then the minimum will be placed in this column. If the thermal

efficiency is greater than the minimum, then the value calculated by the logarithmic equation will be used.

If the parameters in the spreadsheet are changed to reflect a hybrid F-150 pickup (W = 5500 lb, Crr = 0.01, A = 2.77 m², Cd = 0.5, P_{max} = 245 hp, drive eff = 0.90), the truck would get 26.3 mpg in city driving. This would be an excellent fuel economy for such a large vehicle. The spreadsheet developed for this example is shown in Example 11.9.

Example 11.9

The Toyota Prius is a hybrid car. Weight is 3500 lb, Crr is 0.009, drag area is 1.82 m² and the aero drag coefficient is 0.31. The engine is rated at 121 hp and the drive efficiency is about 90%. Develop the spreadsheet and estimate the fuel economy of the Toyota Prius in city driving.

Solution: The model yields a fuel economy of 47.5 mpg for the Toyota Prius, which is very near the EPA city rating for the 2020 version of the car. The first 30 s of the spreadsheet model for the Toyota Prius is shown in Table 11.13. The newest Prius has a more efficient engine than what is modeled using Equation (11.1), and the fuel economy is 60 mpg.

Table 11.13 Urban driving model for a hybrid-electric car.

Car weight	3500	lb		P_{max}	121	hp			Fuel energy	121,000	BTU/gal				
Crr	0.009			Eff	0.9									Min eff	0.15
Area	1.82	m²		Idle	0	W			Econ	47.5	mpg			Regen eff	0.3
Cd	0.31														

Time (s)	Acc (ft/s²)	Speed (ft/s)	Speed (mph)	Dist (ft)	Dist (mi)	Rolling (W)	Aero (W)	Kinetic (W)	Total (W)	Clean (W)	Thermal efficiency	Eff (min)	Fuel (W)	Fuel (BTU/s)	Fuel (gal)
0	0	0	0	0	0	0	0	0	0	0	0.000	0.150	0	0.00	0
1	0	0	0	0	0	0	0	0	0	0	0.000	0.150	0	0.00	0
2	0	0	0	0	0	0	0	0	0	0	0.000	0.150	0	0.00	0
3	0	0	0	0	0	0	0	0	0	0	0.000	0.150	0	0.00	0
4	0	0	0	0	0	0	0	0	0	0	0.000	0.150	0	0.00	0
5	0	0	0	0	0	0	0	0	0	0	0.000	0.150	0	0.00	0
6	0	0	0	0	0	0	0	0	0	0	0.000	0.150	0	0.00	0
7	0	0	0	0	0	0	0	0	0	0	0.000	0.150	0	0.00	0
8	0	0	0	0	0	0	0	0	0	0	0.000	0.150	0	0.00	0
9	0	0	0	0	0	0	0	0	0	0	0.000	0.150	0	0.00	0
10	0	0	0	0	0	0	0	0	0	0	0.000	0.150	0	0.00	0
11	0	0	0	0	0	0	0	0	0	0	0.000	0.150	0	0.00	0
12	0	0	0	0	0	0	0	0	0	0	0.000	0.150	0	0.00	0
13	0	0	0	0	0	0	0	0	0	0	0.000	0.150	0	0.00	0
14	0	0	0	0	0	0	0	0	0	0	0.000	0.150	0	0.00	0
15	0	0	0	0	0	0	0	0	0	0	0.000	0.150	0	0.00	0
16	7	0	0	0	0	0	0	0	0	0	0.000	0.150	0	0.00	0
17	7	7	4.77	3.5	0.001	306	3.3	3620	3929	4366	0.198	0.198	22,036	20.89	0.0002
18	7	14	9.55	14	0.003	627	26.3	10,859	11,512	12,791	0.239	0.239	53,463	50.68	0.0006
19	1	21	14.32	31.5	0.006	961	88.7	18,098	19,148	21,275	0.259	0.259	82,223	77.94	0.0012
20	0	22	15	53	0.010	1010	102.0	3176	4289	4765	0.201	0.201	23,654	22.42	0.0014
21	0	22	15	75	0.014	1010	102.0	0	1112	1236	0.150	0.150	8228	7.80	0.0015
22	0	22	15	97	0.018	1010	102.0	0	1112	1236	0.150	0.150	8228	7.80	0.0015
23	0	22	15	119	0.023	1010	102.0	0	1112	1236	0.150	0.150	8228	7.80	0.0016
24	0	22	15	141	0.027	1010	102.0	0	1112	1236	0.150	0.150	8228	7.80	0.0017
25	0	22	15	163	0.031	1010	102.0	0	1112	1236	0.150	0.150	8228	7.80	0.0017
26	0	22	15	185	0.035	1010	102.0	0	1112	1236	0.150	0.150	8228	7.80	0.0018
27	0	22	15	207	0.039	1010	102.0	0	1112	1236	0.150	0.150	8228	7.80	0.0019
28	0	22	15	229	0.043	1010	102.0	0	1112	1236	0.150	0.150	8228	7.80	0.0019

At this point in time, we are working to develop electric cars that have a long driving range and a fast charging time. The vision is that if all the electricity is produced by renewable energy (wind, solar, hydro), the electric vehicles will have zero emissions. We have made a lot of progress and will probably be able to develop small and medium-sized cars that have an acceptable range. Larger vehicles, especially large trucks, will require many batteries, and it may not be possible to provide an adequate range with lithium-ion batteries. Charging time is a more difficult problem to solve than range. At this point in time, it seems unlikely to me that we will be able to develop a battery system that can be charged in 5 or 10 min, which is typical for how long it takes to fill a vehicle with gasoline. The hybrid vehicle or plug-in hybrid vehicle may be the best overall compromise.

Homework

1. Assume that we do careful fuel economy testing on a light truck that weighs 5400 lb, Crr is 0.012, the aerodynamic drag area is 2.8 m², and Cd is 0.47. Assume the truck is burning 10% ethanol–gasoline blend that has 121,000 BTU/gal.

   (a) Cruising on the level pavement at a steady 30 mph, the truck gets 26.1 mpg. What is the efficiency of the engine/drive system at 30 mph? (Efficiency of the engine/drive system is the product of the thermal efficiency of the engine and the efficiency of the drive system in getting the power to the wheels.) (Answer: 15.58%)

   (b) Cruising on the level pavement at a steady 60 mph, the truck gets 21.9 mpg. What is the efficiency of the engine/drive system at 60 mph? (Answer: 26.02%)

   (c) Assume that the efficiency of the engine/drive system is zero at 0 mph. Plot the efficiency of the engine drive system vs speed over the range of 0 to 60 mph using the data points in (a) and (b) above and the point (0,0). Draw a trend line through the data points and have Excel calculate the slope of the trend line. Please use only the three data points for the plot, 0% efficiency at 0 mph, and the efficiencies you get for 30 and 60 mph.

2. Model an electric car driving on the SAE J1082 suburban cycle. Parameters for the car are: W = 4400 lb, Crr = 0.009, A = 1.9 m², Cd = 0.31, motor/drive efficiency = 88%, battery capacity = 60,000 W-h.

   (a) Assume that there is no regenerative braking on the car and calculate the number of watt-h per mile energy consumption, and the range of the vehicle for suburban driving. (Answer: 270 W-h/mi, 222 mi)

   (b) Assume that regenerative braking is added and that 30% of the braking energy will be recovered and put back into the batteries. Calculate the number of watt-h per mile energy consumption, and the range of the vehicle for suburban driving. (Answer: 245 W-h/mi, 245 mi)

   (c) Evaluate the impact of adding regenerative braking in increasing the range of the car in suburban driving. What is the increase in the range of the car from adding regenerative braking?

   (d) Suppose that by using exotic materials we can reduce the weight of the vehicle by 500 lb. What is the increase in the range of the car from reducing the weight by 500 lb? (Compare to the original case with no regenerative braking and only change the weight of the vehicle.)

(e) Suppose that we can make considerably more luggage room in the hatchback region by squaring off the back of the car like an SUV rather than having it sloped and more aerodynamic. This will increase the drag coefficient to Cd = 0.4. What is the decrease in the range of the car from this modification? (Compare to the original case with no regenerative braking and only change the drag coefficient of the vehicle.)

(f) Suppose we can make the car more pleasant to drive by using wider tires that operate at a lower pressure. This will increase Crr to a value of 0.014. What is the decrease in the range of the car from this modification? (Compare to the original case with no regenerative braking and only change the rolling coefficient of the vehicle.)

3. Model a small gasoline car driving on the SAE J1082 suburban cycle. Parameters for the car are: W = 2800 lb, Crr = 0.010, A = 1.9 m², Cd = 0.32, drive efficiency = 90%. The power rating for the engine is 130 hp and burns the 10% ethanol–gasoline blend that has 121,000 BTU/gal.

(a) Use the method developed in class and calculate the mpg fuel economy of the car driving in a suburban environment. If the car has a 13.2-gal tank, what is the range of the vehicle? (Answer: 40.2 mpg)

(b) Suppose that by using exotic materials we can reduce the weight of the vehicle by 500 lb. What is the increase in mpg of the car from reducing the weight by 500 lb?

(Compare to the original case and only change the weight of the vehicle.)

(c) Suppose that we can make considerably more luggage room in the hatchback region by squaring off the back of the car like an SUV rather than having it sloped and more aerodynamic. This will increase the drag coefficient to Cd = 0.42. What is the decrease in mpg of the car from this modification? (Compare to the original case and only change the drag coefficient.)

(d) Suppose we can improve acceleration and performance by using wider tires that operate at a lower pressure. This will increase Crr to a value of 0.014. What is the decrease in mpg of the car from this modification? (Compare to the original case and only change the rolling coefficient.)

4. [Repeat #3 for the urban cycle] Model a small gasoline car driving on the SAE J1082 urban cycle. Parameters for the car are: W = 2800 lb, Crr = 0.010, A = 1.9 m², Cd = 0.32, drive efficiency = 90%. The power rating for the engine is 130 hp and burns the 10% ethanol–gasoline blend that has 121,000 BTU/gal.

(a) Use the method developed in class and calculate the mpg fuel economy of the car driving in an urban environment. If the car has a 13.2-gal tank, what is the range of the vehicle? (Answer: 39.5 mpg)

(b) Suppose that by using exotic materials we can reduce the weight of the vehicle by 500 lb. What is the increase in mpg of the car from reducing the weight by 500 lb? (Compare to the original case and only change the weight of the vehicle.)

(c) Suppose that we can make considerably more luggage room in the hatchback region by squaring off the back of the car like an SUV rather than having it sloped and more aerodynamic. This will increase the drag coefficient to Cd = 0.42. What is the decrease in mpg of the car from this modification? (Compare to the original case and only change the drag coefficient of the vehicle.)

(d) Suppose we can make the car more pleasant to drive by using wider tires that operate at a lower pressure. This will increase Crr to a value of 0.014. What is the decrease in mpg of the car from this modification? [Compare to the original case and only change the rolling coefficient of the vehicle.]

5. [Repeat #3 for the 70-mph highway cycle] Model a small gasoline car driving on the SAE J1082 highway cycle. Assume that the thermal efficiency of the engine is a constant 30% for this case. Parameters for the car are: W = 2800 lb, Crr = 0.010, A = 1.9 m², Cd = 0.32, drive efficiency = 90%. The power rating for the engine is 130 hp and burns the 10% ethanol–gasoline blend that has 121,000 BTU/gal.

(a) Use the method developed in class and calculate the mpg fuel economy of the car driving on the highway. If the car has a 13.2-gal tank, what is the range of the vehicle? (Answer: 41.0 mpg)

(b) Suppose that by using exotic materials we can reduce the weight of the vehicle by 500 lb. What is the increase in mpg of the car from reducing the weight by 500 lb? (Compare to the original case and only change the weight of the vehicle.)

(c) Suppose that we can make considerably more luggage room in the hatchback region by squaring off the back of the car like an SUV rather than having it sloped and more aerodynamic. This will increase the drag coefficient to Cd = 0.42. What is the decrease in mpg of the car from this modification? (Compare to the original case and only change the drag coefficient of the vehicle.)

(d) Suppose we can make the car more pleasant to drive by using wider tires that operate at a lower pressure. This will increase Crr to a value of 0.014. What is the decrease in mpg of the car from this modification? (Compare to the original case and only change the rolling coefficient of the vehicle.)

6. Model a small hybrid gasoline car driving on the SAE J1082 urban cycle. Parameters for the car are: W = 3100 lb, Crr = 0.010, A = 1.9 m², Cd = 0.32, drive efficiency = 90%. The power rating for the engine is 130 hp and burns the 10% ethanol–gasoline blend that has 121,000 BTU/gal. Assume that the electric motor and battery are designed such that the engine will always operate at 15% thermal efficiency or higher. Also, assume the car has regenerative braking and that 30% of the braking energy will be recovered.

(a) Use the method developed in the book and calculate the mpg fuel economy of the hybrid car driving in an urban environment. If the car has a 13.2-gal tank, what is the range of the vehicle? (Answer: 48.4 mpg)

(b) Suppose that by using exotic materials we can reduce the weight of the vehicle by 500 lb. What is the increase in mpg of the car from reducing the weight by 500 lb? (Compare to the original case and only change the weight of the vehicle.)

(c) Suppose that we can make considerably more luggage room in the hatchback region by squaring off the back of the car like an SUV rather than having it sloped and more aerodynamic. This will increase the drag coefficient to Cd = 0.42. What is the decrease in mpg of the car from this modification? (Compare to the original case and only change the drag coefficient of the vehicle.)

(d) Suppose we can make the car more pleasant to drive by using wider tires that operate at a lower pressure. This will increase Crr to a value of 0.014. What is the decrease in mpg of the car from this modification? (Compare to the original case and only change the rolling coefficient of the vehicle.)

11.5.
Effect of Hills on Fuel Economy

Grades on interstate highways seldom exceed 4% for most of the country. In mountainous areas, it is common to have 6 or 7% grades on the steeper parts of the mountains. With a 4% grade, the road rises 4 ft for every 100 ft traveled, or the sine of the angle is 4/100, which is 2.29°. The angle of the road seems a lot higher than this, but the grades on the roads are very small angles. The power to overcome gravity is equal to the weight of the vehicle multiplied by the upward component of velocity.

Example: Suppose an 80,000 lb truck climbs a 4% grade at 70 mph. What is the horsepower required to overcome gravity? For this example, neglect rolling resistance and aerodynamics and only calculate the portion of horsepower required to climb the hill.

Solution: The upward component of velocity is 70 mph multiplied by 4%. The power to overcome gravity is calculated as follows:

$$\text{Gravity Power} = (80,000\,\text{lb})(70\,\text{mph})(0.04)\left(\frac{5280}{3600}\right) = 328,533\,\text{ft-lb/s} \tag{11.26}$$

Since 1 hp is equal to 550 ft-lb/s, the truck requires 597 hp to overcome the gravity loading. Similar calculations show that 896 hp is required for a 6% grade and 1045 hp is required for a 7% grade. The truck will require an additional 250 hp to overcome rolling resistance and aerodynamics. These are large amounts of horsepower, even for an 18-wheeler truck. The 18-wheelers typically have 400 to 600 hp available to push the truck up the hill. No fully loaded (80,000 lb) truck can climb a 4% grade at 70 mph. If the truck is unloaded so it weighs only 40,000 lb then it is feasible for the truck to climb a 4% grade at 70 mph. Hill climbing requires a substantial amount of horsepower.

Example 11.10

A medium six-cylinder pickup truck is pulling a boat traveling at 80 mph as it approaches a hill with a 7% grade. The weight of the truck and contents is 5000 lb, and the weight of the boat is 3000 lb. The rolling resistance coefficient for the tires is 0.011. The aerodynamic drag coefficient is 0.85 and the drag area is 3.8 m². The truck provides a maximum of 150 hp to the wheels.

(a) Calculate the steady-state speed of the truck up the hill.

(b) Develop a spreadsheet model to show how the velocity decreases as the truck starts up the hill.

Solution: Part (a) can be solved analytically. If we work in watts, the gravity, rolling, and aerodynamic powers can be expressed as a function of speed and we can solve for the steady-state speed.

$$\text{Gravity Power} = \left(8000\right)\left(V\right)\left(\frac{5280}{3600}\right)\left(\frac{7}{100}\right)\left(\frac{745.7}{550}\right) \tag{11.27}$$

$$\text{Rolling Power} = \left(1.989\right)\left(8000\right)\left(V\right)\left(0.011\right)\left(1+\frac{V}{200}\right) \tag{11.28}$$

$$\text{Aero Power} = \left(0.05357\right)\left(V^3\right)\left(3.8\right)\left(0.85\right) \tag{11.29}$$

$$\text{Gravity} + \text{Rolling} + \text{Aero} = \left(150\right)\left(745.7\right) \tag{11.30}$$

The steady-state speed is calculated by solving Equation (11.30), yielding a speed of 58.1 mph. The truck and trailer would be traveling 80 mph when it got to the hill and would gradually slow down to 58.1 mph. It would be able to maintain the 58.1 mph the rest of the way up the hill.

For part (b) of the problem, we will use a spreadsheet approach and develop an approximate model. Kinetic power is the product of mass, velocity, and acceleration, and acceleration is the derivative of velocity with time. In principle, it is possible to set up a first-order differential equation and generate an analytical solution. It is much easier and more straightforward to use the spreadsheet approach to generate an approximate solution.

For this example problem, I chose to step through the process in 0.5-s intervals. The truck starts at 80 mph, and from that number, we calculate the power to overcome gravity, rolling, and aerodynamic resistance. The engine can only provide 150 hp, and at 80 mph, that will not be enough for the truck to maintain speed. The difference in power is made up by a reduction in kinetic energy over the 0.5-s time step. The energy balance at each time step will require a small change in velocity of the truck, which will allow us to find the velocity as a function of time for the truck. Parameters for the vehicle are: W = 8000 lb, Crr = 0.011, A = 3.8 m², Cd = 0.85, power = 150 hp, grade = 7%, and initial speed = 80 mph. The first three steps of the spreadsheet are shown in Table 11.14.

Table 11.14 First three steps of truck pulling boat uphill.

Time (s)	Speed (mph)	Gravity (W)	Rolling (W)	Aero (W)	Engine (W)	Kinetic (J)	$(V_2)^2$ (mph)^2
0	80	89,088	19,600	88,576	111,855	42,705	6282.13
0.5	79.26	88,264	19,367	86,141	111,855	40,959	6169.1
1.0	78.54	87,466	19,143	83,826	111,855	32,290	6060.7

© SAE International.

Gravity, rolling, and aero power are calculated using Equations (11.27), (11.28), and (11.29) using the speed in column 2 of the table. In this model, it is assumed that the engine and drive can provide 150 hp to the wheels, which is 111,855 W. The engine is not capable of providing enough power to overcome the sum of gravity, rolling, and aerodynamic loads, so the difference must be made up with a reduction in kinetic energy. The kinetic energy is the power difference multiplied by 0.5-s time step in Equation (11.31).

$$\text{Kinetic Joules} = (\text{Gravity} + \text{Rolling} + \text{Aero-Engine})(0.5\,\text{s}) \qquad (11.31)$$

If V_1 is the velocity at the beginning of the time step in column 2 and V_2 is the velocity at the end of the time step, the change in kinetic energy is $\frac{1}{2}\,m\,(V_1^2 - V_2^2)$. The speeds are in mph and the weight is given in pounds so there are unit conversions, but the equation can be solved for V_2^2 as shown in Equation (11.32).

$$V_2^2 = V_1^2 - \frac{(22.0811)(\text{Kinetic Joules})}{\text{Weight}} \qquad (11.32)$$

Take the square root of V_2^2 in column 8 to get the value for V_1 in the next row in column 2. The speeds in column 2 are the speed of the truck and trailer at the beginning of each time step. Energy is balanced by the truck and trailer slowing down a little each time step. The reader should be able to duplicate the numbers in the table. Excel results for the first 5 s are shown in Table 11.15. The first 70 s are graphed in Figure 11.5.

In the model the speed of the truck and trailer decreases gradually to the 58.1 mph steady-state speed calculated in part (a) of the problem.

Reality is more complex than is illustrated in the model because the engine will not produce a constant 150 hp over the range of speeds. The engine power will be approximately constant, but the engine RPM will depend on the vehicle speed and engine power varies with RPM. The transmission may downshift, which further complicates the issue. But all things considered, this is a good model to illustrate quantitatively how hill climbing impacts the energy consumption and speed of the vehicle (Figure 11.6).

Table 11.15 Spreadsheet of truck pulling boat uphill.

Vehicle weight	8000	lb				
Roll resistance	0.011					
Drag coeff	0.85					
Drag area	3.8	m²				
Engine power	150	hp				
Hill grade	0.07	grade				
Initial speed	80	mph				
Time steps	**Speed (mph)**	**Gravity (W)**	**Rolling (W)**	**Aero (W)**	**Engine (W)**	**Kinetic (J)**	**$(V_2)^2$**
0	80	89,086.29	19,603.58	88,591.92	111,855	42,713.4	6282.105
0.5	79.25973	88,261.95	19,370.84	86,155.29	111,855	40,966.54	6169.032
1	78.54318	87,464.01	19,146.46	83,839.69	111,855	39,297.59	6060.565
1.5	77.84963	86,691.69	18,930.14	81,638.28	111,855	37,702.56	5956.501
2	77.17837	85,944.19	18,721.58	79,544.65	111,855	36,177.71	5856.645
2.5	76.52872	85,220.75	18,520.48	77,552.82	111,855	34,719.52	5760.815
3	75.90003	84,520.66	18,326.57	75,657.16	111,855	33,324.69	5668.834
3.5	75.29166	83,843.19	18,139.59	73,852.43	111,855	31,990.1	5580.537
4	74.70299	83,187.66	17,959.28	72,133.69	111,855	30,712.81	5495.765
4.5	74.13343	82,553.41	17,785.4	70,496.32	111,855	29,490.06	5414.369
5	73.58239	81,939.78	17,617.71	68,935.97	111,855	28,319.24	5336.204

Figure 11.6 Graph of a truck pulling boat uphill.

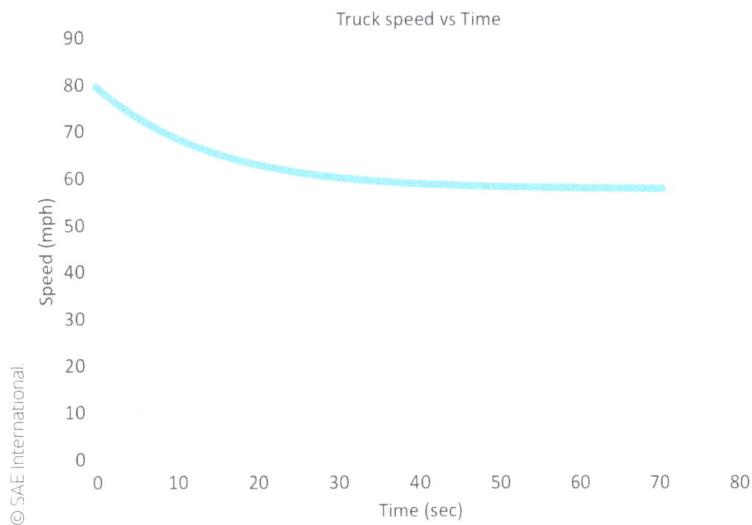

11.6
Effect of Climate Control Systems on Vehicle Energy Consumption

Another vehicle system which has a significant impact on the overall energy consumption of a vehicle is the climate control system. In a vehicle with a conventional internal combustion engine, the main energy consumption impact of climate control system comes when operating in air conditioning (cabin cooling) mode. When cabin heating is needed, there is usually plenty of waste heat available from the engine to accomplish this function, with energy needed to power the blower being minimal by comparison. However, in the case of an electric vehicle, energy must come from the battery for cabin heating in addition to cabin cooling, since there is little or no waste heat available to utilize in heating mode. As a result, operation of the climate control system is generally the second largest consumer of energy in an electric vehicle for most ambient temperature conditions and can have a significant impact on electric vehicle range.

To help understand the operation and energy flow into and out of an air conditioning system in a conventional internal combustion engine vehicle, refer to Figure 11.7. An air conditioning system is able to remove large quantities of heat from the air entering a vehicle's interior (or air recirculating in the cabin, depending on the position of the fresh/recirc door at the blower inlet) by taking advantage of the large amount of latent heat absorbed and released by the refrigerant in the system as it undergoes liquid-to-gas and gas-to-liquid phase changes at the heat exchangers—i.e., the evaporator and the condenser. The air conditioning function itself takes place as energy in the form of heat Q_{EVAP} is removed from the cabin air being cooled by the refrigerant at the evaporator as the air passes through the vehicle's HVAC (heating, ventilation, air conditioning) unit. As liquid refrigerant in a low temperature/low pressure state absorbs heat at the evaporator, it boils off—undergoing a liquid-to-gas phase change. Leaving the evaporator, the low temperature gas-phase refrigerant enters the compressor, where work energy W_{COMP} added to the system, raising the temperature and pressure of the refrigerant. The energy added to the refrigerant at the compressor W_{COMP}, plus the energy absorbed by the refrigerant at the evaporator Q_{EVAP}, is removed from the system at the condenser in the form of heat Q_{COND}—where the refrigerant condenses, undergoing a gas-to-liquid phase change as heat is removed. The high-pressure liquid refrigerant leaving the condenser goes through a throttling process at the expansion valve, which greatly reduces the pressure and temperature of the refrigerant—producing a low temperature mixture of primarily liquid-phase (with some gas-phase) refrigerant at the evaporator inlet, starting the cycle all over again.

Since energy inputs into the system W_{COMP} and Q_{EVAP} must balance out with the energy leaving the system Q_{COND}, if no other losses are present the cooling capacity of the a/c system is expressed as follows:

$$Q_{EVAP} = Q_{COND} - W_{COMP} \text{ (equation for conventional a/c cooling capacity, kW)}$$

Figure 11.7 Schematic of an air conditioning system for an internal combustion engine vehicle.

For cabin heating mode, the internal combustion engine vehicle takes advantage of waste heat from the engine. This waste heat is absorbed by the coolant circulating through the engine, which is pumped from the engine to the HVAC unit through the heater hoses, where heat is transferred from the hot coolant into the cabin air at the heater core. Temperature modulation for air leaving the HVAC unit is achieved by regulating the position of the air mix door at the heater core inlet, producing the desired outlet air temperature through mixing in appropriate proportions the amount of hot air passing through the heater core and cold air bypassing the heater core.

There are two primary differences between an electric vehicle and an internal combustion engine vehicle which must be accounted for when configuring the climate control system. First, there is no accessory drive belt to drive the a/c compressor. To accommodate this difference, an electric vehicle uses an electric-motor driven compressor. Second, and more significantly, there is little or no waste heat available from the electric vehicle powertrain to utilize for cabin heating. Heating in an EV is therefore usually accomplished in one of two ways: by an electric resistance heater (placed either in the air stream where the heater core is normally located, or in the coolant loop, heating up coolant pumped into the heater core) or by a heat pump. Using a heat pump reduces heating mode energy consumption vs. that of an electric resistance heater, since part of the heat the heat pump system transfers into the cabin comes from heat absorbed by the refrigerant from the ambient air. However, a heat pump usually needs supplemental electrical resistance heat and/or advanced technical features for extreme low ambient

temperatures (<–10°C) as the ambient temperature approaches the lowest achievable boiling point of the refrigerant—beyond which the refrigerant can no longer absorb heat from the ambient air.

A heat pump system is more complex than a conventional a/c system and provides both the air conditioning and heating functions for the vehicle. Figure 11.8 provides a schematic of a typical electric vehicle heat pump system. However, being a relatively new technology, significant variations in vehicle heat pump configurations do exist.

Figure 11.8 Schematic of a typical electric vehicle heat pump system.

A. Cabin Heating Mode

B. Cabin Cooling Mode

When in air conditioning mode, the heat pump system functions essentially the same as a conventional air conditioning system, with the dual-function outside heat exchanger (OHX) operating as a condenser, as the OHX solenoid valve is opened, directing high-temperature gas-phase refrigerant leaving the compressor directly to the OHX. Refrigerant leaving the compressor does pass thorough the cabin condenser in the HVAC unit before the OHX; however, no heat transfer takes place there as long as the air-mix door at the cabin condenser inlet is fully closed (the air mix door will open for HVAC outlet air temperature modulation). Similar to the conventional a/c system, cooling capacity for the heat pump system in cabin cooling mode is expressed as:

$$Q_{EVAP} = Q_{OHX} - W_{COMP} \text{ (equation for heat pump cooling capacity, kW)}$$

with the Q_{OHX} term replacing the Q_{COND} term used for conventional a/c.

For cabin heating, instead of a heater core, the heat pump system uses a cabin condenser, where heat is rejected into the HVAC airstream from the high temperature, high pressure refrigerant leaving the compressor, converting the refrigerant into a high-pressure liquid. When in heating mode, the evaporator solenoid valve is opened to prevent refrigerant flow to the evaporator, and the OHX solenoid valve is closed to route refrigerant flow to the OHX expansion valve. In this mode, the OHX functions as an evaporator in the outside air stream, with the high pressure liquid refrigerant leaving the cabin condenser passing though the OHX expansion valve, reducing the refrigerant temperature to below that of the ambient temperature, which allows the refrigerant to absorb heat from the ambient air as the refrigerant in the OHX changes phase from liquid to gas. The low-temperature refrigerant in the OHX can absorb heat from the ambient air as long as the boiling point of the refrigerant is lower that the ambient temperature—even at ambients down to –10°C or below, depending on the minimum boiling point of the refrigerant, and level of heat pump technology in use. This heat energy absorbed from the ambient, combined with the work energy added to the system at the compressor, is then available for cabin heating at the cabin condenser:

$$Q_{CCND} = W_{COMP} + Q_{OHX} \text{ (equation for heat pump heating capacity, kW)}$$

An important concept to grasp when discussing energy consumption of climate control systems is Coefficient of Performance (COP), which is the metric for efficiency for such systems. COP is defined as:

$$COP = \frac{\text{Energy sought}}{\text{Energy that costs}}$$

In the case of air conditioning the *energy sought* is the amount of heat energy being removed from the airstream at the evaporator Q_{EVAP}, and the *energy that costs* is the amount of work energy being put into the system at the compressor, W_{COMP}. COP of a typical a/c system is generally in the range of 2.0 to 2.5, depending on ambient, and the level of technology used—i.e., the system can remove roughly twice as much heat energy from the cabin air as the amount of work energy going in to the system at the compressor.

For a heat pump in heating mode, the *energy that costs* is still the work energy input to the system from the compressor W_{COMP}, but the *energy sought* is the amount of heat energy being added to the cabin air at the cabin condenser, Q_{CCND}. As explained above, the amount of heat energy being provided to the cabin at the cabin condenser is roughly the sum of the heat energy absorbed from the ambient at the OHX and the work energy input at the compressor. The heat energy absorbed from the ambient costs nothing and is therefore is *not* included in the *energy that costs*, and the COP of a heat pump will generally be in the range of 1.5 at around –10°C, up to around 2.0 at 20°C—i.e., twice as much heat energy is being provided to the cabin vs. the amount of work energy added to the system at the compressor. This is not the case with electrical resistance heaters, which will have a maximum COP of 1.0—the electrical energy going into the device is the only source of heat energy in this case. This significant difference in energy consumption when in cabin heating mode, and the resulting improvement in vehicle range, explains why vehicle manufacturers will often opt for the increased complexity and cost of a heat pump system vs. a simpler electrical resistance heater for battery electric vehicles.

SAE J2765, *Procedure for Measuring System COP of a Mobile Air Conditioning System on a Test Bench*, has been developed to quantify the COP of a vehicle's air conditioning system at various standardized operating conditions, and is being updated to include heat pump operating conditions at the time of this writing. SAE J2766, *Life Cycle Analysis to Estimate the CO2-Equivalent Emissions from MAC Operation*, uses data from the J2765 test conditions to annualize an air conditioning system's energy consumption and life-cycle CO_2 emissions produced from operating the system, using climate data from various U.S. cities, as well as cities around the world.

The impact of climate control system operation on the energy consumption of an electric vehicle can be illustrated in the chart shown in Figure 11.9, from a study commissioned by the U.S. Department of Energy [1]. The vehicle used in the study, a production FIAT 500e, used an electric resistance heater, and as shown in Figure 11.3 the energy consumption during heating mode operation is particularly impactful—affecting the vehicle's total energy consumption over a wide range of ambients (–10 to +20°C), representing up to 25% of the vehicle's total energy consumption at –10°C. In a comparable internal combustion engine vehicle, the climate control system energy consumption below 20°C would essentially be zero, since the system utilizes waste heat from the engine that would otherwise be rejected to the ambient through the vehicle's radiator. Fuel used for operation of the air conditioning system in such a comparable internal combustion engine vehicle for ambients above 25°C can be calculated by dividing air conditioning system energy consumption for a given ambient by the engine thermal efficiency and the energy content per gallon of fuel; compressor efficiency would need to be taken into account as well.

Figure 11.9 Energy consumption in an electric vehicle vs. ambient temperature [1, 2] (stock FIAT 500e; 40 km/h auto a/c test).

Courtesy of DENSO INTERNATIONAL AMERICA, INC

Refrigerant selection for use in automotive air conditioning and heat pump systems is based on a variety of factors: safety (flammability and toxicity), ozone depletion potential, global warming potential, and in-vehicle performance being some of the main considerations. Most vehicles currently being produced in the U.S. use R-1234yf refrigerant, which has zero stratospheric ozone depletion potential (ODP) and very low global warming potential (GWP) compared to refrigerants used in the past. Previously used refrigerants included R-12, which was the first refrigerant in use when vehicle air conditioning first began to appear in earnest in the mid-1950s, through when it was phased out in the early 1990s due to high ODP (R-12 has a very high GWP as well). R-134a is a

zero ODP refrigerant that replaced R-12 but was phased out of use in new vehicles in the U.S. in the late 2010s through the early 2020s due to high GWP—though a significant improvement over R-12 in this regard as well. R-134a it is still available to service existing vehicles.

In order to function as a refrigerant in an automotive air conditioning system, a substance needs to have the thermophysical properties necessary to be able transition from a high pressure liquid into a low pressure/low temperature liquid when going through a throttling process, and must have a large latent heat capacity when undergoing gas-to-liquid and liquid-to-gas phase changes within the expected ambient operating temperatures of the vehicle. In order to do so the refrigerant used must be able to condense at a temperature at or above the maximum ambient temperature encountered by the vehicle at the maximum condenser-side pressure the system can tolerate, and if used in a heat pump system it must be able to boil at temperatures below the minimum ambient temperature the vehicle will operate in at the lowest pressure the system can generate at the OHX—or the system must cover the heating needs for extreme low temperature operation with an electrical resistance heater.

One exception to the above condensing temperature rule is found with the transcritical CO_2 cycle, which uses carbon dioxide (R-744) as a refrigerant. The system operates in a "transcritical" manner, meaning that the high pressure side of the system operates at pressures above the "critical point" for the refrigerant, where condensing of the refrigerant in the conventional sense is no longer possible, and there is no longer a clear distinction between the gas and liquid phases—instead of a liquid or a gas, the refrigerant exists in the form of a supercritical fluid at the high-pressure side of the cycle. Instead of condensing the refrigerant in the conventional sense, the "condenser" in this type of system rejects heat as a supercritical fluid cooler—commonly called a gas cooler.

Due to the extremely low boiling point of liquid CO_2, heat pump systems using this refrigerant generally work well down to ambients of −20°C and below. However, in air conditioning mode CO_2 systems have been shown to have limitations in high ambient temperature conditions

(>40°C), such as those encountered in the summer months in the U.S. Southwest, and in the Middle East [3]. As a result, the transcritical CO_2 cycle is currently limited to a few vehicle applications used mainly in Europe, where extreme high ambient conditions are rare.

11.7. Development of a Highway Model

To develop a model of vehicles driving on typical highways, we could survey the route and get accurate survey data. This would take a long time and be expensive. Another way is to drive the route in a vehicle with a Global Positioning System (GPS) connected to a computer and have the GPS collect latitude, longitude, and altitude along the route. The reader should imagine a huge spreadsheet with the first three columns of the spreadsheet being the latitude, longitude, and altitude of points along the highway route. This information can be used to develop a digital model of the highway, and it can be done quickly with relatively inexpensive equipment.

The earth is approximately a spherical polar coordinate system. Assume R is the radius of the earth, θ_{Lat} is the latitude, θ_{Long} is the longitude, and h is the altitude above sea level. The average earth radius is 3959 mi or 6371 km. The radius is smaller at the poles (6353 km) and larger at the equator (6384 km). We will use the average radius in the model. An (x, y, z) cartesian coordinate system can be imposed as follows:

$$r = R + h \qquad (11.33)$$

$$x = r\cos(\theta_{Lat})\cos(\theta_{Long}) \qquad (11.34)$$

$$y = r\cos\left(\theta_{Lat}\right)\sin\left(\theta_{Long}\right)$$ (11.35)

$$z = r\sin\left(\theta_{Lat}\right)$$ (11.36)

The distance between two points in the model is given by the distance formula:

$$\text{Distance} = \left[\left(x_2 - x_1\right)^2 + \left(y_2 - y_1\right)^2 + \left(z_2 - z_1\right)^2\right]^{\frac{1}{2}}$$ (11.37)

The average grade along the distance is the change in height divided by the distance:

$$\text{Grade} = \frac{h_2 - h_1}{\text{Distance}}$$ (11.38)

Table 11.16 shows three hypothetical data points. To be sure and understand the process, the reader should be able to duplicate the numbers in the table.

It is not possible to calculate the distance or slope for the first row in the table because two data points are required to make the calculations. There will be a noise factor in the GPS data, and it may be necessary to do some smoothing to obtain a reasonable digital approximation of the highway. The file used in this book is a model of Highway 63 heading south out of Rolla,

Missouri. It is a road that goes through the Ozark hills. The model can be used to illustrate how hilly terrain impacts the fuel economy and energy consumption of the vehicle.

The model has the distance along the route and the slope of the road in percent grade at each point along the route. It was developed from GPS data as illustrated in Table 11.16. Assuming the weight of the vehicle is in lb and the speed in mph the gravity power in W is given by Equation (11.39).

$$\text{Gravity Power} = \left(1.989\right)W\,V\left(\text{Grade}\right)W$$ (11.39)

Table 11.16 Converting latitude, longitude, and altitude into a digital model of the highway.

Latitude (degrees)	Longitude (degrees)	Altitude (m)	Radius (m)	x-Location (m)	y-Location (m)	z-Location (m)	Distance (m)	Slope grade
37.95143	−91.77127	342.5	6,371,342.5	−155,289.98	−5,021,609.3	3,918,332.646	0	
37.95138	−91.77122	342.2	6,371,342.2	−155,285.696	−5,021,612.6	3,918,328.078	7.0870	−0.04233
37.95133	−91.77125	342.1	6,371,342.1	−155,288.429	−5,021,615.8	3,918,323.632	13.2388	−0.01626

Table 11.17 illustrates how the model is started. The parameters for this example are: W = 3800 lb, Crr = 0.011, A = 1.9 m², Cd = 0.32, and V = 50 mph.

There are 8000 lines of data for the model that gives the distance along the route and the slope of the road in percent grade. The total distance is 175 mi. The maximum grade is less than 4% on this route. A negative grade means it is downhill and a positive grade is uphill. For the example, we will model the vehicle going at a constant speed of 50 mph. The reader should imagine a table 8000 lines in length where the number for the distance along the route and slope are known, and the rest of the table is blank.

We would start at time zero, so the first row under the two-time columns is zero.

The time in hours in column 2 is the distance in column 3 divided by the speed. The time in seconds for column 1 is the time in hours in column 2 multiplied by 3600. Rolling, aero, and gravity power are computed from Equations (11.26), (11.27), and (11.28) and then totaled in column 8. The first eight columns in the spreadsheet are the same whether it is an electric car or a gasoline-powered car. From this information, the reader should be able to duplicate Table 11.18, which is the same as Table 11.17 with the blanks filled in.

Table 11.17 Starting the highway simulation model.

Time (s)	Time (h)	Distance (mi)	Slope grade (%)	Rolling (W)	Aero (W)	Gravity (W)	Total (W)
0	0	0	−3.46				
		0.01352	−2.27				
		0.02844	−1.73				

© SAE International.

Table 11.18 Completion of Table 11.17.

Time (s)	Time (h)	Distance (mi)	Slope grade (%)	Rolling (W)	Aero (W)	Gravity (W)	Total (W)
0	0	0	−3.46	5196.26	4071.32	−13,076	−1142.5
0.97344	0.000270	0.01352	−2.27	5196.26	4071.32	−8579	810
2.04768	0.000569	0.02844	−1.73	5196.26	4071.32	−6538	3211

© SAE International.

For electric vehicles, the total power needs to be "cleaned." If the total power is positive, it must go through the electric motor and drive system, which is typically 85% efficient. If the total power is negative, regenerative braking will be able to recover about 30% of the energy. The energy used or added in each step is subtracted or added to the battery (Table 11.19).

The parameters used in the table are correct for the 2017 Nissan Leaf. In this example, I chose to use the first 50 mi to estimate the W-h/mi energy efficiency and range of the vehicle. There is nothing magical about the 50-mi choice, as long as you choose a distance of at least 50 mi. There are enough hills in the first 50 mi that the average efficiency of the vehicle has been established.

Table 11.19 Electric car on highway example.

Car weight	3800			Battery	30,000		Based on the first 50 mi			
Crr	0.011	lb		Eff	0.85	W-h	W-h/mi	225.1648		
Area	1.9	m²		Speed	50	mph	Range	133.2358	mi	
Cd	0.32			Regen	0.3					
		Distance	**Inclination**							
Time (s)	**Time (h)**	**The route (M)**	**The road (Grade %)**	**Rolling (W)**	**Aero (W)**	**Hill (W)**	**Total power (W)**	**Cleaned (W)**	**Battery (W-h)**	
0.00	0.00000	0.000	−3.50	5196	4071	−13,219	−3952	−1185	30,000	
0.97	0.00027	0.014	−3.27	5196	4071	−12,355	−3087	−926	30,000	
2.05	0.00057	0.028	−2.31	5196	4071	−8715	552	650	30,000	
3.22	0.00089	0.045	−1.77	5196	4071	−6674	2594	3052	29,999	
4.43	0.00123	0.062	−1.43	5196	4071	−5418	3849	4528	29,998	
5.66	0.00157	0.079	−1.15	5196	4071	−4357	4911	5777	29,996	
6.90	0.00192	0.096	−0.48	5196	4071	−1822	7446	8760	29,994	
8.15	0.00226	0.113	−0.17	5196	4071	−638	8630	10,153	29,991	
9.43	0.00262	0.131	0.29	5196	4071	1083	10,351	12,177	29,987	
10.74	0.00298	0.149	0.66	5196	4071	2506	11,774	13,852	29,982	
12.11	0.00336	0.168	1.37	5196	4071	5167	14,434	16,982	29,976	
13.47	0.00374	0.187	1.27	5196	4071	4811	14,079	16,564	29,970	

Choosing a longer distance will yield a slight, but insignificant difference in the results. The model shows that the car would use 225 W-h of energy per mile traveled and have a range of 133 mi. If the electricity cost is $0.12 per kW-h, the energy cost for the car would be $0.02 per mi. The energy cost per mile for the electric car is very low compared to gas cars, but the range of 133 mi would be problematic for most people when traveling on the highway. If the speed is increased to 70 mph the range decreases to 92 mi. That is enough for people to use in daily commutes, but it is not enough for making a long trip on the highway.

The next step is to model a gasoline-powered car driving along the route. For this example, it was assumed that the engine has a constant thermal efficiency of 28%. It is possible to use the logarithmic expression for engine thermal efficiency [Equation (11.21)], but for highway driving this equation underestimates the average thermal efficiency of the engine. Equation (11.21) was developed by fitting a curve through the average thermal efficiency of the engine, which is typical for city driving. For highway driving, the transmission will choose a higher gear and allow the engine to operate closer to the maximum thermal efficiency (Table 11.20).

For comparison, the parameters used in Table 11.20 are for a gasoline car of the same size as the Nissan Leaf. The gas car would be lighter in weight but would have the same Crr, area, and drag coefficient. The engine would produce about 130 hp, and the average thermal efficiency on the highway would be about 28%. Driving at 50 mph this car would get 55.9 mpg, which is an excellent fuel economy, but 50 mph is a slow speed for the highway. If the speed is increased to 70 mph, the fuel economy is 36.9 mpg, which is about right for a small car driving on the highway.

Table 11.20 Gasoline car on highway example.

Car weight	2800	lb		Max power	130	hp						
Crr	0.011			Idle power	193.88	W			Fuel economy	55.9	mpg	
Area	1.9	m²		Drive eff	0.9				Engine eff	0.28		
Cd	0.32			Speed	50	mph						

Time (s)	Time (h)	Distance The route (M)	Inclination The road (Grade %)	Rolling (W)	Aero (W)	Hill (W)	Total power (W)	Clea-ned (W)	Engine Efficiency	Fuel Power (W)	Fuel Power (BTU/s)	Fuel Used (gal)
0.00	0.00000	0.000	−3.50	3829	4071	−9740	−1840	194	0.28	692	0.66	0.00000
0.97	0.00027	0.014	−3.27	3829	4071	−9103	−1203	194	0.28	692	0.66	0.00001
2.05	0.00057	0.028	−2.31	3829	4071	−6422	1478	1643	0.28	5866	5.56	0.00003
3.22	0.00089	0.045	−1.77	3829	4071	−4917	2983	3314	0.28	11,836	11.22	0.00011
4.43	0.00123	0.062	−1.43	3829	4071	−3993	3908	4342	0.28	15,506	14.70	0.00024
5.66	0.00157	0.079	−1.15	3829	4071	−3210	4690	5211	0.28	18,610	17.64	0.00041
6.90	0.00192	0.096	−0.48	3829	4071	−1342	6558	7286	0.28	26,023	24.67	0.00063
8.15	0.00226	0.113	−0.17	3829	4071	−470	7430	8256	0.28	29,485	27.95	0.00090
9.43	0.00262	0.131	0.29	3829	4071	798	8698	9665	0.28	34,516	32.72	0.00122
10.74	0.00298	0.149	0.66	3829	4071	1847	9747	10,830	0.28	38,678	36.66	0.00159
12.11	0.00336	0.168	1.37	3829	4071	3807	11,707	13,008	0.28	46,457	44.04	0.00205
13.47	0.00374	0.187	1.27	3829	4071	3545	11,445	12,717	0.28	45,418	43.05	0.00254

11.8.
Summary of Unit 3

The purpose of Unit 3 was to introduce the reader to methods of using Excel spreadsheets to simulate cars and trucks in urban (city), suburban, and highway driving. The models were developed to estimate the energy consumption and fuel economy of the vehicles. The models are helpful in guiding the design and development of cars and trucks. When proposals are made to reduce weight, lower rolling resistance, or improve the aerodynamics of the vehicle, the models help us assess quantitatively the improvement in energy efficiency or fuel economy that would result from the proposals. The models help us do a cost–benefit analysis of proposed modifications so that we use our design and development resources wisely. Excel is very portable in the sense that almost everyone is comfortable using it. Developing the models in Excel makes it easier to pass the model to someone else. All the spread-sheet models developed for this book are available from the author.

In the first part of Unit 3, we looked at the thermal efficiency of IC engines and electric motors. I showed a method of starting with the engine map and developing an equation that relates the thermal efficiency of an IC engine to the amount of power it produces. The model was developed for a Saturn 1.9 L engine, but the method could be used to develop models for other engines, including diesel engines. The engine will have its best thermal efficiency when producing power at a low RPM, but the car will feel sluggish in city driving if the transmission keeps the engine operating at the RPM for maximum thermal efficiency. For city driving, the transmission will choose a gear so that the engine is operating at an RPM such that the car is more pleasant to drive. For highway driving, the transmission will tend to shift so that the engine is operating at a higher thermal efficiency.

In the middle part of Unit 3, I showed how to use the SAE J1082 specification to develop spreadsheet models for cars and trucks driving in urban (city), suburban, and highway driving. Models were developed for electric and gasoline-powered cars and trucks. The models allow us to quantitatively assess how the parameters (vehicle weight, rolling resistance of tires, aerodynamics, engine efficiency, and drive efficiency) impact the overall energy consumption and fuel economy of the vehicle.

In the last part of Unit 3, I showed how to develop a model for driving on a specific highway or race route. I was an advisor for the solar car project at Missouri S&T for many years, and to optimize the performance of the car on race routes, we developed models as described in this chapter. This was important for the solar car project but less important for production vehicles. For production vehicles, the results of modeling the vehicle traveling on level ground or on an actual highway are essentially the same. The models are to be used to estimate a difference in fuel consumption or electric energy consumption for a proposed modification to the vehicle. The difference will be almost the same if the vehicle is modeled traveling on level pavement or on a more realistic highway or street.

Homework

1. A small four-cylinder pickup truck is pulling a boat traveling at 80 mph as it approaches a hill with a 6% grade. The weight of the truck and contents is 5800 lb, and the weight of the boat is 3200 lb. The rolling resistance coefficient for the tires is 0.012. The aerodynamic drag coefficient is 0.65 and the drag area is 3.8 m². The truck provides a maximum of 140 hp to the wheels.

   (a) Calculate the steady-state speed of the truck up the hill. (Answer: 58.1 mph)

   (b) Develop a spreadsheet model to show how the velocity decreases as the truck starts up the hill.

2. Model an electric car driving on the hilly terrain model. Parameters for the car are: W = 4400 lb, Crr = 0.009, A = 1.9 m2, Cd = 0.31, motor/drive efficiency = 88%, battery capacity = 60,000 W-h. The car drives at a steady 65 mph.

   (a) Assume that there is no regenerative braking on the car and calculate the number of watt-h per mile energy consumption, and the range of the vehicle for driving in hilly terrain. (Answer: 278 W-h/mi, 216 mi)

3. Model a small gasoline car driving on the hilly terrain model. Parameters for the car are: W = 2800 lb, Crr = 0.010, A = 1.9 m², Cd = 0.32, drive efficiency = 90%. Assume that the thermal efficiency of the engine is 29%. The power rating for the engine is 130 hp and burns the 10% ethanol–gasoline blend that has 121,000 BTU/gal. The car drives at a steady 65 mph.

(a) Use the method developed in class and calculate the mpg fuel economy of the car driving in hilly terrain. If the car has a 13.2-gal tank, what is the range of the vehicle? (Answer: 43.7 mpg)

References

1. Chowdhury, S. and Zima, M., Unitary Thermal Energy Management for Propulsion Range Augmentation (UTEMPRA). United States: Fig. 57, 2019. Web. doi:10.2172/1569646.

2. Breakdown of accessory power consumption provided via personal correspondence with Dr. Chowdhry, July 11, 2024.

3. Mathur, G., "Using Natural Refrigerants Propane (R290) & Carbon Dioxide (R744) as Direct & or Indirect Heat Pump Systems to Address PFAS Issue," SAE Technical Paper 2024-01-2876 (2024), doi:https://doi.org/10.4271/2024-01-2876.

The review is a series of example problems. Working through the example problems will give the reader a good review of the material in the book. is an engine map for a small diesel engine produced by Volkswagen to power some of their small cars. When calculating the thermal efficiency of the engine, assume that it takes 69.5 g of diesel fuel to provide 1 kW-h of heat energy.

Figure 12.1 Engine map for a diesel engine.

Example 12.1

The car in question is traveling 70 mph when the transmission is in high gear and the engine is turning 2500 RPM. When the car accelerates from 55 to 85 mph, the horsepower requirements from the engine are shown in Table 12.1. The total hp in the table is the total power to overcome rolling, aero, and a 4 ft/s² acceleration or the vehicle and assumes that the transmission and drive are 90% efficient in transmitting the power to the wheels. It is assumed the car is traveling on level ground. Fill in the table and plot the acceleration line on the engine map in Figure 12.1.

Solution: The first two columns in the table are given and you are to calculate the other two columns. The equations used are Equations (12.1) and (12.2):

$$\text{Angular Velocity} = \frac{(2500)\,V}{70} \tag{12.1}$$

$$\text{Torque} = \frac{(\text{Horsepower})(550)}{\text{RPM}\left(\dfrac{2\pi}{60}\right)} \tag{12.2}$$

Table 12.2 is the completed solution for the table. The numbers in red were added to the table. Because angular velocity and torque are known, the student would plot the points in Figure 12.1 and draw a line through them to complete the solution. The line goes through the most efficient portion of the engine map, so the thermal efficiency of the engine will be very good. If the average fuel consumption for this line is 200 g/kW-h, the average thermal efficiency is 69.5/200 = 34.75%. The line at 85 mph is just outside the engine map curve, so it would not be possible for this engine to produce 228 ft-lb of torque at 3036 RPM.

Table 12.1 Example 12.1.

Speed (mph)	Total horsepower	Angular velocity (RPM)	Torque (ft-lb)
55	74.40		
60	82.84		
65	91.70		
70	101.02		
75	110.83		
80	121.17		
85	132.06		

RPM: Revolution per minute

Table 12.2 Completed Example 12.1.

Speed (mph)	Total horsepower	Angular velocity (RPM)	Torque (ft-lb)
55	74.40	1964	199
60	82.84	2143	203
65	91.70	2321	207
70	101.02	2500	212
75	110.83	2679	217
80	121.17	2857	223
85	132.06	3036	228

RPM: Revolution per minute

© SAE International.

Example 12.2

If we assume a constant engine thermal efficiency of 34.75% during this acceleration, what would you expect the fuel economy to be during this acceleration? Assume 138,700 BTU/gal for diesel fuel. [Note: Fuel economy during hard acceleration is not very good, even for a small diesel-powered car.] (Table 12.3)

Solution: The equations used to complete the table are Equations (12.3), (12.4), and (12.5). The numbers in red in Table 12.4 are the solution.

$$\text{Fuel Power} = \frac{\text{Total Horsepower}}{0.3475} \qquad (12.3)$$

$$\text{Fuel Usage} = \left(\text{Fuel Power}\right)\left(\frac{550}{778.17}\right)\left(\frac{3600}{138,700}\right) \qquad (12.4)$$

$$\text{Economy} = \frac{\text{Speed}}{\text{Fuel Usage}} \qquad (12.5)$$

Table 12.3 Example 12.2.

Speed (mph)	Total horsepower	Fuel power (hp)	Fuel usage (gph)	Economy (mpg)
55	74.40			
60	82.84			
65	91.70			
70	101.02			
75	110.83			
80	121.17			
85	132.06			

© SAE International.

Table 12.4 Completed Example 12.2.

Speed (mph)	Total horsepower	Fuel power (hp)	Fuel usage (gph)	Economy (mpg)
55	74.40	214	3.93	14.00
60	82.84	238	4.37	13.72
65	91.70	264	4.84	13.43
70	101.02	291	5.33	13.13
75	110.83	319	5.85	12.82
80	121.17	349	6.40	12.51
85	132.06	380	6.97	12.19

© SAE International.

Example 12.3

The car in question is traveling 70 mph when the transmission is in high gear and the engine is turning 2500 RPM. When cruising at constant speed, the power requirements are much lower than when accelerating. Assume that the car is cruising at a constant speed with the horsepower requirements shown in Table 12.5. The horsepower shown includes acceleration, rolling, and aero power and assumes that the transmission/drive system is 90% efficient in transmitting the power to the wheels. Plot the speed curve on the engine map in Figure 12.1 and estimate the thermal efficiency.

Solution: The torque numbers in the table are calculated using Equation (12.2). From that, we locate the point on the graph and estimate the grams of fuel required to produce 1 kW-h of useful energy. At 40 mph the estimate used was 254 g/kW-h, and at 70 mph it was estimated as 249 g/kW-h. The numbers in red in Table 12.5 show the calculated torque and thermal efficiency of the engine. Notice that thermal efficiency does not vary much over this range of speeds. This demonstrates why it is reasonable to assume thermal efficiency is constant when driving on the highway.

Table 12.5 Problem and solution for Example 12.3.

Speed (mph)	Total horsepower	Angular velocity (RPM)	Torque (ft-lb)	Efficiency (%)
40	7.13	1429	26.2	27.36
45	8.97	1607	29.3	27.45
50	11.14	1786	32.8	27.54
55	13.67	1964	36.5	27.64
60	16.58	2143	40.6	27.73
65	19.92	2321	45.1	27.82
70	23.72	2500	49.8	27.91

© SAE International

Example 12.4

If the thermal efficiency is as shown in Table 12.6, what would you expect the fuel economy to be during cruising at various speeds? (Diesel fuel has 138,700 BTU/gal)

Solution: The numbers in black in Table 12.6 are the problem statement and the numbers in red are the solution. Equations (12.3), (12.4), and (12.5) are used to calculate the numbers in red in the table. If we had used the average thermal efficiency, the answers would be slightly different. A common question is to ask if it is realistic for the car to get 83.71 mpg at 40 mph. The answer is yes. If you are willing to drive at a steady speed of 40 mph on a level track with the engine turning 1429 RPM in high gear, the car will get approximately 83.71 mpg. The Environmental Protection Agency (EPA) highway estimate for this car is 44 mpg, which is approximately what the model yields at 70 mph.

Table 12.6 Problem and solution for Example 12.4.

Speed (mph)	Total power (hp)	Efficiency (%)	Fuel power (hp)	Fuel usage (gph)	Economy (mpg)
40	7.13	27.36	26.05	0.478	83.71
45	8.97	27.45	32.68	0.600	75.06
50	11.14	27.54	40.45	0.742	67.39
55	13.67	27.64	49.45	0.907	60.62
60	16.58	27.73	59.81	1.097	54.68
65	19.92	27.82	71.62	1.314	49.47
70	23.72	27.91	85.00	1.559	44.89

© SAE International

Example 12.5

The typical acceleration when starting from a stoplight is 7 ft/s². If you are driving a 5400-lb truck, calculate the force (in lb) for the 7 ft/s² acceleration.

Solution: Force is mass times acceleration:

$$\text{Force} = \left(\frac{5400 \text{ lb}}{32.2 \text{ ft} / s^2} \right) \left(7 \text{ ft} / s^2 \right) = 1174 \text{ lb} \tag{2.6}$$

Example 12.6

Assume the 5400 lb truck is to accelerate from rest to 30 mph at a constant 7 ft/s². Fill in Table 12.7 using the units indicated. Consider only kinetic power. Ignore rolling and aero power.

Solution: The speed in mph is the only thing given in the table. Speed in ft/s is a unit conversion from the mph speed. The time column starts at zero, and since it is constant acceleration, the number in the time column is the speed in ft/s divided by the 7 ft/s² acceleration. Power is the 1174-lb force multiplied by the speed in ft/s. The power is converted to watts. The energy used is the average power in watts multiplied by the time step, and it is totaled.

$$\text{Energy}_{i+1} = \text{Energy}_i + \left(\frac{\text{Power}_{i+1} + \text{Power}_i}{2} \right) \frac{\left(\text{Time}_{i+1} - \text{Time}_i \right)}{3600} \qquad (12.7)$$

The calculations are shown in red in the table. For this problem, we could have calculated the kinetic energy at each speed and that would be the energy required in column 6. This works only because we are only considering kinetic energy in the problem. If we had included rolling resistance and aerodynamic losses, this would not have worked, so Equation (12.7) is presented as a more general way to calculate the energy in column 6.

Table 12.7 Problem and solution for Example 12.6.

Time (s)	Speed (mph)	Speed (ft/s)	Power (ft-lb/s)	Power (W)	Energy (W-h)
0	0	0	0	0	0
1.04762	5	7.333	8609	11,672	1.698
2.09524	10	14.667	17,217	23,344	6.793
3.14286	15	22	25,826	35,015	15.285
4.19047	20	29.333	34,435	46,687	27.173
5.23809	25	36.667	43,043	58,359	42.457
6.28571	30	44	51,652	70,031	61.138

Example 12.7

A midsize car has a weight of 4200 lb, Crr is 0.010, the frontal area is 1.68 m², and the drag coefficient is 0.33. The efficiency of the drive system on the car is 90%. A model is to be developed for the car accelerating from rest at 7 ft/s² going up a 3% grade. Fill in the empty boxes in Table 12.8.

Solution: As with previous examples, the numbers in black are given as part of the problem statement and the numbers in red are the solution. In column 3, the speed in ft/s is calculated as the speed at the beginning of the time step plus acceleration times the time step. The speed in column 4 is the speed in ft/s multiplied by (5280/3600).

Equations for rolling, aerodynamic, gravity, and kinetic watts are given previously in the book. For convenience, the equations are listed again in Equations (12.8), (12.9), (12.10), and (12.11). W is in lb, V is in mph, and A is in m².

$$\text{Rolling Power} = \left(1.989\right)\left(W\right)\left(V\right)\left(Crr\right)\left(1 + \frac{V}{200}\right) \tag{12.8}$$

$$\text{Aero Power} = \left(0.05357\right)\left(V^3\right)\left(A\right)\left(Cd\right) \tag{12.9}$$

$$\text{Gravity Power} = \left(1.989\right)\left(W\right)\left(V\right)\left(\text{Grade}\right) \tag{12.10}$$

$$\text{Kinetic Power} = \left(0.0454\right)\frac{\left[V_{i+1}^2 + V_i^2\right]}{T} \tag{12.11}$$

The total power is obtained by adding the rolling, aero, kinetic, and gravity terms and dividing by the drive efficiency.

Table 12.8 Problem and solution for Example 12.7.

Time (s)	Accel (ft/s²)	Speed (ft/s)	Speed (mph)	Rolling (W)	Aero (W)	Kinetic (W)	Gravity (W)	Total (W)
0	7	0	0	0	0	0	0	0
1	7	7	4.773	408.2	3.23	4343.5	1196.1	6612.3
2	7	14	9.545	835.5	25.83	13,030.5	2392.2	18,093.3

Example 12.8

As you move from college to your first job, you are planning to drive your Ford Escape and pull a small U-Haul trailer behind it to move all of your worldly possessions. The Escape will be loaded to a weight of 4800 lb and the trailer will be loaded to 2500 lb. Assume that the rolling coefficient for the Escape and trailer is 0.01, the aerodynamic drag area is 2.8 m², and the drag coefficient is 0.65. The Escape can provide a maximum of 85 hp to the wheels. Calculate the steady-state speed for the Escape and trailer going up a 4% grade and a 7% grade.

Solution: Convert 85 hp to 63,384.5 W. Since it is a steady-state speed, the kinetic power will be zero. Gravity, rolling, and aerodynamic power can be calculated as a function of speed using Equations (12.8), (12.9), and (12.10), respectively. The sum is set equal to the 63,384.5 W power the Escape can provide to the wheels. Plugging in the parameters yields the following equation:

$$(145.2)(V)\left(1+\frac{V}{200}\right)+(0.0957)(V^3)+(580.8)(V)=63,384.5 \qquad (12.12)$$

Solving Equation (12.12) yields a steady-state speed of 57.9 mph for the 4% grade. For a 7% grade, the only change is in the gravity term, and the steady-state speed is 45.4 mph.

Example 12.9

For the Ford Escape of the previous example problem, detail out what happens for the first few time steps as the car comes to a 7% grade driving 75 mph. Fill in Table 12.9.

Solution: As with previous examples, the problem is the black text and the solution is in red. The initial speed is 75 mph. Gravity, rolling, and aero W are calculated using Equations (12.8), (12.9), and (12.10), respectively. The total will be more than the power the engine can provide at the wheels, so the difference must be made up of kinetic energy. Kinetic Joules is calculated as the energy difference divided by the time step of 0.5 s. V_2^2 is calculated using Equation (11.32). This value is then used as the speed in the next row.

Table 12.9 Problem and solution for Example 12.9.

Time (s)	Speed (mph)	Gravity (W)	Rolling (W)	Aero (W)	Engine (W)	Kinetic (J)	$(V_2)^2$ (mph)²
0	75	76,228	14,973	41,132	63,384.5	68,949	5521
0.5	74.30	75,520	14,797	39,996	63,384.5	66,927	5420
1.0	73.62	74,827	14,624	38,904	63,384.5	64,970	5322
1.5	72.95	74,147	14,456	37,853	63,384.5	63,072	5227

© SAE International

12.2.
Review of Units 1 and 2

Example 12.10

We are developing an exercise workout, and part of the workout is to drag a box across the gym floor. The coefficient of friction between the box and the floor is 0.4. We want it to require approximately 100 W power to drag the box at 3 mph. How much should the box weigh? (Figure 12.2)

Solution: The following equations are used:

$$\text{Drag Force} = (0.4)(W) \qquad (12.13)$$

$$\text{Speed} = (3\,\text{mph})\left(\frac{1609}{3600}\right) = 1.34\,\text{m / s} \qquad (12.14)$$

$$100\,\text{W} = (0.4)(W)(1.34) \qquad (12.15)$$

The weight of the box should be 186.5 N, which is approximately 40 lb. A person dragging a 40-lb box across the gym floor at 3 mph will require approximately 100 W of power.

Figure 12.2 Person dragging a box across the gym floor.

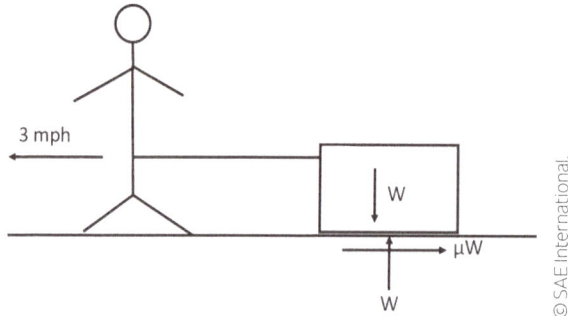

© SAE International.

Example 12.11

Assume 100 W output average during a 30-min workout. Muscles are approximately 20% efficient in converting the carbohydrate fuel into useful work. How many calories will be burned during the 30-min workout? (1 Cal = 4184 J)

Solution: To produce 100 W with muscles that are 20% efficient requires (100/0.2) = 500 W of fuel consumption. The person will consume fuel at a rate of 500 W, which is 500 J/s. Thirty min is 1800 s, so the total calorie consumption is calculated as:

$$\text{Calories Consumed} = \left(500 \text{ J / s}\right)\left(1800 \text{ s}\right)\left(\frac{\text{Cal}}{4184 \text{ J}}\right) = 215 \text{ Cal} \qquad (12.16)$$

Working out at a rate of producing 100 W power for 30 min would be an intense workout for most people. The person would burn about 215 Cal during the 30-min workout. This is the equivalent calories of a small candy bar, and less than the calories in a smoothie the person might purchase as a reward for the intense workout. Working out does not burn as many calories as we would like.

Example 12.12

A bicycle event is to be held where the riders will travel on relatively level pavement for 50 mi. Three classes of riders will enter the event and the planners need to estimate how long it will take for the three classes to finish the race, so the finish line and banquet following the race can be planned. The three classes are:

(a) Casuals—capable of providing 40 W power to the bicycle.

(b) Averages—capable of providing 90 W power to the bicycle.

(c) Athletes—capable of providing 180 W power to the bicycle.

The total drag force on the bicycle is given by the following equation:

$$\text{Drag Force} = \left(4 + \left(0.1\right)V^2\right) \text{N}, \quad \text{with V in mph} \qquad (12.17)$$

Estimate the time required for the three groups to finish the 50-mi event. We will need to have the finish line open and ready when the athletes arrive. The finish line will need to stay open until the casuals have had time to finish the event, and the banquet will need to be planned late enough to allow everyone to get showered and ready.

Solution: Power is drag force multiplied by speed. The drag force equation comes out in N if the speed is in mph for Equation (12.17). To get watts we need to multiply by the speed in m/s, so we need to use the unit conversion of (1609/3600). It is easy to miss that unit conversion in the problem. The average speed for the three classes of riders is calculated as:

(a) $40\ W = (4 + 0.1\ V^2)\ V\ (1609/3600)$ $V = 8.26$ mph for Casuals

(b) $90\ W = (4 + 0.1\ V^2)\ V\ (1609/3600)$ $V = 11.57$ mph for Averages

(c) $180\ W = (4 + 0.1\ V^2)\ V\ (1609/3600)$ $V = 15.07$ mph for Athletes

To travel 50 mi, the Casuals will need 6.05 h, the Averages will need 4.32 h and the Athletes will need 3.32 h to complete the race. Assuming everyone starts the race at the same time, the finish line should open 3 h after the race starts and stay open for 4 h.

Example 12.13

The total drag force on a concrete canoe is 5 lb when traveling through the water at 6 mph. How much paddling power are the students providing in moving the canoe along at 5 mph? Express your answer in ft-lb/s, W, and hp.

$$\text{Power} = (5\ \text{lb})(6\ \text{mph})\left(\frac{5280}{3600}\right) = 44\ \text{ft-lb}\ /\ \text{s} \qquad (12.18)$$

Solution: The power required is 44 ft-lb/s. With unit conversions, this is equivalent to 0.08 hp or 60 W.

Example 12.14

Assume that the drag force on the canoe varies with the square of the speed of the canoe. How much power is required to push the canoe along at 2 mph? Express your answer in ft-lb/s, W, and hp.

Solution: Fluid drag varies with the square of speed in most cases. If the drag force is 5 lb at 6 mph and varies with the square of speed, an equation can be developed for the drag force on the canoe at any speed:

$$\text{Drag Force} = (5\ \text{lb})\left(\frac{V}{6\ \text{mph}}\right)^2 \quad (\text{with V in mph}) \qquad (12.19)$$

At 2 mph the drag force is 0.556 lb and the power is 1.63 ft-lb/s, 0.00296 hp, or 2.21 W. This is a very low power requirement. It would be easy to paddle the canoe along at 2 mph. A similar calculation can be done for 10 mph yielding 203 ft-lb/s or 276.2 W power required. When paddling, the person does not really get to use his or her leg muscles and so the power a person can produce paddling is much less than on a bicycle or rowing with a sliding seat. It is difficult to paddle a canoe at 10 mph. Another way to look at the problem is to note that fluid drag power varies with the cube of speed. If the speed is reduced by a factor of 3, the power will be reduced by a factor of 27. The answers to this example problem are 1/27 of the answers to the previous problem.

Example 12.15

A small airplane cruises at 135 mph using 225 hp from the engine. The density of air is 1.2 kg/m³ at the altitude and temperature in question, and the propeller has a diameter of 1.9 m. Assume that the engine burns aviation fuel that has an energy density of 120,200 BTU/gal, and that the thermal efficiency of the engine is 28%.

  (a) Find the drag force on the airplane.

  (b) Find the fuel economy of the airplane in mpg.

Solution: To solve this problem, we will need to use the propeller problems in Chapter 3, Equations (3.33), (3.34), and (3.35). Substituting in the correct numbers yields the following:

$$V_1 = \left(135\,\text{mph}\right)\left(\frac{1609}{3600}\right) = 60.34\,\text{m / s} \tag{12.20}$$

$$V = \frac{60.34 + V_2}{2} \tag{12.21}$$

$$\text{Thrust Force} = \frac{1}{2}\left(1.2\right)\pi\left(0.95\right)^2\left(V_2^2 - 60.34^2\right) \tag{12.22}$$

$$\text{Power} = \left(\text{Thrust Force}\right)\left(V\right) = \left(225\right)\left(745.7\right) \tag{12.23}$$

The solution is that V_2 = 71.659 m/s. This can be substituted into the thrust force equation to yield a drag force of 2542 lb.

To calculate fuel economy, we start by converting the 225 hp the engine produces into BTU/h, which is 572,528 BTU/h. The thermal efficiency of the engine is 28%, so it will need to burn fuel at a rate of:

$$\text{Fuel Consumption} = \left(\frac{572,528\,\text{BTU / h}}{0.28}\right)\left(\frac{\text{gal}}{120,200\,\text{BTU}}\right) = 17.01\,\text{gph} \tag{12.24}$$

$$\text{Fuel Economy} = \frac{135\,\text{mph}}{17.10\,\text{gph}} = 7.94\,\text{mpg} \tag{12.25}$$

Example 12.16

A jet airplane uses 1 gal of fuel to go approximately 1.45 mi while flying 530 mph. Assuming that the engines are e = 35% efficient, calculate the drag force on the jet. Assume that the engine uses 70 kg of air for each kg of fuel. The energy density of the fuel is E_f = 46.6 MJ/kg. The fuel has a density of 2.88 kg/gal.

Solution: 530 mph is 236.88 m/s. The mass flow rate of the air is given to be 70 times the mass flow rate of the fuel. Letting m_a be mass flow rate of the air and m_f be mass flow rate of the fuel, it follows that m_a = 70 m_f. We calculate the velocity of the combustion products leaving the engine relative to the aircraft V_e from Equation (4.26) as:

$$V_e = \sqrt{\frac{2(0.35)(46.6 \times 10^6)m_f + 70m_f(236.88)^2}{70m_f + m_f}} = 717.5 \, m/s \qquad (12.26)$$

We are given that the aircraft uses 1 gal (2.88 kg) of fuel to go 1.45 mi. Since the speed is 530 mph, we can calculate the mass flow rate of fuel for the aircraft. Once we know the mass flow rate of the fuel, the thrust equation is used to calculate the drag force on the jet.

$$m_f = \left(\frac{2.88 \, kg}{1.45 \, mi}\right)(530 \, mph)\left(\frac{5}{3600 \, s}\right) = 0.424 \, kg/s \qquad (12.27)$$

$$\text{Drag} = \left[0.424 + 70(0.424)\right](717.5) - 70(0.424)(236.88) = 14,569 \, N \qquad (12.28)$$

The thrust of the engine is 14,569 N, and for steady-state flight, this must equal the drag force on the airplane.

Example 12.17

Assume that in the United States, we used 28.3 quads of energy in transportation. Because of the thermal efficiency, about 22.4 quads are rejected to the atmosphere as heat, and 5.9 quads are converted into useful mechanical energy to power the vehicles. Assume that transportation generates 1800 million metric tons of carbon dioxide emissions each year. Based on this information, calculate the number of grams of carbon dioxide generated for each kW-h of useful mechanical energy in transportation.

Solution: First convert the 5.9 quads into kW-h energy.

$$\left(5.9 \times 10^{15} \text{ BTU}\right)\left(\frac{\text{kW-h}}{3412 \text{ BTU}}\right) = 1.729 \times 10^{12} \text{ kW-h} \tag{12.29}$$

1800 million metric tons is 1830×10^{12} g of carbon dioxide. Dividing the numbers:

$$\frac{1800 \times 10^{12} \text{ gCO}_2}{1.729 \times 10^{12} \text{ kW-h}} = 1041 \text{ gCO}_2 \text{ per kW-h} \tag{12.30}$$

Example 12.18

Assume that the atmosphere has a mass of 5.15×10^{18} kg and that the average molecular weight is 29. How many moles of gas are in the atmosphere?

$$\left(5.15 \times 10^{18} \text{ kg}\right)\left(\frac{1000 \text{ g}}{\text{kg}}\right)\left(\frac{\text{mole}}{29 \text{ g}}\right) = 1.776 \times 10^{20} \text{ moles} \tag{12.31}$$

Example 12.19

Assume that the world produces 3.29×10^{13} kg of carbon dioxide annually and that the molecular weight of carbon dioxide is 44. How many moles of carbon dioxide are put into the atmosphere each year?

$$\left(3.29 \times 10^{13} \text{ kg}\right)\left(\frac{1000 \text{g}}{\text{kg}}\right)\left(\frac{\text{mole}}{29 \text{ g}}\right) = 1.134 \times 10^{15} \text{ moles} \tag{12.32}$$

Example 12.20

Based on the two numbers above, how many parts per million does carbon dioxide represent in the atmosphere?

Solution:

$$\frac{1.134 \times 10^{15}}{1.776 \times 10^{20}} = 6.385 \times 10^{-6} = 6.385 \text{ ppm} \tag{12.33}$$

Example 12.21

We are using a liquid diesel fuel that on average has 24 hydrogen atoms for each 12 carbon atoms in the chemical composition. Draw a suitable molecule for $C_{12}H_{24}$.

Solution: There are many possibilities. One possibility is shown in Figure 12.3.

Figure 12.3 $C_{12}H_{24}$ molecule.

$C_{12}H_{24}$

© SAE International.

Example 12.22

The density of the $C_{12}H_{24}$ fuel is 812 g/L and the energy density is 135,000 BTU/gal. Write the chemical reaction for the combustion of the fuel and calculate the number of BTU heat energy we get from the fuel for each gram of carbon dioxide generated.

$$C_{12}H_{24} + 18O_2 \rightarrow 12CO_2 + 12H_2O \qquad (12.34)$$

The chemical balance equation tells us that 1 mole of fuel will produce 12 moles of carbon dioxide (and 12 moles of water). The molecular weights of the fuel and the 12 carbon dioxide atoms are calculated as:

$$C_{12}H_{24} = 12(12) + 24(1) = 168 \text{ g} \qquad (12.35)$$

$$12CO_2 = 12(12 + 2(16)) = 528 \text{ g} \qquad (12.36)$$

The conclusion is that 168 g of fuel will produce 528 g of carbon dioxide. Knowing that the heat value of the fuel is 135,000 BTU/gal and the density is 812 g/L, it is possible to calculate the number of BTU heat energy per gram of fuel.

$$(135,000 \text{ BTU / gal})\left(\frac{0.26417 \text{ gal}}{\text{liter}}\right)\left(\frac{\text{liter}}{812 \text{ g fuel}}\right) = 43.92 \text{ BTU / g fuel} \qquad (12.37)$$

And since 168 g of fuel produces 528 g of carbon dioxide:

$$\left(\frac{43.92 \text{ BTU}}{\text{g fuel}}\right)\left(\frac{168 \text{ g fuel}}{528 \text{ g CO}_2}\right) = 13.97 \text{ BTU / g CO}_2 \qquad (12.38)$$

Example 12.23

A company produces a small wind turbine for home use. The turbine blade is 8 ft in diameter, and the efficiency of the turbine is estimated to be 45%. Assume that the density of air is 1.2 kg/m³. The generator is 90% efficient in converting the energy from the turbine into electric energy. How much power do you estimate the turbine will produce in a 15-mph wind?

Solution: First calculate V_1; the speed of the wind is 6.704 m/s. The radius of the turbine blade is 4 ft, which is converted to 1.219 m. Since efficiency is known to be 45%, the efficiency formula [Equation (7.10)] can be used to calculate V_2.

$$0.45 = \frac{1}{2}\left[1 - \frac{V_2^2}{6.704^2}\right]\left(1 + \frac{V_2}{6.704}\right) \tag{12.39}$$

Solving the equation yields V_2 = 4.571 m/s. The average velocity V is 5.6375 m/s. The turbine power is equal to:

$$\text{Turbine Power} = \frac{1}{2}(1.2)(5.6375)\pi(1.219^2)(6.704^2 - 4.571^2) = 379.75 \text{ W} \tag{12.40}$$

The generator is 90% efficient in converting the turbine power into electricity. The power delivered in a 15-mph wind is 341.8 W.

Example 12.24

Based on a value of $0.13 per kW-h, how long would it take for the wind turbine to produce $1.00 worth of electricity if it operates in a 15-mph wind?

Solution: Assuming the wind blows steady at 15 mph, it takes (1000/341.8) = 2.93 h to produce 1 kW-h of electric energy. At $0.13 per kW-h it takes 7.69 kW-h to be worth $1.00. The answer is that it takes 22.5 h to produce $1.00 worth of electricity.

Putting up the tower and wind turbine and getting the electricity wired would be expensive. At a rate of producing $1.00 worth of electricity every 22.5 h it will produce $3,983.00 worth of electricity in 10 years. To break even, the homeowner could spend about this amount in installing the wind turbine and getting it hooked up.

Index

About the Author

Dr. Douglas R. Carroll

Dr. Douglas R. Carroll received his BS degree from Missouri State University in 1980 and his MS degree from Texas A&M University in 1982. He worked for the Air Force for six years at Tinker AFB in Oklahoma. He returned to academia and received his PhD from Missouri S&T in 1991 and has served as a faculty member at Missouri S&T since 1991.

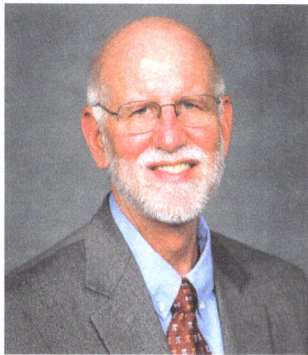

Dr. Carroll led the solar car project at Missouri S&T from 1992 to 2005, working with students designing, building and racing solar powered cars. The team won two national championships in 1999 and 2003. He published *The Winning Solar Car* in 2003, which is a design guide for solar powered cars. Dr. Carroll is an expert in energy efficient transportation and solar power.

In 2008 Dr. Carroll became Director for the Cooperative Engineering program at Missouri S&T, which is a satellite program for the university. Missouri S&T offers degrees in civil engineering, electrical engineering, and mechanical engineering in cooperation with Missouri State University in Springfield, Missouri. Dr. Carroll served as director for the program until 2023. He is currently a Professor in the mechanical engineering program.